In Praise of *Sustaining America's Strategic Advantage*

"Readers looking for insights into how the United States can meet the formidable security challenges it faces should read *Sustaining America's Strategic Advantage*. I know many of the authors, and their lucid writing is grounded firmly in analytical rigor and practical experience. The book's premise, that the United States needs to leverage key advantages—including strategic leader education and a global network of allies and partners—with a greater emphasis on nonmilitary instruments of power, is one that resonated powerfully with me as a retired ambassador who served for 33 years in the U.S. Foreign Service."

—Daniel Shields, Ambassador (ret.), editorial advisor,
Journal of Indo-Pacific Affairs

"This excellent volume sets a multidimensional foundation for American grand strategy that avoids the dangers of either exaggerating or underappreciating the potential of U.S. military power and also avoids the temptation of either overestimating or underestimating the overall strength of the United States at this crucial juncture in world history. Many grand strategies sound good; this one is actually based on a sound diagnosis of world conditions, and a full awareness of the perils as well as the opportunities of the contemporary global order."

—Michael E. O'Hanlon, Phil Knight Chair in Defense
and Strategy, Brookings Institution, and author
of *Military History for the Modern Strategist: America's
Major Wars since 1861*

"We live in dangerous times! For those seeking to understand the complex challenges and challengers facing the United States and how to address them, read *Sustaining America's Strategic Advantage*. The authors bring both scholarly research and practitioner's experience to comprehensively explore the critical security challenges, our competitors' strategies, the benefits and burdens of allies, and the strategic leadership qualities required to leverage these insights and maintain America's competitive advantage. Their call for a grand strategy that is based on American Values, blends the United States' advantages, and embraces our many allies rings true in my experience."

—Curtis M. Scaparrotti, General, U.S. Army (ret.) and former
Supreme Allied Commander Europe (SACEUR)

SUSTAINING AMERICA'S STRATEGIC ADVANTAGE

Joel R. Hillison, Jerad I. Harper, and
Christopher J. Bolan, Editors

Praeger Security International

BLOOMSBURY ACADEMIC
NEW YORK • LONDON • OXFORD • NEW DELHI • SYDNEY

BLOOMSBURY ACADEMIC
Bloomsbury Publishing Inc
1385 Broadway, New York, NY 10018, USA
50 Bedford Square, London, WC1B 3DP, UK
29 Earlsfort Terrace, Dublin 2, Ireland

BLOOMSBURY, BLOOMSBURY ACADEMIC and the Diana logo
are trademarks of Bloomsbury Publishing Plc

First published in the United States of America by ABC-CLIO 2023
Paperback edition published by Bloomsbury Academic 2024

Library of Congress Cataloging-in-Publication Data
Names: Hillison, Joel R., editor. | Harper, Jerad I., editor. |
Bolan, Christopher J., editor.
Title: Sustaining America's strategic advantage / Joel R. Hillison, Jerad I.
Harper, Christopher J. Bolan, editors.
Description: Santa Barbara, California : Praeger, [2023] |
Series: Praeger security international | Includes index.
Identifiers: LCCN 2022042693 (print) | LCCN 2022042694 (ebook) |
ISBN 9781440879920 (cloth) | ISBN 9781440879937 (ebook)
Subjects: LCSH: National security—United States. | United
States—Defenses. | United States—Military relations—21st century.
Classification: LCC UA23 .S89 2023 (print) |
LCC UA23 (ebook) | DDC 355/.033073—dc23/eng/20221025
LC record available at https://lccn.loc.gov/2022042693
LC ebook record available at https://lccn.loc.gov/2022042694

ISBN: HB: 978-1-4408-7992-0
PB: 979-8-7651-3303-3
ePDF: 978-1-4408-7993-7
eBook: 979-8-2161-8351-8

Series: Praeger Security International

To find out more about our authors and books visit www.bloomsbury.com
and sign up for our newsletters.

Contents

Introduction

The United States faces an era of increasing competition and declining material advantage in relation to its adversaries. For a nation that has historically relied upon material superiority, this trend demands a novel approach. To meet the demands of a shrinking competitive advantage, the United States must better employ its strategic advantages: an ability to make sense of the international environment, a global network of allies and partners, and intellectually agile and competent strategic leaders and organizations.

In 2020, the U.S. Joint Chiefs of Staff issued guidance for gaining and sustaining intellectual advantage over our adversaries by reforming how the United States educates those leaders. This book contributes to that effort but goes beyond a narrow focus on education. Rather, it supplies insights and analytical frameworks that policymakers, strategists, and warfighters will need to compete in this challenging environment and, if necessary, to prevail in armed conflict. If the nation is to thrive in the future security environment, the United States will need a better understanding of the environment in which the United States competes, how to collaborate with allies and partners, and how to educate strategic leaders.

This is a timely book given the increased pressures to reduce the U.S. military presence across the globe and to reinvest scarce financial resources at home. At the same time, American allies and partners are clamoring for more U.S. engagement and leadership, especially after the Russian invasion of Ukraine in 2022. With increasing competition for limited national resources, the United States will have to be smarter about how it addresses the challenges and opportunities it faces in the world. Wars in Afghanistan and Iraq, as well as competition below the threshold of war, have shown that earlier ways of engaging in statecraft will not always be enough to achieve our desired strategic outcomes. American foreign policy and military leaders must be able to think forward in time and apply multiple analytical frameworks to understand an emerging international security environment and define the problems facing the country in that environment. They must also understand how to work with a diverse range of allies and partners and use the institutions of national security to achieve our national interests. This edited collection will focus on educating policymakers and warfighters for competition as well as conflict in today's security environment.

The book begins by helping the reader understand the global environment in which the United States is competing. In the first chapter, Dr. Christopher J. Bolan collaborates with Dr. Joel R. Hillison to describe how the United States must adapt its grand strategy to succeed in this changing international landscape. The Cold War strategy of containment is ill-equipped to deal with that landscape. In addition to the challenges posed by China and Russia, there are other threats that will require a unifying vision for future U.S. foreign policy. Yet, grand strategy alone is insufficient to sustain the U.S. strategic advantage.

Leadership was an essential element in the dominance of past empires and states reaching as far back as ancient Greece. Developing and applying leader competencies today will be necessary to improve judgment, solve or mitigate wicked problems, and achieve our national interests. Dr. Craig Morrow begins with an examination of strategic leadership in the twenty-first century. He supplies an array of analytical frameworks and competencies that U.S. leaders will need to embrace if America is to keep its competitive advantage in an era of increasing competition and declining relative strength.

In the following chapter, Dr. John A. Nagl and Commander Thomas J. Newman introduce a different analytical framework, based on concepts from *The Infinite Game*, through which to view U.S. grand strategy. Focusing on the challenge presented by the People's Republic of China to the United States, they articulate a new and recursive approach to grand strategy that will perpetuate U.S. strategic advantages as outlined by Bolan and Hillison. Nagl and Newman reiterate the importance of allies and partners and thoughtfully address the risks of unintended escalation of tensions between the United States and China.

As the amount of coverage in this book shows, nothing has changed the current international landscape more than the rise of China. Dr. James A. Frick helps us to understand China's behavior within international institutions, such as the Bretton Woods organizations, which were set up by the United States at the end of World War II and have been a source of stability and U.S. influence ever since. By better understanding China's economic statecraft, Frick argues that the United States can better leverage its advantages within the liberal international order and keep its strategic advantages in that competition.

While there has been a great deal written about how great powers behave in the international system, Dr. John R. Mowchan sheds light on the behavior of smaller states and proposes a model that better explains how small states align in an increasingly competitive international environment. His study on small state alignment builds upon a recurring theme in this book concerning the importance of international relationships with allies and partners.

The first section of the book ends with a chapter by Dr. Brett D. Weigle examining the role of energy in promoting U.S. national security. In that chapter, he explores opportunities for cooperation and areas of competition among the major energy producers and consumers. In doing so, Weigle redefines notions of energy security and discusses both traditional and nontraditional sources of energy.

The next section of the book examines the importance of allies and partners; these partners represent a significant strategic advantage of the United States. Dr. Kevin J. Weddle and Dr. Joel R. Hillison begin by providing lessons gleaned from our experience fighting with allies in World War II to better frame the importance of contemporary alliances, coalitions, and partnerships. By focusing on the evolution of the U.S. relationship with Britain during that conflict, Weddle and Hillison are able to articulate both the advantages and disadvantages of fighting (and competing) with allies and partners.

Professor Mary F. Foster then teams up with Dr. Joel R. Hillison to look at contemporary allies from the North Atlantic Treaty Organization (NATO) and partners in the European Union (EU) in the context of global competition. Using the instruments of national power as a framework, they analyze the United States' relations with both organizations, identifying both the advantages and disadvantages they entail. Foster and Hillison conclude with four possible futures for those relationships.

Shifting from the European theater back to competition in the Indo-Pacific region, Dr. Jerad I. Harper contrasts the institutionalized and multilateral U.S. alliances and partnerships in Europe to the current hub-and-spoke system found in Asia. He briefly explores the history of those relationships as a baseline to suggest a multitiered and overlapping system of multilateral relationships in the Indo-Pacific. Harper then examines the possibilities for expanding and deepening those relationships from existing bilateral alliances (e.g., Australia, Japan, and the Republic of Korea), multilateral partnerships (e.g., the Quad) and other critical relationships (e.g., Vietnam) toward a potential trilateral U.S.–Australia–Japan defense alliance and expanded "Trilateral-plus" and "Quad-plus" partnerships.

In her chapter examining China's Maritime Silk Road, Colonel Heather Levy draws lessons from China's Belt and Road Initiative to suggest how the United States might better shape its own relationships, articulated by Harper, with partners in the region. Her analysis focuses on U.S. and Chinese access to countries within the Association of Southeast Asian Nations (ASEAN), an often neglected but strategically important organization of states in the region. While they may not always align with the United States, those countries can help the United States deter Chinese aggression and build greater influence in other areas of competition.

One of the ASEAN countries highlighted by both Harper and Levy is Vietnam. Colonel Thomas J. Bouchillon, uses Vietnam as a case study to explore how the United States might expand its partnerships in the region. By methodically describing both the obstacles and opportunities for cooperation between the United States and Vietnam, he reminds us of the importance of these smaller, regional countries in any great power competition. Bouchillon also articulates the tensions between U.S. values and interests that Bolan and Hillison introduced at the beginning of the book in dealing with countries such as Vietnam.

Dr. Christopher J. Bolan adds to this section of the book with his insightful analysis of competition in the Middle East. While the United States has been eager to pivot away from this region, he argues that the United States must remain engaged in the Middle East but should fundamentally change its approach. Building upon the strategic vision he outlined with Hillison in the first chapter, he expands the aperture away from a more centralized and militarized approach to the region to a more holistic approach built upon U.S. soft power.

Dr. Kevin D. Stringer follows this up with a look at an important, yet often overlooked, aspect of warfare: irregular warfare. In this chapter, he delves into the poorly understood definition of irregular warfare and how senior level professional military education covers the topic of irregular warfare. Stringer concludes by suggesting a new, irregular warfare–capable structure for the U.S. military.

In the concluding chapter, Dr. Jerad I. Harper delves into the perilous topic of security force assistance. He draws on U.S. experiences in Vietnam to examine issues surrounding the uses, strengths, and pitfalls of security force assistance in a modern strategic context. U.S. experiences, and shortfalls in Iraq and Afghanistan further support Harper's approach to security force assistance. He argues that such lessons apply not only to weak or fragile states but across the continuum to institutionally stronger states such as Japan. The chapter concludes by suggesting opportunities to better enable such efforts in the future. While the U.S. populace is weary of these types of engagements, they will be essential to the United States keeping its strategic advantages for the future.

As we authored this book, the United States faces unprecedented challenges around the globe. An increasingly assertive China flexes its muscles in the South China Sea and toward Taiwan while exerting its growing economic might internationally through the Belt and Road Initiative and the Global Development Initiative. Russia continues to put pressure on both U.S. allies in NATO and partners in Europe, while demonstrating its own ability to challenge U.S. leadership, whether in the Arctic, the Caucasus region, or in the Middle East. Smaller states, such as Iran and North Korea, continue to flout international rules underpinned by U.S. leadership and power. In addition, challenges such as COVID, climate change, migration, democratic backsliding, and pervasive economic imbalances weaken the power of not only the United States but also those allies and partners who share the same values and interests as the United States. While it will be impossible to return to the very brief unipolar moment at the end of the Cold War, the insights contained in this book can help guide policymakers and warfighters as they confront the challenges ahead to U.S. leadership and enable them to sustain America's strategic advantage.

PART I

Understanding the International Environment

CHAPTER 1

Competitive Strategy in a Changing International Landscape

Christopher J. Bolan and Joel R. Hillison

Great power competition is now the central organizing principal of American security strategy. Both President Donald Trump's 2017 *National Security Strategy* and President Joe Biden's *Interim National Security Strategic Guidance* view a resurgent Russia and rising China as the priority threats challenging American interests.[1] The 2022 National Defense Strategy goes further, describing China as the pacing challenge and Russia as an acute threat to U.S. interests requiring a stronger deterrence.[2] This shift from a focus on global terrorism and globalization to renewed concern over great power competition suggests a return to the zero-sum game mentality of the Cold War. This sentiment was reinforced in the wake of Russia's brutal and unprovoked invasion of Ukraine in March 2022; many pundits and former policymakers have concluded that the world is now in the throes of a new Cold War with the potential to end up in another World War.[3]

However, the international context of today's competition with Beijing and Moscow is fundamentally different from that of the twentieth century. Power, in all its manifestations, is much more diffuse, giving nonstate actors and smaller regional states newfound leverage in influencing outcomes. This diffusion of power makes collective action infinitely more complex, even within established alliances and international institutions. Adding to the complexity of the current environment, climate change, infectious diseases, and other emerging transnational security challenges cannot be addressed successfully without the assistance of China and Russia as well as our traditional allies and partners. In addition, the

American and Chinese economies are interdependent in ways unimagina-ble during the Cold War. Open military conflict between great powers or even escalating tit-for-tat economic retaliations now place global economic growth at risk, as we see with the tightening of Western sanctions target-ing Russia in response to Moscow's invasion of Ukraine. Fourth, growing competition with traditional allies in nonmilitary fields such as arms sales, commercial access to emerging markets, and technological innovation coupled with emerging nontraditional security challenges will necessitate that Washington adapt its grand strategy and modernize its already exten-sive global alliances and partnerships. To deal with these challenges and remain focused on vital U.S. interests, the United States will have to estab-lish a strategic vision that sets the parameters of its future foreign policy.

But what should that vision look like? In his famous "House Divided" speech in 1858, Abraham Lincoln delivered his vision to the Republican State Convention for how the United States should deal with slavery and why it was essential to do so. His speech opened with "If we could first know where we are, and whither we are tending, we could then bet-ter judge what to do, and how to do it."[4] This chapter follows a similar approach to describing the grand strategy needed today. It first describes where we are and explores the implications of the major shifts in the global security environment for American national security strategy. It then out-lines the ideological struggle between the United States and its two major competitors: Russia and China. That competition over ideals and national purpose will define how we compete and when we cooperate. Finally, the chapter lays out the foundation of a new grand strategy better suited to the security environment faced by the United States and its allies. That grand strategy is built upon the foundations of the liberal international order that the United States and its allies built in the aftermath of World War II. Shared liberal values and democratic institutions represent the house which currently faces the tempest of a growing diffusion of power, a host of transnational challenges, greater global interdependence, and increasing competition. Just as Lincoln argued that "A house divided against itself cannot stand," so too we argue that the countries and institu-tions of the liberal international order cannot stand divided in the face of these challenges.[5]

The traditional realist approach to growing great power competition that informed containment during the Cold War is ill-suited to the new strategic reality. The liberal approach that underpinned the unipolar moment, after the fall of the Soviet Union, is also insufficient to main-taining that liberal order. Similarly, a constructive approach focused nar-rowly on values that are interpreted differently even among close partner nations does not provide a single coherent alternative. Rather, it is a practi-cal blend of all these insights from international relations theory that offers the most realistic prospect for framing an effective U.S. foreign policy in today's complex and shifting environment. Certainly, where vital security interests are at stake, U.S. policymakers will need to harness the traditional elements of hard military and economic power. Yet, urgent global challenges require taking advantage of the transparency and cooperative mechanisms provided by existing and newly created global, regional, nongovernmental, and private sector organizations. Finally, U.S. policymakers cannot afford to ignore the soft power that derives from America's image (even flawed) as a promoter of basic human rights and representative governance. Achieving the right mix of these insights will not be easy and will require constant evaluation, reappraisal, and redirection. But this is the goal that should preoccupy U.S. policymakers in adapting to the changing circumstances and environments of this emerging new security environment.

WHERE WE ARE: SHIFTS IN THE GLOBAL SECURITY ENVIRONMENT

The Cold War between the United States and the Soviet Union occurred in a vastly different strategic context from what the United States now confronts with China and Russia. One critical distinction is the diffusion of power throughout the global system. The Cold War was principally a competition between two ideological, economic, and military heavyweights, reflecting a largely bipolar distribution of global power. This pitted Western-style democracy and free-market economies led by private sectors against Soviet-style communism and state-controlled economies. Within this prism, international relations were largely viewed as a relatively simplistic game of zero-sum politics—a gain for the United States represented a loss for the USSR and vice versa. The relative balance of global power thus essentially divided the world into two opposite camps with Washington and Moscow competing for friends and allies in this two-sided competition. While there are certainly echoes of this Cold War competition reflected in today's challenge of a rising China and resurgent Russia, there are also fundamental differences and distinctions that should be reflected in U.S. strategies and policies.[6]

In his 2008 *Foreign Affairs* article Richard Haas (president of the Council on Foreign Relations) argued that the world was neither unipolar, bipolar, nor multipolar but rather "nonpolar."[7]

. . . one of the cardinal features of the contemporary international system is that nation-states have lost their monopoly on power and in some domains their preeminence as well. States are being challenged from above, by regional and global organizations; from below, by militias; and from the side, by a variety of nongovernmental organizations (NGOs) and corporations. *Power is now found in many hands and in many places* (italics added for emphasis).[8]

In a similar vein, Thomas Friedman has written several books detailing his central thesis that globalization has empowered individuals, activists, communities, innovators, entrepreneurs, and others at the expense of more traditional hierarchical structures of state and industrial power.[9] Meanwhile Moisés Naím has taken this argument further, suggesting that power is not only more diffuse but is rather decaying, that "power no longer buys as much as it did in the past . . . power is easier to get, harder to use . . . and easier to lose."[10] Globalization has enabled the three revolutions that are central to Naím's theme of decaying power.[11] The "more" revolution (the growth in global population and greater resource availability) has challenged the state's ability to control "people, resources, and land." The "mobility" revolution (the greater mobility of people, goods, and ideas) has both increased global standards of living and increased political instability. Finally, the "mentality" revolution (manifested in the rapid spread of technology, the proliferation and vulnerability of information, and the interconnectedness of ideas) has increased the pace of change and challenged the ability of existing power structures to react.[12]

Contemporary examples of this diffusion of power are abundant and painfully evident in the inability of overwhelming U.S. conventional power to achieve strategic success in either Iraq or Afghanistan despite twenty years of massive diplomatic, economic, and military investments. Stiff Ukrainian resistance against a numerically and technologically superior Russia in 2022 also reflects that diffusion of power. The spread of advanced military and civilian technologies in missiles, drones, and cyber warfare have allowed far weaker competitors (as measured by traditional means of power) to threaten U.S. interests both globally and regionally. Past assumptions of U.S. technological superiority are therefore suspect.[13] Examples include Chinese economic espionage[14] and cyberattacks on U.S. private[15] and defense technology companies,[16] Russian, Chinese, and Iranian hacking of the 2016 US election campaign,[17] attacks by Iranian and Russian-backed militia on U.S. forces in Syria,[18] and ongoing drone[19] and ballistic missile strikes[20] on U.S. forces in Iraq by Iran and Iranian-backed forces. U.S. conventional military superiority alone is simply incapable of easily delivering strategic success against opponents taking advantage of this diffusion of power and employing these asymmetric means of competition. Of course, these same dynamics can also work against China and Russia. In Ukraine, demonstrably smaller and relatively poorly equipped Ukrainian forces have (at least for now) not only stalled[21] Russia's massive invasion but have begun to undertake offensive counterattacks[22] to retake lost territory in certain areas. U.S. political and military leaders are certainly watching these developments closely and drawing lessons that might be useful in bolstering the defensive capabilities of Taiwan, thereby deterring a Chinese invasion.[23]

Yet the diffusion of power impacts the United States differently in ways that will influence the competition with China and Russia. The United States is more economically and culturally connected to the world than either China or Russia. This can yield greater benefits but also present challenges. As a democracy, the United States may react more slowly to challenges as they arise than our autocratic competitors. However, the United States, and its democratic allies, are more adaptable over time and better able to identify problems, and change harmful policies. To take advantage of this position, America must husband its military resources so as not to get caught up in endless and frequent conflicts that do not threat vital U.S. interests. It must also better develop and leverage its diplomatic and informational instruments of power to get ahead of its competitors in the information space and leverage its intelligence assets to support its narratives surrounding critical events. For instance, in advance of Russia's 2022 invasion of Ukraine, the United States smartly highlighted intelligence exposing Russia's false flag operations and consequently preempted Russia's preferred narrative justifying the attack. U.S. policymakers similarly harnessed public statements by celebrities such as China's tennis star Peng Shuai to draw international condemnation of China's treatment of the Uighurs.[24] Gaining an edge in the informational space will require the United States to improve the strategic thinking skills of its civilian and military leaders and to become more proficient in the gray zone competition that both China and Russia are quite good at.

SOMETIMES, COOPERATION WILL BE POSSIBLE: TRANSNATIONAL ISSUES[25]

In addition to the complications imposed by this diffusion of power throughout the international system, new and emerging security challenges will also place a premium on active cooperation among state and nonstate actors to address other transnational security concerns. For example, U.S.–Chinese global cooperation will be necessary to any effort to effectively combat climate change[26] as these two countries are the highest emitters of greenhouse gases, accounting for nearly 40% of total global emissions.[27] As the spread of COVID-19 demonstrated, U.S.–Chinese bilateral cooperation on issues of health security will also be critical in protecting the lives and economic prosperity of Americans at home and of the entire world.[28] Limited cooperation in such areas as nuclear nonproliferation is also essential. For example, Chinese and Russian cooperation in pressing Iranian leaders to resume adherence to limitations on its civilian nuclear program will be essential if Iranian nuclear capabilities are to be contained short of armed military action. Moreover, U.S.–Chinese–Russian cooperation in the sensitive field of cyberspace could increase transparency and help establish international rules of the road in these vital emerging areas upon which international commerce is increasingly dependent and to which critical domestic infrastructures are increasingly vulnerable.[29]

Additionally, even in the military and economic space where competition might well be more prominent, focused cooperation on narrow issues might prove to be mutually beneficial. For instance, while U.S. policymakers resist the spread of Chinese communications firms like Huawei both at home and abroad, the United States could welcome other targeted Chinese investments in Africa[30] and elsewhere in the developing world where the need exceeds the ability and willingness of U.S. firms to invest. Additionally, U.S. policymakers could seek to encourage Chinese naval participation to bolster international security efforts in the Arabian Gulf and antipiracy operations off the horn of Africa.[31] Doing so would ensure China begins to assume a growing share of the burden of providing the stable international security environment from which Beijing has profited so handsomely. This would also give China a larger stake in preserving that stability. In all of these areas, though, negotiations themselves are not the desired end state—in other words, talking for talking sake. Rather, cooperation should be pursued as a way to achieve U.S. interests without seriously compromising the larger geostrategic position of the United States and its allies in other vital areas.

COMPETITION IN A COMMUNITY OF NATIONS: INTERDEPENDENCE AND COMPETITION[32]

There are significant differences between the U.S. competition with China and the competition with the USSR during the Cold War. First, China and the United States have significant economic ties that would be difficult to untangle without a major disruption of the global economy. While both the United States and China would suffer from such a retrenchment, many non-European allies and partners would suffer even more as they are much more dependent upon trade and investment from China. Second, unlike the USSR, China is a participant in many international institutions and has much to lose from starting a conflict with the United States. Third, China has a robust economy that is gaining advantages such as artificial intelligence, green technology, and 5G communications, whereas the Soviet Union was (and Russia remains) heavily dependent upon fossil fuels. Thus, the liberal international order the United States and its allies built after World War II is facing a more serious challenge from a more capable competitor.[33] U.S.–China relations will certainly shape the post-pandemic order just as U.S.–USSR relations did during the Cold War.[34] Nowhere is that relationship more intertwined than in trade.

China is one of the largest trading partners of the United States, after Canada, Mexico, and the European Union (EU).[35] China also holds significant amounts of U.S. debt, only second to Japan.[36] Yet, Chinese investment in the United States, and vice versa, is not as significant when compared with other investing partners. While the United States and China have extensive trade relations, our European allies are even more entangled with Chinese trade and investment.[37] Similarly, while the United States is now a net energy exporter, its European allies are dependent upon Russia for energy. Russia is also a major supplier of key commodities, such as nickel, that are needed for clean energy technology.[38] The developing world (especially sub-Saharan Africa) is also heavily dependent upon Russia for agricultural products, which might limit their willingness to side too closely with the United States.[39] On the other hand, China is also dependent upon Russia and the Middle East for its energy. While U.S. close relations with Israel and the Arab Gulf could give the United States some leverage in deterring Chinese aggression elsewhere, a broader collective approach is needed to address issues such as the conflict in Ukraine and prevent global famine and instability.

While more interdependent with China than it ever was with the Soviet Union, the United States faces rigorous competition from China.[40] Taiwan is a great example of that competition as it is a long-time democratic ally of the United States but is considered by China to be an integral part of its country. In a 2021 speech commemorating the 100th anniversary of the Chinese Communist Party, Xi Jinping stated that "solving the Taiwan question and realizing the complete reunification of the motherland are the unswerving historical tasks of the Chinese Communist Party and the common aspiration of all Chinese people."[41] Meanwhile the United States has an ambiguous commitment to defend Taiwan,[42] giving Beijing a natural advantage of escalation dominance given the asymmetry of interests involved. Nonetheless, if the United States failed to support Taiwan in the event of a major attack or invasion, it would suffer a potential significant setback to its influence in the region and the democratic ideals for which it stands.[43] Such an attack could also represent a severe blow to the United States and global economy.[44]

While China is the major long-term challenge facing the United States, the United States is still vulnerable to competition from a declining yet remarkably assertive and nuclear-armed Russia. The invasion of Ukraine in 2022 exemplifies that competition. At the time of writing, the United States confronts a very realistic prospect of being drawn into unwanted conflicts in two critical areas at the same time: Ukraine and Taiwan. Of course, the U.S. commitment to Ukraine is slightly more ambiguous as Ukraine is not yet a NATO member.[45] While aggressively seeking to bolster Ukrainian military capabilities through the provision of weapons and material, the United States and its allies have been extremely careful to avoid any actions (like establishing a No-Fly Zone) that could lead to a direct U.S. or NATO confrontation with Russian forces. Such a confrontation could spread horizontally (beyond Ukraine) and/or escalate vertically (including use of chemical, biological, or nuclear weapons).[46]

Prevailing in this environment will require the United States to both compete and, where possible, cooperate with China and Russia.[47] Over the past three decades, China has significantly narrowed the power differentials with the United States. By some measures, China now has the world's largest economy, though it is still a middle-income state on a per capita basis. China also has extensive investments abroad, through its Belt and Road Initiative, which extends its political and economic influence while also providing the potential to enhance its ability to project military power.[48] Where the United States retains a significant advantage over both Russia and China is in its global network of allies and partners.

BUILDING UPON STRATEGIC ADVANTAGES: ALLIES AND PARTNERS[49]

The United States retains close to sixty treaty allies worldwide and partnerships with many more states.[50] In spite of foreign policy missteps over the past two decades, many countries still view the United States as their top ally according to a 2019 Pew Research survey.[51] This wide array of partners and allies represents a significant competitive advantage for the United States compared to the relatively weak and unstable allies of China (e.g., Pakistan) and Russia (e.g., Syria and Belarus). Given the increasing military capabilities of China[52] and the challenges posed by Russia's invasion of Ukraine, the United States must carefully nurture these relationships without allowing them to draw it into unwanted conflict.

In Europe, this means supporting more robust EU military capabilities without diminishing the role of NATO for collective defense. This will be easier said than done as the United States must overcome its concerns over Europe's quest for strategic autonomy.[53] In a common refrain, European (and other) allies will have to pick up more of the slack for collective defense.[54] While there is certainly room for improvement in bolstering defense capabilities, a deeper analysis demonstrates that many are already increasing their spending on defense and national security programs.[55] Also, aggressive actions by Russia have prompted Germany—the most powerful economy in Europe—to pledge a massive increase in defense spending.[56] With that increased capability and contributions to collective defense and security, European nations will demand a greater voice.

The United States and the EU will continue to compete in certain sectors (e.g., airlines) and to disagree over other issues such as genetically modified organisms and digital services.[57] However, the United States and EU must work together to protect intellectual property and reduce vulnerabilities in key infrastructure and technologies where China has made inroads. One example of how the United States is trying to coordinate with these contentious issues is the U.S.–EU Trade and Technology Council.[58] The United States also competes with both allies and competitors in the arms industry. The largest arms exporters in the world are the United States, Russia, France, Germany, and China.[59]

In the Middle East, the United States must manage fraught relations and diverging interests with Iran, Saudi Arabia, Turkey, and Israel.[60] Stability in the region requires reducing tensions with Iran so as to avoid a major regional armed confrontation in the immediate term while helping partners transition to a post-oil economy in the longer term.[61] Unexpectedly, the 2022 Russian invasion of Ukraine increased the tensions and mistrust between the United States and both Saudi Arabia and the United Arab Emirates.[62] In addition, many traditional U.S. allies, and hopeful partners, invest in arms from America's competitors such as Russia. For example, both Turkey (a NATO ally)[63] and Saudi Arabia[64] purchased or agreed to purchase S-400 air defense missiles from Russia, straining relations with the United States.

In the Indo-Pacific, leveraging U.S. advantages means modernizing bilateral ties with traditional allies (such as Japan[65] and the Republic of Korea)[66] and solidifying or expanding multilateral relations (such as the security pact with the United Kingdom and Australia)[67] and the Quadrilateral Security Dialogue, called the Quad (which includes Australia, India, Japan, and the United States). The United States must also encourage allies to strengthen their own bilateral security cooperation (e.g., Australia and Japan).[68] Yet, as the nuclear submarine deal with Australia demonstrated, sometimes U.S. cooperation with its allies can hurt relations with other allies.[69] Improving relationships in the region also means closer cooperation with allies, such as the Philippines, and partners like India, Vietnam, and other nations where interests overlap without pressuring them into binding defense commitments that they are not yet ready to assume. For example, one of the key members of the Quad, India, also signed a major trade and weapons deal with Russia in 2021, which complicates U.S. aspirations for closer security ties.[70] Clearly, the United States will have to determine where it can or cannot compromise with its allies and partners in competing with both China and Russia.

Understanding and adapting to these major shifts in the global security environment is essential to developing a competitive strategy for the United States. While some elements of the current strategic environment mirror those of the Cold War, others are clearly different and require a new U.S. grand strategy or strategic vision. The diffusion of power, the proliferation of truly global issues requiring effective and broad international cooperation, deepening economic interdependence, and the need to capitalize on an extensive U.S. network of allies and partners present both challenges and opportunities for the United States. At the same time, the United States cannot wish away competition from Russia and China. Just as NSC-68 outlined a grand strategy to contend with the Cold War, this chapter seeks to establish the foundations of a new grand strategy for the United States in a time of increasing competition, especially from Russia and China, as well as the need for cooperation with those same two states to address some of the challenges outlined previously.[71]

WHITHER WE ARE TRENDING: THE BATTLE OF IDENTITY AND IDEALS

A productive grand strategy begins with a set of "first principles" to guide the difficult decisions and trade-offs required to implement that strategy.[72] These principles, or focal points, can best be found in the self-identification of the United States and its competitors. The United States has always seen itself as an experiment in democracy with the unalienable human rights of life, liberty, and the pursuit of happiness enshrined in the Declaration of Independence. Those core interests have informed U.S. foreign policy ever since. The preamble to the U.S. Constitution begins with the words, "We the people," outlining the purpose of the United States to, among other things, establish justice and promote the general welfare on behalf of and with the consent of the governed. This ideal was evident in President Abraham Lincoln's Gettysburg Address, some sixty years after the Constitution, which ended by describing a government "of the people, by the people, for the people."[73] Almost one hundred years later, NSC-68 articulated the national purpose of the United States as "to assure the integrity and vitality of our free society, which is founded upon the dignity and worth of the individual."[74] That document went on to say that the intention of the United States was to "foster a world environment in which the American system can survive and flourish."[75] President Ronald Reagan later echoed Lincoln's words and NSC-68's national purpose to portray the United States as an example to the whole world when he described the United States as a "shining city on a hill" in his January 25, 1988, State of the Union Address.

The United States still sees itself as the leader of the free world. President Trump, in his "America First" National Security Strategy, claimed U.S. international leadership: "America is leading again on the world stage."[76] He went on to say that "the NATO alliance of free and sovereign states is one of our great advantages over our competitors, and the United States remains committed to Article 5 of the Washington Treaty,"[77] even though he had on other occasions disparaged the alliance and linked U.S. commitment to defend them to whether those nations had "fulfilled their obligations to us."[78] In his *Interim National Security Strategic Guidance*, President Biden "committed to engage with the world once again, not to meet yesterday's challenges, but today's and tomorrow's."[79] Surveys by the Pew Research service demonstrate that support for international organizations is still strong in the United States with support for NATO at 52%[80] and support for the United Nations at 59%.[81] Surveys also demonstrate an increase in favorable ratings of the United States by democratic allies in 2021, with the election of President Biden.[82]

KNOWING YOUR COMPETITION: NATIONAL PURPOSE AND STRATEGY OF RUSSIA

NSC 68 described the USSR's fundamental design as, first and foremost, the survival of the government, requiring "the complete subversion or forcible destruction of the machinery of government and structure of society in the . . . non-Soviet world."[83] Instead of individual liberty and prosperity, the Soviet government valued collective stability, which required extensive control by the government. The document then went on to describe the purpose of the USSR or "the slave state (is) to eliminate the challenge of freedom."[84] Russia's weakening of institutional checks and balances and harsh crackdown on dissent similarly attack the very notion of individual freedom.

Russia, under President Putin, has translated this national purpose into three core objectives: regime survival, end of U.S. hegemony, and restoration of Russia's place in the world.[85] Regime survival is the primary consideration for Putin and his oligarchic allies. Russian leaders seek to rule over their people and exploit the natural resources of the country for their personal benefit and, at their discretion, for the people of Russia. While it is an exaggeration to call Russia a slave state, as NSC-68 did with the Soviet Union, it is certainly an authoritarian state with severe limits on freedom of expression, checks and balances, and other rights such as the right to assemble. As an *Economist* article pointed out, "truth is a risky business" in Russia.[86] According to Garry Kasparov, former world chess champion and Russian opposition leader, Russia's ideology has been "Let's steal together."[87]

To understand Russia's stance on U.S. hegemony, it is important to understand Russia's strategic culture. In her 2016 article, "From Stalin to Putin: Russian Strategic Culture in the XXI Century, Its Continuity, and Change," Polina Sinovets describes Russia's strategic culture:

In order to understand Russia's strategic culture, (it) is necessary to underline that Russia has a mentality of a great state, used to power projection and having a high feeling of vulnerability. To some extent, both categories are interdependent, as the necessity to surround itself with a belt of the buffer states derives from Russian high feeling of insecurity, which gives a good soil for aggravation of its "besieged fortress complex."[88]

From the view of this complex, Russia sees U.S. hegemony as detrimental to Russian interests and ultimate survival. Russia sees itself as a great power, whose role is necessary to maintain a stable, multipolar global order, but not the global order set up by the United States and its allies in the aftermath of World War II. In fact, Russia believes that the United States is in decline and is a cause of international instability.[89] According to President Putin, "The United States has overstepped its national borders in every way."[90] NATO and EU expansion is viewed as another effort by the West to remove Russia's buffer states and thus weaken or destroy Russia and impose its own interests on the rest of the world. U.S. and NATO military intervention in Kosovo, Iraq, Libya, and now Ukraine reinforces this perception. The color revolutions, which Putin blamed on U.S. instigation, also suggest that popular democratic uprisings on Russia's border could cause instability at home.[91] The invasions of Ukraine in 2014 and again in 2022 reinforced Russia's view that limiting the sovereignty of Ukraine is a vital interest and demonstrates the necessity of using force to keep Ukraine, and potentially other states, in its orbit. In Russia's view, only the great powers truly have sovereignty, which is consistent with the view under the Soviet Union.

Russia's grand strategy seeks to regain the status of the USSR and return its near abroad to a subservient position. This goal entails reclaiming much if not all of Ukraine, keeping Belarus as a puppet state, and dissolving or weakening NATO. By maintaining or increasing energy dependence in Europe, especially Germany, Russia hopes to weaken NATO and EU resolve to stand up to Russia. It also seeks to leverage friendly states, such as Hungary, to keep those organizations from reacting too harshly to Russian provocation or decoupling from Russian energy. Russia also seeks to build military and economic ties with Turkey to draw it further away from the United States and NATO. Promoting greater economic and financial ties with China is another way to mitigate vulnerability to U.S., EU, and G-7 sanctions on the Russian economy. This includes granting favorable energy deals to China, agreeing to currency swaps between the ruble and the *renminbi*, and building alternative structures such as the Eurasian Economic Union and the Shanghai Cooperation Organization. Russia's significant nuclear arsenal, large (though less-capable) conventional forces, and vast natural resources such as oil, gas, and nickel provide the means for this strategy.

KNOWING YOUR COMPETITION: NATIONAL PURPOSE AND STRATEGY OF CHINA

China has often been referred to as the Middle Kingdom.[92] This term reflects how current Chinese leaders see their country as the center of the world and thus deserving of certain prerogatives with the states in its gravitational field.[93] After a century of humiliation, in which foreign powers carved up and dominated China, China sees itself as returning to its rightful place in the world. As Odd Arne Westad argued in a 2019 *Foreign Policy* article, "in the party's internal communications, the line is always that the United States is planning to undermine China's rise through external aggression and internal subversion."[94] In addition to these malign intentions attributed to the United States, America's seemingly endless string of economic and foreign policy missteps leads China to believe that it will soon replace the United States as the global hegemon. In that pursuit, China seeks to fully modernize its economy and become a global leader the next two decades.[95]

Like Russia, Chinese leaders are primarily focused on the survival of the Communist Party at home. Also like Putin, Xi has been consolidating power during his tenure. However, Xi has done so while continuing economic growth, increasing military capability, and cracking down on corruption within the Communist Party.[96] In pursuit of the China Dream, Xi seeks to achieve technical dominance in key industries such as artificial intelligence, green energy, and other fields.[97] Through the Belt and Road Initiative, he also seeks to build global interconnectedness with China and maximize the use of excess Chinese capital and construction capacity. Xi has waged a war of ideas both internally, by stoking Chinese nationalism, and externally through a "Wolf Warrior" imagery as captured in films of the same name.[98] And, opposed to the US values of individualism, Xi seeks to promote Chinese values of collective respect, prosperity, and stability over individual political and economic freedoms.

China, like Russia, believes the world should be dominated by the great powers, each with their own sphere of influence. It too would like to maintain a stable, multipolar global order but is more willing to work within the existing institutions of the international order, so long as the rules of those institutions support China's interests. For example, China is seeking greater influence in international economic organizations such as the International Monetary Fund (IMF) and World Bank. At the same time, China is also looking to set up parallel economic organizations that are more in line with their view of political and economic relations, such as the New Development Bank and the Asian Infrastructure Investment Bank. This is consistent with China's vision of itself as a leader of the "global south."[99] Like Russia, China believes in great powers' prerogatives in relationship to their immediate neighbors. In January 2021, President Xi spoke to the World Economic Forum in Switzerland.[100] In that speech, he rejected the notion of universal values and the danger of trying to foist foreign values and systems on others. In other words, great powers should not interfere in the domestic affairs of other great powers.

China also believes that the United States is in decline and has been a cause of international instability, especially when interfering in domestic issues such as the Uighurs, Hong Kong, and Taiwan. China's objective of reunification with Taiwan continues to be a source of tension with the

United States and U.S. allies in the region. China will pursue its political goals through a robust and competitive economic growth strategy and an increasingly assertive foreign policy in the East and South China Sea regions. It is also willing to threaten or use military force in pursuit of its vital interests, specifically the reunification of Taiwan. China's military modernization efforts and growing economy are designed to provide the resources needed for this strategy to succeed. Given its size and growing economic strength, China represents a more formidable competitor to the United States, but also one more interdependent with it and the liberal international order, than Russia or the Soviet Union. Any U.S. grand strategy must address the unique competition and interdependence between China and the United States.

FRAMEWORK FOR A NEW COMPETITIVE GRAND STRATEGY

In their seminal article, "Competing Visions for U.S. Grand Strategy," Barry Posen and Andrew Ross compared four different approaches to grand strategy in the immediate, post–Cold War world.[101] In order to deal with the contemporary security environment, the United States will have to develop a grand strategy that sits somewhere between what Posen and Ross call cooperative security (founded on the school of liberalism) and selective engagement (with roots in realism), as outlined by Posen and Ross. The United States will have to create a hybrid grand strategy of cooperating with authoritarian regimes where important interests coincide (as in Posen and Ross's cooperative security grand strategy) and containing authoritarian regimes where vital interests diverge (as in their selective engagement approach). To maintain its competitive advantage, the United States will also have to work more closely with its allies and partners to maintain and strengthen the liberal international order, embodied in the "rules, norms, and institutions that govern relations among the key players," founded on political (e.g., democratic) and economic (e.g., free market) liberalism.[102]

One of the key features of the global environment is a battle of ideas between individual freedom and state-centric rights or cultural prerogatives. The United States should pursue a grand strategy based on shared values, which promotes individual freedoms and free-market economies; and shared interests, which protect collective security and shared prosperity. This cannot be a purely pragmatic approach where the ends justify the means. The United States and its allies should identify which of its values cannot be compromised and determine the extent to which there is room for trade-offs to promote the greater good. For example, the democratic values (free speech, free press, elections, adherence to the rule of law, and protection of minority rights) that prevailed during the Cold War remain attractive to people everywhere, if not to authoritarian elites. These values should not be readily compromised in the absence of some tangible and overriding contribution to U.S. security interest. However, there are times when the United States will have to deal with some lesser authoritarian regimes in order to check the actions of a more dangerous regime or to deal with a shared security challenge. For example, Western efforts to sanction Russia for its February 2022 invasion of Ukraine included cooperation from authoritarian or authoritarian-leaning governments as well as democratic ones.

Illiberal forces already espouse a competing narrative based on the inviolability of state's rights to sovereignty within their borders and the dominance of collective security and prosperity over individual freedom. Though Russia's invasion of Ukraine puts lie the to these professed values. Yet, these illiberal state and nonstate actors will band together under this narrative to justify and protect their mutually reinforcing spheres of influence.

Building upon this ideological foundation, the United States must continue to grow, modernize, and strengthen its network of allies and partners, as well as its open, democratic political system. This network remains a tremendous long-term competitive advantage for the United States over its adversaries. The United States and its allies must be prepared to balance against the autocratic powers to preserve the liberal international order and offset declining U.S. relative advantage in material strength. A key to this will be transparency in the West and publicly exposing the lies of the two largest autocratic systems (China and Russia) both internationally and to their publics. The public disclosure of Russian aims and false flag narratives in the lead up to and conduct of the invasion of Ukraine provide a perfect example.[103] If democracies can band together, they can outlast the autocracies. As Steven Heffington, a professor of national security strategy at the National War College, wrote in his article about great power competition, "the ability to leverage vibrant and robust governmental and civilian analysis and public debate is an asymmetric advantage of our liberal society."[104]

To build this network of allies and partners, the United States will need to help democratic allies and partners strengthen democratic institutions, improve military capabilities, and promote economic development that increases prosperity and the agency of individuals. These efforts improve U.S. security and the security of those allies and partners over the long term. The democratic peace theory has demonstrated that, overall, mature democracies are less likely to go to war with each other, especially the stronger and more resilient those democracies are.[105] Accepting this theory, the United States should promote a stronger and more autonomous Europe as well as more capable partners in both Japan and the Republic of Korea. Some degree of strategic autonomy is beneficial to both the United States and its allies. The United States also must accept that some of its partners will demand a less comprehensive relationship with it than the NATO allies. Therefore, allies and partners that share democratic values and key interests should not be shunned for pursuing their own path on less important interests. For example, a stronger and more autonomous India, outside of a formal defense arrangement with the United States, is not necessarily a threat to U.S. interests so long as India remains a democracy espousing rights and protections cherished here in the United States and is supportive of vital U.S. interests.

Yet, competition extends beyond the ideological and political space to economics. The United States political system and its more open economy are long-term competitive advantages. What was formerly known as the Washington Consensus still provides an advantage in the long run over more state-centered approaches to economic development.[106] Private ownership, private companies, and largely market forces historically result in a better allocation of resources. The market economies are better able to adjust to new changes and direct resources where needed most. Yet economic policy must be flexible enough to correct for market distortions and inefficiencies along the lines of Keynesian fiscal policies in response to depressions and liquidity traps.

Despite this competition across the political, informational, military, and economic spheres, it is possible to accommodate a rising China and declining Russia without direct conflict with the United States. However, that possibility depends in great deal on the actions and priorities of the governments in those two countries. Cooperation requires the United States to consider both identity, history, and culture in addition to power and interests when dealing with both China and Russia. As Thucydides identified over two millennia ago, fear, honor, and interests play a role in whether states go to war with each other. Understanding China and Russia's perception of self and their identity and history enables us to identify and mitigate actions that instill fear or compromise their honor. Each state sees itself having a privileged place in the world and fears internal chaos as much as any external security concern. Where possible, U.S. policy should consider those perspectives and have a more empathetic view of their possible reactions. The United States should also identify where China and Russia's interests are not necessarily aligned and work to widen any fissures that might bring one of the two more closely in line with U.S. values and vital interests. For example, Russia may bristle at being the junior partner of China and may one day be open to a more benign security relationship with Europe and the United States.

REIMAGINING THE TOOLS OF NATIONAL POWER

As in NSC-68, the new grand strategy must take into account the tools of national power and leverage those where the United States has a competitive advantage. As Joseph Nye argues, power is getting others to do what you want, or preventing them from doing something you do not want, through a combination of persuasion (soft power), coercion, and deterrence (hard power).[107] The United States must also protect its key source of power, the collective will of the United States and its allies and partners to uphold the liberal international order it helped to establish at the end of World War II. To do this, the United States must become more agile in its use of its instruments of national power and invest more heavily and strategically in the nonmilitary instruments of power.

Diplomatic and Informational Power

The United States retains significant advantages in its diplomatic and informational power. Much of the former is represented in the institutions of global governance. The United States, and its allies, make up a commanding majority of the UN Security Council, and the location of the United Nations in the United States gives the U.S. government easy access to the other nations of the world. In addition, representatives of those governments have unfettered access to U.S. civil society. That level of transparency gives the United States a great deal of credibility in dealing with other countries even as it can expose U.S. politics to meddling from outside actors such as China, Russia, and Iran.

As Stacie E. Goddard observed in a 2022 *Foreign Affairs* article about the international system, for all their shortcomings, international institutions dominated by the United States remain a "tool to manage power politics . . . by providing channels of communication, forums for negotiation, and clear rules about what counts as appropriate behavior."[108] In the specific case of Ukraine, the existence of these international and regional institutions such as the UN, the International Criminal Court, EU, and NATO allowed the United States and its allies to rapidly coordinate a range of diplomatic, economic, and military actions to bolster Ukraine's ability to resist Russia's naked aggression. Moreover, even as competition with Russia and China intensifies over the short term, addressing major global security challenges such as climate change, transnational crime, nonproliferation, and the spread of infectious disease will require more than a modicum of global and regional cooperation. Forums provided by international and regional institutions can facilitate this needed cooperation and provide multiple transparent means of exchanging information that could prevent miscalculation and mitigate against further escalation in the event of a crisis.

U.S. soft power also represents an advantage for the United States as represented in informational power. The ubiquitous presence of U.S. culture such as music and movies in addition to U.S. dominance of the Internet represents a strength in democratic societies and in authoritarian societies when their citizens gain access to that information. U.S. release of classified information about Russian intentions, prior to the invasion of Ukraine, provides just one example of how the United States can leverage its capabilities. Still the United States needs to increase its capacity in this area and develop capacity in areas that once were covered by the U.S. Information Agency.

Military Power

The United States retains a significant advantage in the military instrument of power, though that relative advantage may be shrinking. The United States needs to keep its military strong and continue to focus efforts on crisis management and cooperative security. However, it needs to be judicious in the use of military power in combat operations that sap resources, including political will, and tarnish the U.S. image abroad. To be successful, U.S allies will have to increase their own capabilities to internally balance against China and Russia. Russia's invasion of Ukraine appears to have spurred Germany to increase its military spending and capability and has likewise increased dialogue in Japan for the military to take a more active role. In addition, the United States should consider a new Truman Plan for critical emerging and threatened democracies such as Ukraine and Taiwan, contingent upon their own efforts at self-defense and democratic reforms. That assistance should focus on making them resilient enough to resist Chinese and Russian coercion and intimidation.

Moreover, while seeking to safeguard its current advantages in global basing access, combined arms training, logistics, and warfighting experience, the U.S. military will also need to devote greater attention to competition in nontraditional fields such as artificial intelligence, space, drones, cyber warfare, and improving the capabilities of our partners to contribute to global security.[109]

Economic Power

In the economic realm, the United States again retains significant advantages despite the rise of China. First, the United States must work actively and responsibly to promote the dollar as the international currency. The frivolous use of sanctions represents a threat to dollar dominance, as does fiscal irresponsibility in the U.S. budget. The United States needs to continue to advocate for open markets and free trade, but with mechanisms to protect intellectual property, protect critical industries from compromise, and to ensure that the United States and its allies have access to energy. Part of this requires reengaging with the Comprehensive and Progressive Agreement for Trans-Pacific Partnership (CPTPP) and developing an even more comprehensive trade agreement with the EU and democratic nations such as the United Kingdom.

The United States also dominates global financial institutions such as the World Bank, which is historically led by an American, and the IMF, which has been historically led by a European. In addition, the United States has a de facto veto over any major changes to either the IMF or the World Bank, locking in its institutional advantages. The G-7 consists of entirely of U.S. allies, and democracies and the United States allies also represent a majority in the G-20. This combined with the political heft of the European Union provides the United States with a clear advantage over both China and Russia. The United States needs to invest in its parallel domestic institutions to strengthen its ability to use this institutional power. The Departments of Commerce and State, as well as the U.S. Trade Representative's Office, need to have sufficient resources and support to enable diplomacy to move to the forefront in U.S. foreign policy.

RISK ASSESSMENT

A new U.S. strategy will also require a fundamental reassessment of strategic risks. One of the most immediate challenges will be to ensure that any new approach is sustainable. In the wake of the collapse of the Soviet Union, America enjoyed a "unipolar moment" that gave its leaders an unusual cushion for miscalculation and strategic overreach. That margin of error will inevitable grow smaller as competition with China and Russia intensifies and uncertainties about the future persist. As a result, U.S. policymakers will need to be more strategic in setting priorities, allocating limited resources, and seeking targeted contributions from allies and partners. This process should begin with an acknowledgment that great power competition will necessarily prioritize America's attention to both Asia as the emerging geopolitical center of gravity and to Europe as a vital trading partner with whom we have deeply shared democratic values. This does not mean the abandonment of the Middle East, Africa, or Latin America to the predations of others. It does, however, mean that these regions will in effect become secondary theaters of competition.

Making American strategy sustainable also means fostering more mature and genuine partnerships with friends and allies across the globe aimed at achieving a better balance in burden sharing. In the case of Europe, the United States should encourage European initiatives to build military capabilities that complement U.S. strengths and that can be deployed independently without reliance on U.S. logical, intelligence, or other support. In Asia, it means joining agreements like the CPTPP to extend U.S. economic influence and promote U.S. trade and investment abroad. In the Middle East, the United States should promote budding diplomatic engagements aimed at healing regional rifts that fuel civil war and undermine regional integration. For instance, the Abraham Accords normalizing relations between Israel and four Arab states represent tentative steps in easing Arab–Israeli tensions and offer potential venues for coordinating regional security efforts. Similarly, contacts between Iranian and Saudi diplomats hold the potential to ease Sunni–Shi'a sectarian rifts dividing the region and prevent an open military confrontation with Iran in which all parties would suffer greatly.

CONCLUSION

While the challenges are many in the competition between the United States and both China and Russia, the United States remains well-positioned to meet those challenges. By promoting a positive and confident vision of America, both at home and abroad, as a champion of democratic values, the rule of law, and of economic freedom, the United States can continue to successfully compete with both state and nonstate actors. However, it must do so in a nuanced way, incorporating and listening to its many allies and partners. As Hal Brands acknowledged in his 2014 book, *What Is Good Grand Strategy*, U.S. officials will need to "work through problems systematically, to understand the limits and possibilities of U.S. power, and to keep an eye on the long term in dealing with the immediate."[110] This is easier said than done in a democracy, especially one that is politically polarized, especially when it is reliant upon other, equally polarized democracies. However, the United States was able to prevail during the Cold War through the skillful application of both soft and hard power, leveraging the global institutions and extensive network of partners and allies it was able to establish after World War II. Now is not the time to lose heart. As Lincoln stated in the conclusion of his "House Divided" speech, "The result is not doubtful. We shall not fail—if we stand firm, we shall not fail."[111]

ACKNOWLEDGMENT

This first portion of this chapter is based on (and contains significant portions taken verbatim from) an earlier article by the authors: Christopher J. Bolan and Joel R. Hillison, "Managing the Twin Challenges of Competition and Cooperation," *Modern Warfare Institute*, April 19, 2022, and is published with the permission of the Modern Warfare Institute. We would also like to thank Dr. Paul R. Kan and Dr. Jerome Sibayan for their substantial inputs.

NOTES

1. Joseph R. Biden, Jr., *Interim National Security Strategic Guidance* (Washington, DC: The White House, March 2021).

2. *Fact Sheet: National Security Strategy* (Washington, DC: Department of Defense, March 28, 2022).

3. Nathaniel Lee, "The Second Cold War Is Already Beginning, Experts Say, and Many of the Battles Are Being Fought with Economic Weapons," CNBC, March 25, 2022. https://www.cnbc.com/2022/03/25/here-is-how-a-new-cold-war-may-impact-the-us-economy.html; and Elliott Abrams, "The New Cold War," National Review, March 3, 2022. https://www.nationalreview.com/magazine/2022/03/21/the-new-cold-war/.

4. Abraham Lincoln, "House Divided Speech," National Park Service, accessed June 27, 2022. https://www.nps.gov/liho/learn/historyculture/housedivided.htm#:~:text="A%20house%20divided%20against%20itself,thing%2C%20or%20all%20the%20other.

5. *Ibid.*

6. Hal Brands and John Lewis Gaddis, "The New Cold War: America, China, and the Echoes of History," *Foreign Affairs*, November/December 2021, 10–20.

7. Christopher J. Bolan and Joel R. Hillison, "Managing the Twin Challenges of Competition and Cooperation," *Modern Warfare Institute*, April 19, 2022.

8. Richard N. Haass, "The Age of Nonpolarity: What Will Follow U.S. Dominance," *Foreign Affairs*, May/June 2008, 44–56.

9. See for example Thomas L. Friedman, *Thank You for Being Late* (New York: Picador, Farrar, Straus, and Giroux, 2017); *That Used To Be Us* (New York: Farrar, Straus, and Giroux, 2011); *The World Is Flat* (New York: Picador/Farrar, Straus, and Giroux, 2005); *Hot, Flat, and Crowded* (New York: Picador/Farrar, Straus, and Giroux, 2008); and *The Lexus and the Olive Tree* (New York: Picador/Farrar, Straus, and Giroux, 1999).

10. Moisés Naím, *The End of Power* (New York: Basic Books, 2013), 2.

11. Joel Hillison, "Strategic Landpower: Application at the Nexus of Deviant Globalization and Nonstate Actors," in *Landpower in the Long War*, ed. Jason W. Warren (Lexington: The University of Press of Kentucky, 2019), 44–60.

12. Moisés Naím, *The End of Power* (New York: Basic Books, 2013), 53–62.

13. Con Crane, "The Danger of Technological Surprise: Expect the Unexpected or Suffer the Consequences," *The War Room*, January 6, 2022. https://warroom.armywarcollege.edu/articles/tech-surprise.

14. Remarks by FBI Director Christopher Wray, "Responding Effectively to the Chinese Economic Espionage Threat," Washington DC, February 6, 2020. https://www.fbi.gov/news/responding-effectively-to-the-chinese-economic-espionage-threat.

15. Emily Feng, "The White House Blamed China for Hacking Microsoft," NPR, July 20, 2021. https://www.npr.org/2021/07/201018283149/china-blames-united-states-for-cyberattacks.

16. Sean Lyngass, "Suspected Chinese Hackers Breach More US Defense and Tech Firms," CNN, December 3, 2021. https://www.cnn.com/2021/12/02/politics/china-hackers-espionage-defense-contractors/index.html.

17. Kevin Collier, "Russia, China and Iran Launched Cyberattackes on Presidential Campaigns, Microsoft Says," NBC News, September 10, 2020. https://www.nbcnews.com/tech/security/russian-china-iran-launched-cyberattacks-presidential-campaigns-microsoft-says-n1239803.

18. Robert Burns, "US Military Coalition in Syria Takes Out Rocket Launch Sites," *Military Times*, January 4, 2022. https://www.militarytimes.com/news/pentagon-congress/2022/01/04/us-military-coalition-in-syria-takes-out-rocket-launch-sites/.

19. Rachel Nostrant, "Second Drone Strike in Two Days Attempted on US-led Coalition in Iraq," *Military Times*, January 4, 2022. https://www.militarytimes.com/flashpoints/2022/01/04/second-drone-strike-in-two-days-attempted-on-us-led-coalition-in-iraq-report/.

20. Bill Chappell, "What We Know: Iran's Missile Strike Against the US in Iraq," NPR, January 8, 2020. https://www.npr.org/2020/01/08/794501068/what-we-know-irans-missile-strike-against-the-u-s-in-iraq.

21. Anton Troianovski and Michael Schwirtz, "As Russia Stalls in Ukraine, Dissent Brews Over Putin's Leadership," *The New York Times*, March 22, 2022. https://www.nytimes.com/2022/03/22/world/europe/putin-russia-military-planning.html.

22. Yaroslav Trfimov and Mauro Orry, "Ukraine Launches Counteroffensive to Disrupt Russian Supply Lines," *The Wall Street Journal*, May 15, 2022. https://www.wsj.com/articles/ukraine-launches-counteroffensive-to-disrupt-russian-supply-lines-11652531731.

23. Jeffrey W. Hornung, "Ukraine's Lessons for Taiwan," *War on the Rocks*, March 17, 2022. https://warontherocks.com/2022/03/ukraines-lessons-for-taiwan/.

24. Helen Regan, "Peng Shuai Accused a Retired Chinese Communist Party Leader of Sexual Assault," CNN, November 22, 2021. https://www.cnn.com/2021/11/19/tennis/peng-shuai-china-explainer-intl-hnk/index.html.

25. Christopher J. Bolan and Joel R. Hillison, "Managing the Twin Challenges of Competition and Cooperation," *Modern Warfare Institute*, April 19, 2022.

26. U.S. Department of State, "U.S.–China Joint Glasgow Declaration on Enhancing Climate Action in the 2020's," Washington, DC, November 10, 2021. https://www.state.gov/u-s-china-joint-glasgow-declaration-on-enhancing-climate-action-in-the-2020s/.

27. Helen Regan and Carlotta Dotto, "US vs China: How the World's Two Biggest Emitters Stack Up on Climate," CNN.com, October 28, 2021. https://www.cnn.com/2021/10/28/world/china-us-climate-cop26-intl-hnk/index.html.

28. Scott Kennedy, "Advancing U.S.–China Health Security Cooperation in an Era of Strategic Competition," Center for Strategic and International Studies, Washington, DC, December 1, 2021. https://www.csis.org/analysis/advancing-us-china-health-security-cooperation-era-strategic-competition.

29. See for example, Adam Segal, "Project on Cyberspace and U.S.-China Relations," Council on Foreign Relations. https://www.cfr.org/project/cyberspace-and-us-china-relations.

30. Earl Carr, "The US Versus Chinese Investment in Africa," Forbes.com, September 4, 2020. https://www.forbes.com/sites/earlcarr/2020/09/04/the-us-versus-chinese-investment-in-africa/?sh=7ecb505165d4.

31. Matthew G. Minot-Scheuermann, "Chinese Anti-Piracy and the Global Maritime Commons," *The Diplomat*, February 25, 2016. https://thediplomat.com/2016/02/chinas-anti-piracy-mission-and-the-global-maritime-commons/.

32. Christopher J. Bolan and Joel R. Hillison, "Managing the Twin Challenges of Competition and Cooperation," *Modern Warfare Institute*, April 19, 2022.

33. "The World that the West Built after Pearl Harbor Is Cracking," *The Economist*, December 11, 2021. https://www.economist.com/briefing/what-will-america-fight-for/21806660.

34. See for example, Interview with Fareed Zakaria, The Foreign Correspondents' Club, April 14, 2021. https://www.fcchk.org/fareed-zakaria-on-u-s-china-relations-and-the-post-pandemic-world/.

35. U.S. Trade Representative, "The People's Republic of China: Facts," Washington, DC, accessed May 16, 2022. https://ustr.gov/countries-regions/china-mongolia-taiwan/peoples-republic-china.

36. "Which Countries Hold the Most US Debt?" *Fox Business*, June 18, 2020. https://www.foxbusiness.com/markets/which-countries-hold-the-most-us-debt.

37. Silvia Amaro, "China Overtakes US as Europe's Main Trading Partner for the First Time," CNBC.com, February 16, 2021. https://www.cnbc.com/2021/02/16/china-overtakes-us-as-europes-main-trade-partner.html.

38. "The Energy Superpowers," *The Economist*, March 26, 2022, 67–69.

39. "The Food Catastrophe," *The Economist*, May 21, 2022, 11.

40. William Giannetti, "Two Things to Know About the US–China Competition," *War Room*, December 16, 2021. https://warroom.armywarcollege.edu/articles/two-things.

41. Yew Lun Tian and Yimou Lee, "China's Xi Pledges 'Reunification' with Taiwan, Gets Stern Rebuke," Reuters, July 1, 2021.

42. Kevin Liptak, "Biden Vows to Protect Taiwan in Event of Chinese Attack," CNN.com, October 22, 2021. https://www.cnn.com/2021/10/21/politics/taiwan-china-biden-town-hall/index.html.

43. Blake Herzinger, "Abandoning Taiwan Make Zero Moral or Strategic Sense," Foreign Policy.com, May 3, 2021. https://foreignpolicy.com/2021/05/03/taiwan-policy-us-china-abandon/.

44. Keoni Everington, "China's Intent to Seize Taiwan Threatens US Economic Security," Taiwan News, November 5, 2021. https://www.taiwannews.com.tw/en/news/4335857.

45. "US Committed to Ukraine's Sovereignty, Territorial Integrity," Reuters, December 9, 2021. https://www.reuters.com/world/europe/us-committed-ukraines-sovereignty-territorial-integrity-white-house-2021-12-09/.

46. Tracy Wilkinson, "Ukraine Wants a No-Fly Zone. Why Do the US and NATO Reject the Idea?" *The Los Angeles Times*, March 8, 2022. https://www.latimes.com/politics/story/2022-03-08/no-fly-zone-ukraine-russia-biden-nato.

47. Christopher J. Bolan and Joel R. Hillison, "Compete and Cooperate to Make US National Security Strategies Great Again," Foreign Policy Research Institute, March 2, 2018. https://www.fpri.org/2018/03/compete-cooperate-make-u-s-national-security-strategies-great/.

48. Andrew Chatzky and James McBride, "Backgrounder: China's Massive Belt and Road Initiative," Council on Foreign Relations, January 28, 2020. https://www.cfr.org/backgrounder/chinas-massive-belt-and-road-initiative.

49. Christopher J. Bolan and Joel R. Hillison, "Managing the Twin Challenges of Competition and Cooperation," *Modern Warfare Institute*, April 19, 2022.

50. Office of Legal Adviser, "Treaties in Force," U.S. Department of State TIF-Supplement-2022.pdf (state.gov).

51. Laura Silver, "US Is Seen as a Top Ally in Many Countries—But Others View It as a Threat," Pew Research Center, December 5, 2019. https://www.pewresearch.org/fact-tank/2019/12/05/u-s-is-seen-as-a-top-ally-in-many-countries-but-others-view-it-as-a-threat/.

52. Alex Marquardt and Oren Liebermann, "Senior US General Warns China's Military Progress Is 'Stunning' as US Is Hampered by 'Brutal' Bureaucracy," CNN.com, October 28, 2021. https://www.cnn.com/2021/10/28/politics/hyten-stunning-china-military-progress/index.html.

53. Stefano Graziosi and James Jay Carafano, "Europe's Strategic Autonomy Fallacy," Heritage Institute, October 14, 2021. https://www.heritage.org/europe/commentary/europes-strategic-autonomy-fallacy.

54. Lindsey W. Ford and James Goldgeier, "Who Are America's Allies and Are They Paying Their Fair Share of Defense?" Brookings Institute, December 17, 2019. https://www.brookings.edu/policy2020/votervital/who-are-americas-allies-and-are-they-paying-their-fair-share-of-defense/.

55. Marc Selinger, "NATO Allies 'Stepping Up' on Defence Spending, Stoltenberg Says," Janes.com, June 16, 2021. https://www.janes.com/defence-news/news-detail/nato-allies-stepping-up-on-defence-spending-stoltenberg-says.

56. Christian Molling and Torben Schutz, "Unpacking Germany's Billion-Dollar Spending Question," *Defense News*, March 11, 2022. https://www.defensenews.com/opinion/commentary/2022/03/11/unpacking-germanys-billion-dollar-spending-question/.

57. Kati Suominen, "On the Rise: Europe's Competition Policy Challenges to Technology Companies," Center for Strategic & International Studies (Washington, DC, October 26, 2020). https://www.csis.org/analysis/rise-europes-competition-policy-challenges-technology-companies.

58. "U.S.-E.U. Trade and Technology Council (TTC)," Office of the United States Trade Representative, accessed May 25, 2022. https://ustr.gov/useuttc.

59. "SIPRI: Biggest Arms Exporters and Importers," *European Security & Defence*, March 12, 2020. https://euro-sd.com/2020/03/news/16688/sipri-biggest-arms-exporters-and-importers/.

60. Christopher J. Bolan, Jerad I. Harper, and Joel R. Hillison, "Diverging Interests: US Strategy in the Middle East," *Parameters* 50, no. 4 (2020). https://press.armywarcollege.edu/parameters/vol50/iss4/10/.

61. Nader Kabbani and Nejla Ben Mimoune, "Economic Diversification in the Gulf: Time to Redouble Efforts," Brookings Institute, January 31, 2021. https://www.brookings.edu/research/economic-diversification-in-the-gulf-time-to-redouble-efforts/.

62. Karen DeYoung and Missy Ryan, "Ukraine Has Widened the Breach between the U.S. and Persian Gulf Countries," *The Washington Post*, March 30, 2022.

63. Amanda Macias, "Biden Unable to Reach Agreement with Turkey's Erdogan over Russian Missile System Deal during NATO Summit," CNBC, June 17, 2021. https://www.cnbc.com/2021/06/17/us-unable-to-reach-agreement-with-turkey-over-russian-s-400.html.

64. Stephen Bryen, "Why Did Saudi Arabia Want Russia's S-400?" The National Interest, January 17, 2020. https://nationalinterest.org/blog/buzz/why-did-saudi-arabia-want-russias-s-400-114711.

65. Nike Ching, "US Japan Agree to Bolster Security Alliance on Sidelines of G-7," Voice of America News, December 12, 2021. https://www.voanews.com/a/us-japan-agree-to-bolster-security-alliance-on-sidelines-of-g-7/6350778.html.

66. "U.S. South Korea Plan to Deepen Economic, Security Ties," Reuters, May 21, 2021. https://www.usnews.com/news/world/articles/2021-05-21/us -south-korea-plan-to-deepen-economic-security-ties.

67. "Aukus: UK, US and Australia Launch Pact to Counter China," BBC News, September 16, 2021. https://www.bbc.com/news/world-58564837.

68. "Japan, Australia Sign Defence Pact for Closer Cooperation," Reuters, January 6, 2022. https://www.reuters.com/world/asia-pacific/japan-australia-sign -defence-cooperation-pact-2022-01-06/.

69. Sylvie Corbet and Zeke Miller, "Biden Tells Macron US 'Clumsy' in Australia Submarine Deal," AP News, October 29, 2021. https://apnews.com/article/joe-biden -france-paris-australia-emmanuel-macron-cc6b8d85e39e7a66021372313e380b6c.

70. "India Signs Trade and Arms Deals with Russia during Putin's Visit to New Delhi," CNN, December 6, 2021. https://www.cnn.com/2021/12/06/india/india -russia-arms-deal-putin-modi-intl-hnk/index.html.

71. The Executive Secretary, "NSC-68: A Report to the National Security Council," *Naval War College Review* 27 (May–June 1975): 51–60, 87–108.

72. Hal Brands, *What Good Is Grand Strategy?: Power and Purpose in American Statecraft from Harry S. Truman to George W. Bush* (Ithaca: Cornell University Press, 2014), 193. https://doi.org/10.7591/9780801470288.

73. Abraham Lincoln, "Gettysburg Address," November 19, 1863.

74. "NSC-68: A Report to the National Security Council," *Naval War College Review* 28, no. 3 (1975), 54.

75. "NSC-68: A Report to the National Security Council," *Naval War College Review* 28, no. 3 (1975), 68.

76. Donald J. Trump, *National Security Strategy* (Washington, DC: The White House, December, 2017), I.

77. Donald J. Trump, *National Security Strategy* (Washington, DC: The White House, December, 2017), 48.

78. David E. Sanger and Maggie Haberman, "Trump Plays Down U.S. Commitment to NATO and Role Abroad," *The New York Times*, July 21, 2019, A1.

79. Joseph R. Biden, Jr., *Interim National Security Strategic Guidance* (Washington, DC: The White House, March 21, 2021).

80. Moira Fagan and Jacob Poushter, "NATO Seen Favorably across Member States," Pew Research, February 9, 2020.

81. Moira Fagan and Christine Huang, "UN Gets Mostly Positive Marks from People around the World," Pew Research, September 23, 2019.

82. Richard Wike, Jacob Poushter, Laura Silver, Janell Fetterolf, and Mara Mordecai, "America's Image Abroad Rebounds with Transition from Trump to Biden," Pew Research, June 10, 2021.

83. "NSC-68: A Report to the National Security Council," *Naval War College Review* 28, no. 3 (1975), 54.

84. "NSC-68: A Report to the National Security Council," *Naval War College Review* 28, no. 3 (1975), 55.

85. Nataliya Bugayova, "How We Got Here with Russia: The Kremlin's Worldview," Institute for the Study of War, March 2019, 23.

86. "Welcome to the Putin Show," *The Economist*, May 21, 2022, 56–58.

87. Garry Kasparov, "Winter Is Coming," *Public Affairs*, 2015, XXIII.

88. Polina Sinovets, "From Stalin to Putin: Russian Strategic Culture in the XXI Century, Its Continuity, and Change," *Philosophy Study* 6, no. 7 (July 2016), 422.

89. Michael Kofman and Andrea Kendall-Taylor, "The Myth of Russian Decline: Why Moscow Will Be a Persistent Power," *Foreign Affairs* 100, no. 6 (Nov, 2021): 142–152.

90. Putin, "Speech and Following Discussion at the Munich Security Conference," February 10, 2007, in Nataliya Bugayova, "How We Got Here with Russia: The Kremlin's Worldview," Institute for the Study of War, March 2019, 19.

91. Putin, "Speech and Following Discussion at the Munich Security Conference," February 10, 2007, in Nataliya Bugayova, "How We Got Here with Russia: The Kremlin's Worldview," Institute for the Study of War, March 2019, 16.

92. Joseph Esherick, "How the Qing Became China," in *Empire to Nation: Historical Perspectives on the Making of the Modern World*, Rowman & Littlefield, 2006, 232–233.

93. "Xi Jinping and the 'Chinese Dream,'" *Deutsch Weld*, May 7, 2018.

94. Odd Arne Westad, "The Sources of Chinese Conduct: Are Washington and Beijing Fighting a New Cold War?" *Foreign Affairs* (September 2019), 86–95. https://www.proquest.com/magazines/sources-chinese-conduct-are-washington-beijing/docview/2275092890/se-2?accountid=4444.

95. Kevin Rudd, "Xi Jinping and China's Global Ambitions," *Financial Times*, October 22, 2017.

96. Kevin Rudd, "Xi Jinping and China's Global Ambitions," *Financial Times*, October 22, 2017.

97. Central Committee of the CPC, *The 14th Five-Year Plan for Economic and Social Development of the PRC: 2021–2025*, 8, 11.

98. Zhiqun Zhu, "PacNet #26—Interpreting China's 'Wolf Warrior Diplomacy,'" Pacific Forum, May 14, 2020.

99. Stephen M. Walt, "Xi Tells the World What He Really Wants," *Foreign Policy*, January 29, 2021.

100. Stephen M. Walt, "Xi Tells the World What He Really Wants," *Foreign Policy*, January 29, 2021.

101. Barry R. Posen and Andrew L. Ross, "Competing Visions for U.S. Grand Strategy," *International Security* 21, no. 3 (Winter 1996/1997), pp. 5–53.

102. Hal Brands, *American Grand Strategy and the Liberal Order: Continuity, Change, and Options for the Future*. Rand Corporation, 2016, 2.

103. Ken Dilanian et al., "In a Break with the Past, U.S. Is Using Intel to Fight an Info War with Russia, Even When the Intel Isn't Rock Solid," NBC News, April 6, 2022. https://www.nbcnews.com/politics/national-security/us-using-declassified-intel-fight-info-war-russia-even-intel-isnt-rock-rcna23014.

104. Steven Heffington, "Channeling the Legacy of Kennan: Theory of Success in Great Power Competition," *Modern War Institute*, February 8, 2022.

105. See for example, Edward D. Mansfield and Jack Snyder, "Democratization and the Danger of War," *International Security* 20, no.1 (1995).

106. Theodore H. Cohn, *Global Political Economy* (New York: Routledge, 2016), 419.

107. Joseph S. Nye, Jr., *Power in the Global Information Age: From Realism to Globalization* (New York: Taylor & Francis, 2004), 5.

108. Stacie E. Goddars, "The Outsiders: How the International System Can Still Check China and Russia," *Foreign Affairs* (May/June 2022). https://www

.foreignaffairs.com/articles/ukraine/2022-04-06/china-russia-ukraine
-international-system-outsiders.

109. Gregory C. Allen, "Across Drones, AI, and Space, Commercial Tech Is Flexing Military Muscle in Ukraine," Center for Strategic & International Studies (Washington, DC, May 13, 2022). https://www.csis.org/analysis/across-drones-ai
-and-space-commercial-tech-flexing-military-muscle-ukraine.

110. Hal Brands, *What Good Is Grand Strategy? Power and Purpose in American Statecraft from Harry S. Truman to George W. Bush* (Ithaca: Cornell University Press, 2014), 194. https://doi.org/10.7591/9780801470288.

111. Abraham Lincoln, "House Divided Speech," National Park Service, accessed June 27, 2022. https://www.nps.gov/liho/learn/historyculture/housedivided
.htm#:~:text="A%20house%20divided%20against%20itself,thing%2C%20or%20
all%20the%20other.

CHAPTER 2

Strategic Leadership for the Twenty-First Century

Craig Morrow

INTRODUCTION

Sustaining America's competitive advantage requires leaders capable of effectively identifying risks and opportunities, then leveraging the instruments of national power to mitigate those risks and maximize those opportunities. The skills necessary to accomplish this are generally grouped under the rubric of strategic leadership. Although some leadership principles are applicable at all levels, there are several unique aspects of leadership at the strategic level. Although this chapter often focuses on strategic leaders within the U.S. military, the concepts involved are applicable across the entire U.S. government.

At every level, leadership is founded upon a basic set of underlying principles. Among these principles are the need to provide dignity and respect for all people, empowering people to do the right thing and the continuing development of subordinates. Although relevant at all levels, these elements are most applicable to the face-to-face work of first-line supervisors. These direct leaders are task-oriented and work to build cohesion and motivation in teams, with the goal of achieving a specified objective. As leaders progress through their careers, some will rise to higher levels of responsibility, transitioning from direct leadership into the realm of organizational leadership.

Leadership at the organizational level differs from direct leadership in that the leader is transitioning to a less-structured environment in which their responsibilities are less well-defined and their focus shifts from the near-term to the longer term. There are levels of continuity as well as change in the transition to organizational leadership. As with direct leadership, the organizational leader's focus remains primarily internal to the organization. While those in a direct leadership role focus on the execution of assigned tasks, the organizational leader assumes a greater role in planning and directing the organization and thereby determining which tasks will be accomplished. The organizational leader continues to foster unit cohesion and moral but executes this through a focus on organizational climate rather than individual motivation.

As leaders in the government of the United States approach the pinnacle of their careers, a select few will transition from organizational leadership to strategic leadership. These strategic leaders will operate in an environment of increased uncertainty and complexity, one in which there are often no clear solutions and problems are often "managed" rather than solved. Strategic leaders also differ from leaders at lower levels in that they are primarily oriented on the environment external to the organization. While leaders at the direct and organizational levels are charged to accomplish their organizational mission, using resources resident within their organization, the strategic leader operates in a more ambiguous environment where their mandate comes from outside their organization or institution and often requires resources and expertise that are also outside their institution. The strategic leader is future-focused and provides the vision for the organization, looking to ensure the organization is properly aligned with the competitive environment and therefore able to leverage opportunities and cope with challenges the future will present.

Within the U.S. Department of Defense (DoD), the Joint Chiefs of Staff (JCS) published their vision and guidance for professional military education (PME) and talent management to ensure the military of the United States had the leaders necessary to sustain a competitive advantage in the new millennium. Their vision includes a PME system that "identifies, develops, and utilizes strategically minded, critically thinking, and creative joint warfighters skilled in the art of war and the practical and ethical application of lethal military power."[1] Consistent with that guidance, the emerging strategic leaders within the DoD—and, indeed, across the U.S. government—must be self-aware, ethically grounded, strategic thinkers who are effective communicators capable of leading organizational change. This represents the five fundamental competencies required of strategic leaders: strategic thinking, self-awareness, ethics, change leadership, and effective communication. These competencies do not function in isolation; rather, each operates in combination with the others. They both depend upon and reinforce each other: self-awareness facilitates both strategic thinking and ethical reasoning; strategic thinking and effective communications are both necessary for effective change management.

STRATEGIC THINKING

We cannot solve our problems with the same thinking we used when we created them.

—attributed to *Albert Einstein*

Ingrid Bonn has noted that strategic thinking is essential to remaining competitive in the twenty-first century, and organizations that success-fully integrate strategic thinking into their leadership processes create an enduring competitive advantage.[2] Although every strategic leadership competency relates to each of the other competencies, strategic thinking is perhaps more central to effective leadership at the strategic level than any of the other competencies. The challenges faced by strategic leaders tend to be much more intractable than those confronted by leaders at lower levels. The simple reality is that problems with a clear "right" answer tend be solved by leaders at lower levels of an organization. The "wicked prob-lems" faced by strategic leaders require a different approach than the more clear-cut, black-and-white issues these leaders dealt with during the first decades of their careers. The different nature of the issues at the strategic level requires a different type of thinking. Strategic thinking has the goal of facilitating good judgment to inform decision-making and the develop-ment of innovative strategies.[3]

Within the field of strategic leadership, there is general agreement that strategic thinking is crucial.[4] However, there is no widely accepted defi-nition of exactly what constitutes strategic thinking.[5] Several leadership scholars have put forth various perspectives about the nature of strategic thinking, its component pieces, and ways to measure it.[6] There is, how-ever, general agreement that strategic planning is not strategic thinking. The primarily analytical nature of strategic planning contrasts with the nature of strategic thinking, which—in addition to analysis—requires creativity and synthesis. Jeanne Liedtka of the Darden School of Business has noted: "The literature draws a sharp dichotomy between the creative and analytic aspects of strategy-making, when both are clearly needed in any thoughtful strategy-making process."[7] Similarly, Doug Waters, writ-ing in *Joint Forces Quarterly*, suggests that although critical thinking is an essential competency for military officers at every level, at the strategic level critical thinking alone is insufficient. He argues that "creativity and the ability to use systems thinking to holistically assess all aspects of an organization's internal and external key factors are what truly empower effective strategic thinking."[8] Strategic thinking differs from other types of thinking in that it is both an analytical and a creative process that incorpo-rates both a systems perspective and "thinking in time."

As noted at the start of this chapter, a key distinction between strate-gic leadership and leadership at the direct and organizational levels is the outward focus of the strategic leader. Ross Harrison, in *Strategic Thinking in 3D*, notes that effective strategic thinking requires the leader to have a clear understanding of the environment beyond their organization.[9] Olson and Simerson suggest that systems thinking can expand the range of factors a decision maker considers when "identifying options and prioritizing actions."[10] Part of fully understanding the external environment is recognizing the systems nature of the environment and how the relevant systems can be affected to achieve (or, more likely, merely advance toward) a more desirable state of equilibrium. A full understanding of the strategic environment also involves "thinking in time," an appreciation that enhanced decision-making results from the deliberate analysis of relevant history.[11] Strategic thinking requires an understanding of the present situation, a vision for a changed future situation, and an understanding of the past that informs the development of strategy.

Waters has offered a strategic thinking framework that incorporates the substantial complexity of strategic thinking. Waters' model shows the integration of the past, the present, and the future as the strategic thinker seeks to apply historical knowledge to move the organization from its present state to a desired future state. At its very core, strategic thinking begins with thinking in time, seeking to answer the question: "Having seen the future that we want to create, what must we keep from our past, lose from the past, and create in our present, to get there?"[12] The model also depicts a process in which critical thinking and creative thinking are integrated with systems thinking to develop an understanding of the complex environment and the factors that influence their organization and the larger environment.[13] Heracleous has suggested that strategic thinking involves creative, divergent strategies emerging from strategic thinking which are then fused with convergent, analytical thought.[14] The strategic thinking framework depicts this iterative process of creative thinking generating new possibilities and critical thinking analyzing that information to support further creative thinking before ultimately converging on the opportunities likely to produce the most favorable outcomes. Linking back to the need for self-awareness among strategic leaders is the requirement for strategic thinkers to be cognizant of their personal biases and assumptions. Waters' model acknowledges this reality by depicting a strategic thinking model that rests upon a foundation of self-awareness, awareness of cultural influences on thinking, the considerations of ethics and values, and openness to discourse and reflection.

In the development of any national strategy, leaders will apply each element of the strategic thinking framework. Strategy formulation often begins with an analysis of the current situation. This requires an understanding of complex adaptive systems and the ability to recognize the strategic environment as a system, with the United States—itself a system composed of multiple systems (e.g., the DoD)—being just one element of that larger environment. Defining the desired "ends" for the strategy also requires the interplay of creative thinking (envisioning alternate futures) and critical thinking to evaluate the feasibility, acceptability, and suitability of the options generated as a result of that creative thinking. Figuring out the ways to accomplish identified ends requires the same interplay of creative and critical thinking. Deciding the requisite means necessary to implement the various ways might be seen as a strictly analytical endeavor (critical thinking), but this process is also enhanced when creative thinking and thinking in time are incorporated into the process. Assessing the risks associated with a strategy is another element of the model, requiring both creative and critical thinking.

In his 2011 book *Thinking Fast and Slow*, Daniel Kahneman popularized the concept of "system 1" and "system 2" thinking.[15] Although not apparent in Waters' framework, awareness of these different systems of thinking facilitates the effective exercise of strategic thinking. System 1 thinking represents the instantaneous, reflexive thoughts and emotions that come to us without significant cognitive effort (thinking fast). System 2 thinking (thinking slow) is deliberate, effortful thinking that requires cognitive strain.

The value of thinking fast is obvious in many situations, and leaders at all levels tend to employ system 1 most of the time. System 1 thinking allows us to complete the routine tasks of our lives, both personal and professional, without deliberate thought. System 1 thinking also makes routine, bureaucratic tasks and decisions more automatic and less taxing. Additionally, our emotional responses are the result of system 1 thinking. As noted later in this chapter, emotional awareness is a key aspect of effective leadership; a lack of empathy limits the value of any leader. Norris and Epstein have found that system 1 thinking is positively associated with empathy. Norris and Epstein have also found system 1 thinking to be associated with greater creativity—a central element of Waters' strategic thinking framework.[16] One critical weakness of system 1 thinking is its vulnerability to bias. We can think reflexively because system 1 thinking relies exclusively on information already stored in our memory. The data we draw on in that instant may seem relevant, but further reflection might reveal key differences in context that make the information irrelevant. Despite being prone to ill effects of biases and heuristics, system 1 thinking can be a valuable attribute at lower levels of leadership.

System 1 thinking is employed by all of us, every day, regardless of the level where we sit in our institutional hierarchy; however, the value of this reflexive thought diminishes as a leader rises to higher levels. It is the effortful thinking of system 2 that is essential for effective critical thinking and systems thinking—central aspects of strategic thinking. As stated earlier, a rising leader will encounter increasingly complex problems during their career; the number of relevant variables (known and unknown) will increase and the degree of control over many of those variables will diminish. Fortunately, the "wicked problems" encountered at the strategic level typically, although not always, have a longer time horizon which allows the leader to engage in the effortful thinking required.

Mindful of the dual-process nature of thinking, successful strategic leaders can leverage the strategic thinking framework to effectively advance American interests in the twenty-first century. This is the outcome sought by the JCS to develop leaders "who think critically and can creatively apply military power to inform national strategy, conduct globally integrated operations, and fight under conditions of disruptive change."[17]

SELF-AWARENESS

If you know the enemy and know yourself, you need not fear the result of a hundred battles.

—Sun Tzu, *The Art of War*

According to Daniel Goleman, the very best leaders are aware of their own emotional makeup and the emotions of those around them; this competency enhances their ability to accomplish their organization's mission.[18] Many theorists have suggested self-awareness is the essential foundation of emotional intelligence.[19] Although awareness of your own emotions is a key aspect of emotional intelligence, senior leaders must develop not only individual self-awareness but competence in the larger concept of emotional intelligence. Goleman's research suggests that, while technical skills, IQ, and emotional intelligence are all related to excellent performance, "emotional intelligence proved to be twice as important as the others for jobs at all levels."[20] Accordingly, the Officer Professional Military Education Policy (OPMEP) published by the JCS within the Department of Defense lists the production of joint officers with refined interpersonal skills and effective verbal communications skill among the desired outcomes from the military's PME system.[21] The *JCS Vision and Guidance for Professional Military Education & Talent Management* reinforces this requirement, directing PME and joint (JPME) programs to produce graduates who "possess . . . emotional intelligence, and effective written, verbal, and visual communications skills."[22] Appropriate self-awareness underpins each of these larger competencies.[23]

General Ulysses S. Grant used self-awareness and awareness of his soldiers to successfully fight against an enemy that seemed unstoppable in the early years of the U.S. Civil War. Aware of his personal shortcomings, Grant made his friend, John Rawlins, his top advisor to help keep him on track and away from drinking during his stressful battles during the war.[24] Knowing the emotional state of his soldiers before the siege of Vicksburg, he put them to work digging canals to keep them fit and mentally occupied prior to investing the formidable town.[25] Grant's self-awareness improved his decision-making abilities and fortified him for the long and grueling struggle.

Importantly, Goleman's model of emotional intelligence suggests emotional competencies are learned skills that may be developed and improved.[26,27] However, a review of the literature discussing the malleability of emotional intelligence found the results to be equivocal.[28] That review by the RAND Corporation did, however, reveal one element of emotional intelligence, metacognition, for which the literature does appear to support the idea of malleability.[29] Metacognition, defined by RAND as a person's awareness of their own cognitive and problem-solving processes, is a key component of self-awareness.[30] Army Doctrine Publication 6–22 notes "[s]elf-aware Army leaders build a personal frame of reference from schooling, experience, self-study, and assessment while reflecting on current events, history, and geography."[31] The development of self-awareness among the nascent senior leaders in the U.S. military (war college students) is accomplished through individual assessment as well classroom instruction. For example, students at the Air War College are offered an Executive Leadership Feedback Program that provides 360-degree assessment—comparing students' self-ratings with ratings from peers, superiors, and subordinates. This program also provides a comparison of how each student's ratings compare to their fellow students.[32] Similarly, the Army War College's Strategic Leadership Feedback Program provides students a comprehensive assessment of their leadership style and behaviors to foster self-awareness and, consequently, leadership effectiveness.[33] Both programs provide students a one-on-one leadership feedback session with a highly trained staff member. The education of strategic leaders for the armed forces has included various forms of personality assessments over the years; the Myers–Briggs Type Indicator® was long the assessment of choice; however, assessments based on the "Big Five" are now becoming more prominent.[34]

In the classroom, students are presented with relevant theories (e.g., the work of Daniel Goleman) and ways to enhance their individual self-awareness and better focus their development. In their writings on emotional intelligence, Goleman and his colleagues refer to resonant leaders and dissonant leaders. Dissonant leaders are those who fail to connect with others, are insensitive to others' reactions, and create distance between themselves and followers; ultimately resulting in poor performance. Conversely, the resonant leader is attuned to others' feelings, inspiring and uplifting them, creating a more productive work environment.[35] Other research has suggested leaders with greater self-awareness more effectively foster self-empowerment, as well as the empowerment of their colleagues and their organizations.[36] Leaders who are viewed as empowering have been linked to higher levels of employee retention.[37] Emotional intelligence may also enhance decision-making processes, another fundamental element of strategic leadership education.[38]

ETHICS

All of us in the Department of Defense enjoy the highest trust and confidence of the American people because we live by core values grounded in duty and honor that influence how we think and act.

     —former Secretary of Defense Mark Esper, *SECDEF Memorandum*

Ethics relate to leader actions and their relationship to individual and societal values. Values and ideas matter, especially for democracies. America's image as a "city on a hill" or a role model for democratic governance can be a powerful asset in attracting like-minded allies and partners. Alliances, such as the North Atlantic Treaty Organization (NATO), have gone beyond transitory arrangements based on convenience to organizations with shared values, especially under the leadership of the United States. This is one reason countries were clamoring to join NATO at the end of the Cold War, while they were fleeing NATO's counterpart in the Warsaw Pact. Ethical leadership is essential to maintaining those alliances and partnerships that continue to provide a competitive advantage for the United States.

Strategic leaders determine the direction for the institutions they lead. For that reason, it is critical that leaders at the highest levels be guided by the mores, norms, and principles of the society within which they operate. As representatives of the American people, strategic leaders in the U.S. government are guided by a collection of foundational principles, some codified in our Constitution; other, more latent principles, are embedded in the social fabric of the nation. Within the Department of Defense, the various guiding principles are understood to be the basis of what is referred to as "the profession of arms."

The American people empower strategic leaders to act on their behalf. This relationship requires our strategic leaders to earn (and retain) the trust of the American people. That trust-based relationship requires that the moral and ethical conduct of strategic leaders be beyond reproach. Relatedly, within the Department of Defense, the OPMEP provides the JCS vision for JPME being a system that develops joint warfighters "skilled in the . . . *ethical* application of lethal military power [emphasis added]."[39] Ethical leadership at the strategic level involves three interrelated topics: ethical decision-making, personal ethics, and civil–military relations.

Ethical Decision-Making

Societies grant professions a degree of autonomy to be self-policing. The professions determine who is admitted, who is promoted, and who should be removed from the profession. In the case of the military, another obvious example of the autonomy granted by society is the Uniform Code of Military Justice, through which the military services are permitted, by

Congress, to exercise prosecutorial and judicial functions in order to maintain good order and discipline. This autonomy, however, can be revoked. The autonomy the military enjoys requires that society perceives the profession to be competent in the exercise of its expertise and operating in accordance with the values of the profession. Although the U.S. military is routinely ranked among America's most trusted institutions, that trust must be earned continuously.[40] Potential threats to the military's professional autonomy are typified by the perception within the U.S. Congress in recent years that the military had not responded effectively or appropriately to the issue of sexual assault and harassment. This produced an erosion of trust within Congress and resulted in increased Congressional oversight in the form of more restrictive legislation.[41] This also erodes the standing of the United States abroad as a champion of human rights and democratic values.

All leaders must contend with the ethical implications related to their decisions, but leaders at the strategic level are confronted with decisions that are less clearly "right" or "wrong" than those faced by leaders at lower echelons, wrestling with gray-zone issues that might involve choosing between multiple "right" answers or choosing the least bad of multiple bad alternatives. Unfortunately, a small number of strategic leaders in the DoD continue to choose poorly. As noted by Banks, Eckley, and Stackle, "substantiated ethical violations continue to plague the most senior ranks" in the DoD.[42] In 2017 and 2018, forty-nine strategic leaders in the DoD (generals, admirals, and members of the civilian Senior Executive Service) had substantiated ethical violations.[43] While these numbers represent only about 2% of the strategic leadership in the Defense Department, the high visibility of these leaders and the impact of their decisions makes even this small number significant.

Although the DoD continues to produce strategic leaders who demonstrate ethical failings, the department also continues its efforts to better equip its senior leaders for success in this domain. Many variables that have been linked to the exercise of moral judgement are nonmalleable (e.g., sex, nationality, religiosity).[44] However, education (i.e., ethics within a graduate-level curriculum) has been demonstrated to impact ethical perceptions of various situations.[45] Indeed, even a single course of instruction can have a measurable impact on ethical judgement.[46] The education of war college students is intended to foster successful decision-making in the "gray zone" where these future strategic leaders will operate. In making decisions, strategic leaders consider multiple factors (e.g., legal and policy implications, national interests, regional issues, resource constraints); these considerations often overlap with ethical considerations. Although ethics remains a distinct domain within the overall decision-making processes, ethical considerations are considered—not to the exclusion of policy, national interests, or other concerns, but in addition to them.

One means of fostering ethical decision-making at the strategic level is to have leaders examine decision-making through three ethical lenses: consequences (teleology), principles (deontology), and the actor's moral character (virtue-based ethics). In seminar sessions at the war colleges, discussion of the three lenses helps students understand the advantages and disadvantages of each lens. This also serves to foster individual self-awareness of their own tendencies to reason principally with one lens or another. The application of these principles to case studies (e.g., the use of chemical weapons, nuclear weapons, or lethal autonomous weapons systems) serves to further develop the students' understanding.

Personal Ethics

Strategic leaders make decisions—with ethical implications—on behalf of their institutions, but they also make more personal decisions with ethical implications. In teaching personal ethics, both the Army War College and the Naval War College require students to read "The Bathsheba Syndrome" by Ludwig and Longenecker.[47] This article suggests that ethical failure among senior leaders is often the byproduct of their individual success, which they are often unprepared to cope with. A career marked by sustained success affords strategic leaders a degree of freedom not enjoyed by those who are less successful or further down in the organizational hierarchy. Strategic leaders often enjoy relatively unrestrained access to organizational resources; they are also provided with privileged access to people and information. Their pattern of success over many years can lead to complacency and a loss of focus; it can also create an inflated sense of their personal ability to manipulate outcomes. Taken together, all these byproducts of success combine to increase the likelihood that strategic leaders, even those with a highly developed sense of right and wrong, will be tempted by the opportunities presented to them. Understanding of this phenomenon equips the fledgling strategic leader with an awareness of the structures, procedures, and practices which enhance the likelihood of senior leaders falling prey to the temptations they may face. This awareness among war college graduates, many of whom become close advisors of strategic leaders in the Department of Defense shortly after graduation, as well the strategic leaders themselves, helps to reduce overall susceptibility to "The Bathsheba Syndrome."

Civil–Military Relations

Although the military is only one of the many instruments of national power available to the strategic leadership of the United States, it is often seen as a tool of first resort, as evidenced by its frequent employment for both international and, in the case of the National Guard, domestic crises. For this reason, it is important for strategic leaders in the United States to understand the nature of appropriate civil–military relations (CMR). Carl von Clausewitz 's assertion that "War is nothing but a continuation of the political intercourse of governments, with a mixture of other means" makes clear the intimate relationship between the province of the military professional and that of the political leader.[48] The nature of this relationship requires all strategic leaders in the U.S. government—civilian as well as uniformed leaders—to understand what is appropriate within the realm of CMR. As military officers move through their careers, the relationship between the political leadership of the nation and the uniformed leadership of the military transitions from conceptual to practical, and ultimately personal.

The president, as commander-in-chief, has the power to determine when, where, and how the military will be employed. Similarly, Congress retains the constitutional power to declare war, although they rarely use that power. If the military is not seen as being effective in promoting the nation's interests (e.g., Afghanistan), or is not performing in an ethical manner (e.g., Abu Ghraib), both the president and the Congress can act to impose greater control. Both the president and the legislature also have the power to influence strategic leaders on a personal level. Senior leaders in the U.S. government (military and civilian) are appointed by the president, with the advice and consent of the Senate. When the military has had ethical failings as an institution, or these senior leaders have personal ethical failings, both the president and Congress have the power to impose a variety of sanctions. The fundamental nature of CMR in the United States is something that is taught to military officers prior to receiving their initial commission in the military.[49]

Discussion of CMR in the United States during the twentieth century was largely rooted in the ideas expressed by Samuel Huntington in *The Soldier and the State* (1957). While Clausewitz saw a natural mingling of the political and military spheres, Huntington saw a clearer distinction between the two domains. At the outset of the Cold War, with the Soviet Union posing an existential threat to the United States, Huntington felt that the pattern of CMR that best ensured the security of the state was the most appropriate; this led to the theory of what he called "objective civilian control." The objective control theory focused on a professional military—a military that was respected by the civilian leadership (and given the requisite autonomy)—but also a military profession that recognized and respected civilian authority and control. The less desirable alternative envisioned by Huntington would be "subjective civilian control," with the civilian leadership of the state micromanaging the military. In essence, the civilian leadership would decide the objectives and the military professionals would determine and implement the military means to achieve those objectives (or inform the civilian leaders if military means cannot, in fact, achieve the desired objectives).

In *The Professional Soldier* (1960), Morris Janowitz countered Huntington with a theory that emphasized the narrowing gap between the culture of the U.S. military and the society from which it is formed. Janowitz noted that, among other factors, the population of the military (particularly the officer corps) was becoming more representative of the general population and that the nature of modern warfare meant that relatively few of the technical tasks that soldiers performed were exclusive to the military. Janowitz also noted that the exponential increase in the destructive power of warfare necessitated more, not less, political involvement in traditionally military actions. Janowitz's perspective harkens back to the views of Clausewitz and presaged a general turn away from Huntington's theory of objective civilian control toward a greater acceptance of the *need* for the political leadership to be involved in many aspects of military affairs.

Today's military professionals are grounded in a more Clausewitzian view of CMR, as typified by Eliot Cohen in his book *Supreme Command*. Cohen refers to the Huntingtonian theory of CMR as the "normal" perspective, suggesting this is the most broadly accepted view. Examining the relationship between military leaders and their civilian masters during large-scale wars in the nineteenth and twentieth centuries, Cohen shows that the civilian leaders were continually and deeply involved in military affairs. By engaging in what Cohen refers to as an "unequal dialogue," with their military commanders, the political leaders provided a venue for an ongoing discussion of national strategy.[50] It is this deep involvement by the political leadership, Cohen argues, that produced positive outcomes for Abraham Lincoln, Georges Clemenceau, Winston Churchill, David Ben-Gurion, and the nations they each led during the various wars. Informed by an understanding of the military as a profession and senior military leaders as stewards of that profession, strategic leaders appreciate that civil control of the military is not merely a fact derived from the U.S. Constitution—and certainly not a restriction of military autonomy—but a valuable process that produces more favorable strategic outcomes for the people of the United States.

CHANGE LEADERSHIP

If you don't like change, you're going to like irrelevance even less.
            —General Eric Shinseki, former U.S. Army Chief of Staff[51]

While leaders at tactical and operational levels generally focus on the personnel and the missions of their organization ("down and in"), strategic-level leaders spend the majority of their efforts focusing on the external environment and constituents ("up and out"). Indeed, some argue that the distinguishing characteristic of strategic leadership is the requirement to monitor, interpret, and, where possible, influence the external environment. That environment—the competitive context outside of the organization's formal boundaries and, therefore, beyond one's direct control—can significantly affect both current and future choices available to senior leaders and their organizations. This external orientation highlights an organization's struggle for influence, acquisition of scarce resources, and even survival.[52] Another important driver of the senior leader's external focus is the need for the strategic leader to assess the compatibility of their organization with the strategic environment. When a senior leader foresees an incompatibility (e.g., inability to cope effectively with emerging threats), they provide the vision that guides the organization to a new state that restores compatibility. Leaders at lower echelons may *manage change*, but the strategic leader must *lead change*.

In his book *Leading Change: Why Transformation Efforts Fail*, John Kotter notes that today's organizations must continuously change and adapt to stay competitive in the rapidly transforming global environment. Organizational change requires skilled leadership; mere management will not accomplish lasting, systemic change.[53] Consistent with these ideas, the OPMEP lists "recognize change and lead transitions" as an attribute which PME supports by equipping graduates with the ability to "anticipate and lead rapid adaptation and innovation during a dynamic period of acceleration in the rate of change in warfare under the conditions of great power competition and disruptive technology."[54]

Evaluating many organizations undergoing change, Kotter noted areas of commonality among failed change efforts and developed an eight-step model for effective organizational change. Since Kotter first published his treatise on leading change in 1995, his framework for successfully leading organizational change has become the dominant wisdom in this arena.[55] As is done in business schools around the globe, the war colleges have brought Kotter's change model into their curricula. Applying Kotter's model to a military case study, students at the Army War College examine various historical case studies where senior leaders have led successful change efforts. Students may also take part in a change leadership game that tests their ability to apply the concepts of Kotter's change model and highlights some cost–benefit trade-offs that might need to be considered in leading change in the organization.

Consistent with Kotter's change model, Tichy and Devanna have suggested change leadership requires the strategic leader to create a positive vision of the future organization, while also supplying effective emotional support through the change process.[56] Allen and Hill have argued that developing and promulgating a vision for the organization is the most important task of a strategic leader.[57] They note:

Vision provides a sense of ultimate purpose, direction, and motivation for all members and activities within an enterprise. It provides an overarching concept that serves to initiate and then specify goals, plans, and programs. . . . The vision helps to identify what in the environment is important, what requires action, and what action should be taken.[58]

As noted previously, the strategic leader tends to be more outwardly focused than leaders at lower levels. The outward focus of the strategic leader allows them to assess the alignment of the organization with potential threats and opportunities in the competitive environment. It is this environmental scanning and assessment that generates a change vision.

Kotter's change model provides strategic leaders with an evidence-based framework for leading organizational change; it also serves as an example of the intertwining nature of each of the strategic leadership skills covered in this chapter. Kotter's first step (creating a sense of urgency) requires the leader to employ strategic thinking to identify potential threats and opportunities and then effective communication to relay the perceived threat or opportunity to the relevant constituents. The second step (creating a guiding coalition) requires an understanding of the organization as a system (strategic thinking) to find the right people within that system, with the requisite influence (emotional intelligence) to serve as the guiding coalition. The development of a change vision and strategy (step 3 of the Kotter model) also requires effective strategic thinking. After developing the change vision, the strategic leader will need to convincingly communicate that vision (the fourth step of Kotter's model). As Kotter notes, without credible communications "employees' hearts and mind are never captured."[59] Effective communications, important throughout the process, come to the fore here. The next step in the process involves empowering action (step 5). Again, strategic thinking is required to understand which organizational processes and structures need to be influenced (i.e., aligned with the change vision) in order to influence the larger system. This includes strategic communications to stakeholders outside of the military, including Congress, the American people, and the allies and partners of the United States. The iterative nature of strategic thinking becomes patent as the leader continuously checks for barriers or resistance to change within the system.

In addition to strategic thinking, identifying and overcoming resistance to change also requires self-awareness. The effective leader of change must empathize with the position of those in opposition to fully understand the resistance; this is essential to gaining the long-term support of the change vision from all constituents. Self-awareness (emotional intelligence) is also important to the sixth stage of Kotter's model, involving the generation of "short-term wins." By creating wins early in the change process, leaders give their subordinates a feel of victory in the preliminary stages of change.

The penultimate stage of Kotter's model focuses on consolidating gains and producing more change. Self-aware leaders leverage the increased credibility generated by the initial change effort to influence additional nodes of the system. The final stage of the change process looks to anchor the change in the culture of the organization. Effective communication is critical to ensuring the change becomes embedded in organizational culture.

One particular type of change effort relates to organizational culture itself. Wong and Gerras note that organizational culture "provides the underlying foundation for decisions in strategy, planning, organization, training, and operations."[60] The fundamental importance of culture is also reflected in *The Army People Strategy*, which states that "Strategy-culture misalignment results in mission failure."[61] Strategic leaders must be effective in shaping organizational culture to set conditions inside the organization that facilitate strategic success. Strategic leaders lead organizational change whenever the organization is deemed to be ill-suited to the competitive environment. Similarly, if the organizational culture is not aligned with organizational values and/or the values of the larger society, the leader must correct that misalignment. By effectively employing strategic thinking, a self-aware leader can effectively communicate the need for change and produce the organizational change(s) needed to ensure that their organization is properly aligned to exploit opportunities and mitigate threats to the organization (internal as well as external).

EFFECTIVE COMMUNICATIONS

Communications aren't owned by the communication department. You have to have good executives who can and will communicate.

—Jean-Pierre Garnier, CEO of GlaxoSmithKline[62]

Understanding the various elements of effective strategic leadership described thus far can enhance strategic leaders' effectiveness but, to varying degrees, the successful application of any of these skill sets rests upon the leader's ability to effectively communicate, both within the organization and with external constituencies. As President Obama noted in the National Framework for Strategic Communication: "effective strategic communications are essential to sustaining global legitimacy and supporting our policy aims."[63] It is important to clarify some terms relating to strategic communications. One accepted definition for strategic communications, that used within the Defense Department, is useful in this regard:

Focused United States Government (USG) efforts to understand and engage key audiences in order to create, strengthen or preserve conditions favorable for the advancement of USG interests, policies, and objectives through the use of coordinated programs, plans, themes, messages, and products synchronized with the actions of all instruments of national power.[64]

When discussing the various aspects of strategic leadership, the word "strategic" typically describes the elevated level at which the action is taking place. You will notice, however, that the level at which the communication takes place is not an integral part of this definition. Strategic communications are—simply put—communications used to advance a strategy. Looking at strategic communication in more depth, the Principal Deputy Assistant Secretary of Defense for Public Affairs offered nine principles of effective strategic communication: credible (perception of truthfulness and respect), dialogue (multifaceted exchange of ideas), unity of effort (integrated and coordinated), responsive (right audience, message, time, and place), understanding (deep comprehension of others), pervasive (every action sends a message), results-based (tied to desired end state), continuous (analysis, planning, execution, assessment), leadership-driven (leaders must lead communication process).[65]

The "leadership-driven" principle of strategic communications puts senior leaders at the very center of the process. Similarly, the epigraph at the start of this section reminds us that strategic leaders are the focal point of any effective communications strategy. The strategic leader is directly impacted by the "pervasive" nature of strategic communications. Strategic leaders, by virtue of their position, are constantly communicating. Everything that is said or done by a senior leader is a form of strategic communication. This is a double-edged sword that serves as a powerful tool but also places a substantial burden on all senior leaders. Senior leaders must always be cognizant of the fact that everything they say, or fail to say, can be used against them; everything they do, or fail to do, can similarly be used against them.

Beyond their personal actions and statements, senior leaders must coordinate the actions of the various elements of their strategic communications "orchestra." This is necessary to ensure unity of effort. The Office of the Undersecretary for Public Diplomacy and Public Affairs, within the Department of State, is the lead agency for strategic communications within the U.S. government.[66] As such, this office ensures the national strategic communications effort is integrated and coordinated to achieve synchronization across all instruments of national power. This approach must be balanced against the need for proactive and rapid messaging. Coordination takes time, and opportunities to influence the debate early in a crisis can be missed.

To be effective, strategic communications should be an exchange of ideas, seeking mutual understanding. Similarly, the assessment of current messaging impacts the planning and execution of subsequent messaging. The resulting feedback loops represent the dialogue nature of effective strategic communications. The dialogue builds and strengthens relationships that can be leveraged to advance our national interests. The understanding that results in part from dialogue is not only valuable in its own right but also supports the strategic communications principle of credibility. As stated in the National Framework for Strategic Communication, "understanding the attitudes, opinions, grievances, and concerns of peoples . . . is critical to allow us to convey credible, consistent messages, develop effective plans and to better understand how our actions will be perceived."[67] The feedback loops are also essential to support the need for communications to be responsive—communicating the right message to the right audience at the right time and place. All these activities, properly orchestrated by the strategic leader, ensure a results-based effort that moves the institution toward the desired end state.

CONCLUSION

If America is to retain its competitive advantage in an era of increasing competition and declining relative strength, superior leadership will be essential. International competition will require our current strategic leaders, as well as future generations, to effectively lead change and communicate a clear and compelling vision to both domestic and international audiences. Fortunately, the United States has developed the institutions needed to create strategic leaders who both understand and can apply the concepts underlying self-awareness, strategic thinking, ethical decision-making, change leadership, and effective communication. A continuing focus on effective leader development throughout a career of public service is essential to ensure these institutions continue to produce what is arguably America's greatest resource: effective strategic leaders.

While nations can seldom determine the genius of the commander, an all-important component in winning wars according to Clausewitz, they can better certainly take steps to better prepare their leaders to compete and win in the nonlinear and volatile environment of the twenty-first century.

NOTES

1. U.S. Joint Chiefs of Staff, *Developing Today's Joint Officers for Tomorrow's Ways of War: The Joint Chiefs of Staff Vision and Guidance for Professional Military Education & Talent Management* (Washington, DC: Joint Chiefs of Staff, 2020), 2.

2. Ingrid Bonn, "Developing Strategic Thinking as a Core Competency," *Management Decision*, 39, 1 (2001), 63–71.

3. Charles Allen and Stephen Gerras, "Developing Creative and Critical Thinkers," *Military Review* (November–December 2009), 77.

4. Fiona Graetz, "Strategic Thinking Versus Strategic Planning: Towards Understanding the Complementarities," *Management Decision*, 40, 5 (2002), 456–462; Henry Mintzberg, *The Rise and Fall of Strategic Planning* (New York: The Free Press, 2000); Iraj Tavakoli and Judith Lawton, "Strategic Thinking and Knowledge Management," *Handbook of Business Strategy*, 6, 1 (2005), 155–160.

5. Bonn, 63–71.

6. See, for example, Mintzberg; Bonn; Graetz.

7. Jeanne Liedtka, "Strategic Thinking: Can It Be Taught?" *Long Range Planning* 31, 1 (1998), 121.

8. Douglas E. Waters, "Understanding Strategic Thinking and Developing Strategic Thinkers," *Joint Force Quarterly* 63, 4 (2011), 115.

9. Ross Harrison, *Strategic Thinking in 3D: A Guide for National Security, Foreign Policy, and Business Professionals* (Lincoln, NE: University of Nebraska Press, 2013).

10. Aaron K. Olson and B. Keith Simerson, *Leading with Strategic Thinking: Four Ways Effective Leaders Gain Insight, Drive Change, and Get Results* (John Wiley & Sons, 2015), 22.

11. Richard E. Neustadt and Ernest R. May, *Thinking in Time: The Uses of History for Decision Makers* (New York: Free Press, 1988).

12. Liedtka, 123.

13. Waters, 115.

14. Loizos Heracleous, "Strategic Thinking or Strategic Planning?" *Long Range Planning*, 31, 3, (1998), 481–487.

15. Daniel Kahneman, *Thinking, Fast and Slow* (New York: Farrar, Straus, and Giroux, 2011).

16. Paul Norris and Seymour Epstein, "An Experiential Thinking Style: Its Facets and Relations with Objective and Subjective Criterion Measures," *Journal of Personality*, 79, 5 (2011), 1043–1080.

17. Chairman of the Joint Chiefs of Staff, *Chairman of the Joint Chiefs of Staff Instruction (CJCSI) 1800.01F* (Washington, DC: U.S. Joint Chiefs of Staff, 2020), 1.

18. Daniel Goleman, "What Makes a Leader?" *Harvard Business Review* (1998), 93–102.

19. See, for example, Richard E. Boyatzis and Daniel Goleman, *Emotional Competency Inventory* (Boston, MA: Hay Group, 2007); Richard E. Boyatzis and Annie McKee, *Resonant Leadership* (Boston, MA: Harvard Business School, 2005); Cary Cherniss and Daniel Goleman, *The Emotionally Intelligent Workplace* (San Francisco, CA: Jossey–Bass, 2001); Daniel Goleman, *Emotional Intelligence: Why It Can Matter More Than IQ* (New York, NY: Bantam, 1995); Karen F. Osterman and Madeline M. Hafner, "Curriculum in Leadership Preparation," in Michelle D. Young, Gary M. Crow, Joseph Murphey, and Rodney T. Ogawa (Eds.), *Handbook of Research on the Education of School Leaders* (New York, NY: Routledge, 2009), 269–318.

20. Goleman, "What Makes a Leader?" 94.

21. Chairman of the Joint Chiefs of Staff, A–2.

22. U.S. Joint Chiefs of Staff, 4.

23. See, for example, Daniel R. Ames and Abbie S Wazlawek, "Pushing in the Dark: Causes and Consequences of Limited Self–Awareness for Interpersonal Assertiveness," *Personality & Social Psychology Bulletin* 40, 6 (2014), 775–790; John W. Keltner, *Elements of Interpersonal Communication* (Belmont, CA: Wadsworth Publishing Company, 1973).

24. Allen J. Ottens, *General John A. Rawlins: No Ordinary Man* (Bloomington, IN: Indiana University Press, 2021), viii.

25. Bob Higgins, "Viewing the Civil War Through a Geological Window," *Cultural Resource Management*, 25, 4 (2002), 22.

26. Daniel Goleman, *Emotional Intelligence*.

27. Daniel Goleman, "What Makes a Leader?" 8.

28. Susan G. Straus et al., *Malleability and Measurement of Army Leader Attributes: Personnel Development in the U.S. Army*, Research Report, RR–1583–A (Santa Monica, CA: RAND, 2018), 35, as cited in Stephen Banks, David Eckley, and Mark Stackle, "Strategic Leadership Meta-Competencies: Malleability and Measurement" (Integrated Research Project, U.S. Army War College, 2020), 40.

29. Straus et al., 24.

30. Straus et al., 21.

31. U.S. Department of the Army, *Army Doctrine Publication 6–22: Army Leadership and the Profession* (Washington, DC: Department of the Army), 10–5.

32. Air University, *Air University Catalog: Academic Year 2018–2019* (Maxwell AFB, AL: Air University, 1 October 2018), 61. https://www.airuniversity.af.edu /Portals/10/AcademicAffairs/documents/2018–2019_AU_Catalog_20190115.pdf.

33. U.S. Army War College, *Academic Program Guide* (Carlisle Barracks, PA: Army War College), 31. https://www.armywarcollege.edu/documents /Academic%2520Program%2520Guide.pdf.

34. Stephen J. Gerras and Leonard Wong, "Moving Beyond the MBTI: The Big Five and Leader Development," *Military Review* (March–April 2016), 54–57. As of 2020, the MBTI is still in use at the Naval War College.

35. James D. Hess and Arnold C. Bacigalupo, "Applying Emotional Intelligence Skills to Leadership and Decision Making in Non–Profit Organizations," *Administrative Sciences* 3.4 (2013), 202–220.

36. Cam Caldwell and Linda A. Hayes, "Self-Efficacy and Self-Awareness: Moral Insights to Increased Leader Effectiveness," *Journal of Management Development*, 35, 9 (2016), 1163–1173. https://doi.org/10.1108/JMD–01–2016–0011.

37. Janie Bester, Marius W. Stander, and Llewellyn E. van Zyl, *Leadership Empowering Behaviour, Psychological Empowerment, Organisational Citizenship Behaviours and Turnover Intention in a Manufacturing Division. South African Journal of Industrial Psychology,* 41, 1, (2015), 1–14.

38. Hess and Bacigalupo, 202–220.

39. Chairman of the Joint Chiefs of Staff, 1.

40. Gallup Inc., "Confidence in Institutions," *Gallup.com*, June 22, 2007, https:// news.gallup.com/poll/1597/Confidence–Institutions.aspx.

41. John Vermeesch, "Trust Erosion and Identity Corrosion," *Military Review* 93, 5 (September–October 2013), 2–10.

42. Banks, Eckley and Stackle, 57.

43. Inspector General, U.S. Department of Defense, *FY19 Top DOD Management Challenges* (October 15, 2018), 56–66.

44. See, for example, Terry W. Loe, Linda Ferrell, and Phylis Mansfield, "A Review of Empirical Studies Assessing Ethical Decision Making in Business," *Journal of Business Ethics* 25, 3 (2000), 185–204.

45. Yvette P. Lopez, Paula L. Rechner, and Julie B. Olson-Buchanan, "Shaping Ethical Perceptions: An Empirical Assessment of the Influence of Business Education, Culture, and Demographic Factors," *Journal of Business Ethics*, 60, 4 (2005), 347.

46. Peggy Cloninger and T. T. Selvarajan, "Can Ethics Education Improve Ethical Judgment? An Empirical Study," *S.A.M. Advanced Management Journal* 75, 4 (Autumn, 2010), 9.

47. Dean C. Ludwig and Clinton O. Longenecker, "The Bathsheba Syndrome: The Ethical Failure of Successful Leaders," *Journal of Business Ethics* 12 (April 1993), 265–273.

48. Carl von Clausewitz, *On War* (United Kingdom: K. Paul, Trench, Trübner & Company, Limited, 1908), 121.

49. See, for example, paragraph 2–2. "Pre–Commissioning Training and Education" in U. S. Army Cadet Command Regulation 145–3, *Army Senior Reserve Officers' Training Corps On–Campus Training and Leadership Development* (June 18, 2019), 13.

50. Eliot A. Cohen, *Supreme Command: Soldiers, Statesmen, and Leadership in Wartime* (New York: Anchor Books, 2002), 208–224.

51. Fast Company, *The Rules of Business: 55 Essential Ideas to Help Smart People (and Organizations) Perform at Their Best* (New York, NY: Currency/Doubleday, 2005), 7.

52. R. Craig Bullis, "The Competitive Environment," in *Strategic Leadership Course Directive* (Carlisle Barracks, PA: U.S. Army War College, September 10, 2020), 35.

53. John P. Kotter, *Leading Change* (Boston: Harvard Business Review Press, 2012), vii, 3.

54. Chairman of the Joint Chiefs of Staff, *CJCSI 1800.01F*, A–2.

55. Kotter, *Leading Change*, 59–67.

56. Noel M. Tichy and Mary Anne Devanna, *Transformational Leaders* (New York: John Wiley & Sons, 1986).

57. Charles D. Allen and Andrew A. Hill, "Vision," Faculty Paper (Carlisle Barracks, PA: U.S. Army War College, 2012), 1.

58. Allen and Hill, 2.

59. Kotter, *Leading Change*, 9.

60. Leonard Wong and Stephen J. Gerras, "Culture and Military Organizations," in *The Culture of Military Organizations*, ed. Peter R. Mansoor and Williamson Murray (Cambridge, UK: Cambridge University Press, 2019).

61. U.S. Department of the Army, *The Army People Strategy* (October 2019).

62. Paul A. Argenti, Robert A. Howell, and Karen A. Beck, "The Strategic Communication Imperative" in *MIT Sloan Management Review, Summer Edition* (2015), 88.

63. Barack H. Obama, *National Framework for Strategic Communication* (Washington, DC: The White House, 2010), 1.

64. Joint Warfighting Center, *Commander's Handbook for Strategic Communication and Communication Strategy*, Version 3.0 (Suffolk, VA: US Joint Forces Command, 24 June 2010), I–2.

65. Robert T. Hasting, "Principles of Strategic Communication" (Washington, DC: U.S. Department of Defense, August 15, 2008).

66. Joint Warfighting Center, II–1.

67. Obama, 1.

CHAPTER 3

Does Winning Matter?

John A. Nagl and Thomas P. Newman

If the contest is based on interests, tyranny wins.
If the contest is based on values, democracy wins.
　　　　—former Chinese official cited by Matthew Pottinger in *Foreign Affairs*[1]

In August 2021, students shuffled into seminar rooms of Root Hall at the U.S. Army War College to begin a new academic year. They were greeted with live images on classroom televisions of the fall of Kabul to the Taliban and the chaotic yet valiant evacuation of Americans and partners that marked a strategic failure many had spent a significant portion of their careers fighting to avoid. The War College student body consists of senior officers from all U.S. military branches, international fellows from allied and partner militaries, and U.S. government interagency civilians. The curriculum encourages deep personal and professional reflection of its students before they assume positions of strategic responsibility in the national security sphere. The fall of Afghanistan intensified that reflection and reinforced the importance of developing strategic competence.[2]

　　A few weeks following the Taliban victory, a visiting Army general addressed the student body. A student asked the general what he could say to those who had served in Afghanistan and were seeking closure.[3] His answer was not as interesting as the nature of the question itself, which was the natural desire for an end to the conflict. The current and previous U.S. presidential administrations campaigned to end "endless" and "forever" wars. There is an underlying and natural aversion to the idea of indefinite troop deployments into harm's way and political expediency in opposing them. However, that aversion provides future adversaries a solid blueprint for strategic advantage should the U.S. engage in war again—impose costs in blood and treasure while outlasting finite domestic political will.

In his book *The Infinite Game,* Simon Sinek argues this was the case in the Vietnam War, noting that the North Vietnamese committed to an infinite strategy that provided a distinct advantage over an opponent with finite will and resources. He defines two types of games—finite and infinite. Finite games have known players, defined rules, and set timeframes to determine a clear winner, while infinite games have known and unknown players and no rules (or rule-breaking), with limitless time horizons.[4]

In the wake of the Afghan war, America has new opponents or old opponents that have regained prominence. The return to great power competition has come into sharp focus with malign activities by China and Russia against the United States, our allies and partners, and the rules-based international system. The characteristics of the competition align with an infinite game. This chapter examines how concepts from *The Infinite Game* can be applied to grand strategy—the orchestration of all elements of national power to accomplish national objectives.[5] It argues that adopting infinite game principles within U.S. grand strategy will better preserve American security and prosperity.

THE GRAND STRATEGIC IMPERATIVE

The People's Republic of China (PRC) is America's chief competitor and serves as the primary actor for analysis.[6] China employs a grand strategy to achieve its "Great Rejuvenation" centenary goal to overcome what it calls "the century of humiliation" by 2049.[7] To achieve this end, China shifted to blunting U.S. hegemony in the early 1990s following a trifecta of key events that altered China's threat perceptions: the collapse of the Soviet Union, the Tiananmen Square massacre, and the 1991 Gulf War.[8] China grew more aggressive in the wake of the 2008 global financial crisis, which it viewed as a sign of Western vulnerability.[9] Chinese President Hu Jintao remarked in 2009, "the international financial crisis has caused . . . the world economic governance structure to receive a major shock; the prospects for global multipolarity have grown clearer."[10] People's Liberation Army (PLA) Senior Colonel Liu Mingfu, while a professor at China's National Defense University, authored the *China Dream* in 2010, calling for a more aggressive militarized strategy of "displacing" the United States, by war if necessary, making waves inside and outside of China.[11] President Xi Jinping embraced the view openly in 2017, stating, "Let us work together to create a military force for realizing the Chinese dream."[12] Bogged down and largely distracted by conflict in the Middle East over the last twenty years, the United States was strategically slow to adapt to China's goal of displacing U.S. power and influence in the global order.

America has not faced such a formidable rival since the Soviet Union. During the Cold War, the United States outlasted its opponent through a grand strategy heavily informed by *NSC-68: United States Objectives and*

Programs for National Security. NSC-68 recognized the infinite nature of
the contest: "The seeds of conflicts will inevitably exist or will come into
being . . . Not to acknowledge it can be fatally dangerous in a world in
which there are no final solutions."[13] Winning in combat always matters for those involved in the fighting—pursuit of anything less would
be unethical and break trust with the American people. However, at the
grand strategic level, NSC-68 recognized that a military victory by itself
would be futile because of the nature of the contest:

Resort to war is not only a last resort for a free society, but it is also an act which
cannot definitively end the fundamental conflict in the realm of ideas. The idea of
slavery can only be overcome by the timely and persistent demonstration of the
superiority of the idea of freedom. Military victory alone would only partially and
perhaps only temporarily affect the fundamental conflict.[14]

Sinek views the U.S. declaration of victory in the Cold War as a strategic
misstep that blinded the nation to the emergence of new players in a game
that does not end, writing, "hubris increases the chance that any weaknesses
our organization may have are left open to exploitation by other players . . .
which contributes to the draining of will and resources we need to stay in
the game."[15] Despite the lack of a U.S. peer competitor, Al Qaeda shattered
America's sense of invulnerability on 9/11, triggering two costly decades
of protracted conflict while China watched and gained power. China was
deft at concealing a long game that caught the United States flat-footed, but
long games should not be confused with infinite ones. An infinite mindset
perpetuates strategic advantage by dismissing the noble but naïve idea of
"winning the peace," rejecting the very notion of finish lines.

DEFINING AN INFINITE MINDSET

The infinite mindset must be defined before examining how an infinite
approach to grand strategy can better serve American interests. Sinek
compares and contrasts the characteristics of infinite versus finite thinking of various leaders and the impact of both approaches on the long-term
health of organizations. Finite-minded leaders seek to "win," often for
personal benefit, with little regard for second- and third-order effects on
the organization or nation they lead.[16] Infinite-minded leaders perceive
competition as unending with no triumphalist victory; instead, they focus
on leaving the nation on better ground for the next generation.[17] Infinite-
minded leaders view their primary role as stewards and are less prone
to "ethical fading" in the pursuit of strategic objectives.[18] Additionally,
finite-minded strategic leaders often become overly focused on what they
oppose rather than placing strategic effort on what they are for—what
Sinek terms a "just cause."[19]

In geopolitical terms, the just cause is the unifying purpose of the nation, the overriding strategic objective that should inform ways and means. An effective just cause is inspiring, challenging, and ultimately aspirational—an endless pursuit, not a final destination.[20] It must be powerful enough to sustain will and commitment of resources across generations. For those in government, it is worthy of career-long effort by national security professionals. America already has a just cause based on the most powerful idea in the world, and authoritarian rivals fear it.[21]

The just cause is articulated in the U.S. Declaration of Independence: "We hold these truths to be self-evident, that all men are created equal, that they are endowed by their Creator with certain unalienable Rights, that among these are Life, Liberty, and the pursuit of Happiness."[22] Cynics may argue that lofty idealism has no place in serious grand strategic design, but there is solid precedent for the direct linkage. NSC-68 aptly noted, "In essence, the fundamental purpose is to assure the integrity and vitality of our free society, which is founded upon the dignity and worth of the individual."[23] American values include freedom, democracy, and the rule of law. Interests include national wealth, security, power, and influence. A truly just cause elevates enduring values over interests, or more correctly, better secures national interests through the primacy of values.

Failure to live up to our founding values is part of the American story, and rivals often seize on those events. Foreign malign actors increasingly exploit lingering American political and cultural fault lines through information operations, taking advantage of the nature of a free and open society to destabilize the nation from within.[24] Infinite-minded leaders contribute to national resiliency through tireless efforts to advance the just cause. Democratic institutions, enshrined in the U.S. Constitution, provide checks and balances for national resilience in periods of democratic backsliding when finite-minded leaders aggressively pursue personal or national interests ahead of values.[25] As former Vice President Mike Pence reflected, "if we lose faith in the Constitution . . . we'll lose our country."[26] Malign activity to undermine democracy is the most insidious threat to American survival. It will continue because the just cause is also America's grand strategic center of gravity—a "game over" target for determined rivals.

Infinite-minded players also embrace a concept Sinek terms "worthy rival." A worthy rival is a formidable competitor, friendly or adversarial, in the game that warrants careful study. The goal is not to win a finite victory over them, but to achieve or sustain relative advantage.[27] The quality of the just cause combined with the selection of worthy rivals informs strategic design. Sinek notes, "We get to choose our worthy rivals and we would be wise to select them strategically . . . because there is something about them that reveals our weaknesses and pushes us to constantly improve."[28] The United States has no shortage of rivals; while Russia meets "worthy rival" criteria, it pales in comparison to a China that outpaces all others in terms of the scale and scope of challenge to U.S. global hegemony, a relationship made more complex through deep economic interdependence.[29] China and the United States share common interests, but our national values are diametrically opposed. Managed well, the bilateral relationship can extend the post–World War II period of relative peace and prosperity in the Asia–Pacific region; managed poorly, the relationship will deteriorate from strategic competition into armed conflict with devastating consequences for both nations and the world. An infinite mindset deeply rooted in the advantage of American values holds the key to securing the former outcome rather than the latter.

BENEFITS OF AN INFINITE-BASED STRATEGIC APPROACH

The Russian invasion of Ukraine brought the enduring contest between democracy and autocracy back into view on the world stage and unified Europe, the North Atlantic Treaty Organization (NATO), and the United States around a just cause. A joint statement by the United States and European Union read, "we are united in our resolve to defend our shared values, including democracy, respect for human rights, global peace and stability, and the rules-based international order."[30] The benefits of an infinite-based approach are playing out in real time through the unified Western response. Russian President Vladimir Putin, an archetype of finite-minded leadership, largely miscalculated his ability to exploit European divisions and underestimated Europe's willingness to endure hardship in defense of the free world. The unified Western response, combined with Ukraine's resolve to fight against Putin's expansionist ambitions, turned Russia into a weakened and isolated pariah state almost overnight.

In contrast to Western unity and resolve, China's "no limits" friendship with Russia has to date largely failed, demonstrating Chinese reluctance to fully ally with Russia at the expense of lucrative trade relations with both Europe and the United States.[31] The invasion also shone a spotlight on continued foreign malign influence operations.[32] Pro-Kremlin disinformation, mainstreamed to millions of Americans by some U.S. media outlets, fed a growing domestic extremist nexus of QAnon followers, Capitol attack supporters, and Putin apologists.[33] In this multidomain contest between democracy and autocracy, an infinite-based grand strategy preserves our comparative strategic advantages, expands our will and resources, and ensures national resilience.

NSC-68 was clear-eyed on the stakes of the Cold War and rooted strategy in infinite concepts. But it focused solely on the Soviet Union. This approach is ill-suited for a modern threat environment that includes state and nonstate actors with increasing existential reach and capability. Still, China stands out head and shoulders above all. In 1999, the PLA published a book entitled *Unrestricted Warfare* by Colonels Qiao Liang and Wang Xiangsui that detailed the changing character of the contest:

As the arena of war has expanded, encompassing the political, economic, cultural, and psychological spheres, in addition to the land, sea, air, space, and electronic spheres, the interactions among all factors have made it difficult for the military sphere to serve as the automatic dominant sphere in every war. War will be conducted in nonwar spheres. . . . If we want to have victory in future wars, we must be fully prepared intellectually for this scenario.[34]

Two decades later, the head of U.S. domestic law enforcement, FBI Director Christopher Wray, observed the success of the Chinese approach: "There is no country that presents a broader, more severe threat to our innovation, our ideas, and our economic security than China."[35] China masterfully pursued its strategy of nonwar aggression and has expanded its global reach through its robust Belt and Road Initiative. But the superiority of American democracy over China's closed authoritarian regime is evidenced by China's adoption of quasi-capitalism and reliance upon the theft of American intellectual property and the free access to the global commons provided by American security forces to fuel an ascendancy that would not have been possible otherwise.

While formidable, China is not invincible. It faces its own set of internal challenges that strain its path to rejuvenation politically, economically, demographically, and geographically.[36] Instead of remaining fixated on rival strengths, the infinite approach pivots to building upon America's considerable comparative strategic advantages rooted in a functioning democracy.[37] America's advantages over China include the free flow of ideas and capital in open and civil society, unfettered institutions of higher learning that feed technological innovation, and a free-market economy (and U.S.-dollar-based world reserve currency) that secures economic prosperity. The Chinese communist system cannot match U.S. innovation and growth, and China rejects the just cause that sustains it.

Therefore, all threats to American democracy pose a risk to vital interests. The just cause is a shared lens for multidomain threat recognition across the diplomatic, informational, military, economic, financial, intelligence, and law enforcement (DIME-FIL) spheres of national power. The democratic values central to the just cause provide a grand strategic frame of reference to accurately assess that death threats designed to intimidate honest election workers at home pose as great a risk to vital U.S. interests as militarized Chinese outposts in the South China Sea.[38] The infinite approach safeguards and advances the sources of enduring American power. As Ryan Hass, former China director on the National Security Council staff, warns,

In recent years, the United States has been squandering its sources of strength: a strong global network of alliances, international prestige, and national purpose in addressing America's own shortcomings. By inconsistently standing up for American values and individual liberties, frequently showing sympathy for autocratic behaviors abroad, picking fights with erstwhile allies . . . and deepening social divisions at home, the United States has been undermining its own advantages in its competition with China.[39]

The second benefit of an infinite approach is the expansion of will and resources through the attraction of allies and partners. The global threat environment requires that America stand with her allies around the globe. Commanding General of U.S. Army Pacific Charles Flynn articulated this well:

The future fight will be global, it'll be multidirectional, it'll be multidimensional, and it'll be multi-domain. If we fight the next war domain on domain, we may not like the outcome. But if we fight across all domains with our allies and partners . . . demonstrating our ability to make that problem more complex and harder every day, that is the core of integrated deterrence, and there is no adversary . . . that can match that teamwork.[40]

The Western response to Russian aggression demonstrates the power of unified and expanding teams of teams (state, nonstate, treaty alliances, private sector, and individuals) working in concert and autonomously across all domains to support Ukraine and impose costs on Russia.[41] The just cause provides a North Star—an organizing principle for disparate benevolent actors addressing the threat of autocracy at home and abroad.[42] Credible combat power postured forward remains essential to conflict deterrence, but it is one element of many.[43] It is too early to determine how the unified response to Ukraine will impact China's thinking on the Taiwan question.[44] However, Russia's loss of face, domestic instability, and cratered economy should give pause to a Chinese Communist Party (CCP) leadership in pursuit of national rejuvenation and obsessed with regime survival.

In the infinite approach, *how* the game is played matters. Just cause principles inform and expand strategic ways and means. Rush Doshi, founding director of the Brookings China Strategy Initiative, notes: "American openness attracts the allies that sustain the global liberal order . . . US soft power flows from the country's open society and civic creed."[45] When America stays true to its values at home and abroad, efforts to build a free and open Indo-Pacific region with allies and partners gain traction—not only as an anti-China bloc based on the finite thinking of *opposition*, but as a coalition of like-minded nations standing together to *advance* mutual peace and prosperity. China is also a benefactor of regional stability when it behaves as a responsible power within the international order, but it shows no signs of doing so.

China's aggressive actions in the South China Sea and toxic style of "wolf diplomacy" are damaging its global image.[46] Even before the current Ukraine invasion, former Secretary of State Madeleine Albright argued that global despots were losing ground: "They blew it . . . unflattering views of China are at a historic high, and a median of 74 percent of those polled reported that they had no confidence in Russian President Vladimir Putin to do the right thing in world affairs."[47] Putin's finite approach spurred the expansion of NATO with the addition of Finland and Sweden, longtime independents who abhorred Russian violation of international norms and saw value in joining a coalition of like-minded partners. While they have different forms of authoritarian systems, China and Russia play the game in similar ways that engender global distrust. China has contracts, Russia has oil, and the United States has allies who share our values; the difference should not be underestimated in the infinite game.

The third benefit of adopting the infinite approach is national resiliency through resistance to what Sinek terms "ethical fading."[48] Ethical fading occurs whenever the nation fails to live up to its values at home or abroad. Periods of ethical fading weaken the nation and provide exploitable threat vectors to adversaries, whether they call attention to formerly sanctioned racial discrimination policies or photos from the mistreatment of prisoners at Abu Ghraib. There have been many periods of ethical fading in American history, and they should be studied, but not in a vacuum. Contrasting America's failures with its work to overcome them demonstrates the power of the just cause intended by the Founders to continually work to form a more perfect union.[49]

For example, the evils of slavery gave way to a Civil War to end that atrocity, followed by the Jim Crow era, which gave way to the Civil Rights movement. In contrast, China's Cultural Revolution gave way to the Tiananmen Square massacre, followed by the betrayal and repression of the citizens of Hong Kong and ethnic Uyghur genocide in Xinjiang, which may be followed by the slaughter of countless civilians in Taiwan. America is resilient because it can change and adapt for the betterment of all its citizens by advancing the just cause, willingly placing power in the hands of the many, not the few. Yet, a closed authoritarian regime like China remains in *stasis*, pursuing the narrow self-interests of regime survival at the expense of human freedom and dignity. It is a finite approach to an infinite game. Imprisoned Chinese citizen and Hong Kong democracy activist Joshua Wong sums up this commonality of autocracies well:

Their motivation is singular: self-perpetuation. To consolidate and maintain power domestically these regimes have shown no compunction about crushing dissenters, crippling civil society and removing other obstacles that stand in their way. Outside their borders they flex their military muscles to make a show of strength abroad . . . to impress their home audiences . . . this two-front strategy is the only way for them to retain power.[50]

America is only as strong as its civil society and the institutions that underpin democracy. Democracy can be messy, and no leader is perfect. Ethical fading is rooted in the self-delusion of doing what one believes to be right at the time, often through rationalization.[51] Ethical fading weakens a nation strategically. A modern example was the decision to invade Iraq pre-emptively in 2003. It squandered global goodwill from a broad range of allies and partners in the wake of 9/11, diverted desperately needed resources from the war in Afghanistan, and hampered America's ability to modernize the military for conflict against near peer competitors. It also provided Russia and other autocratic regimes an ongoing propaganda opportunity to justify violating the territorial integrity of sovereign nations under false pretexts.[52]

Also damaging are periods when a finite-based strategy doggedly pursues national interests at the expense of values for sustained periods, weakening American strategic advantage. These periods are marked by retrenchment and retreat from global leadership; they may also include extorting, bullying, or abandoning allies and partners. Will and resources are reduced during these periods by putting America at risk of having to act alone in a crisis or conflict. However, the hope of historical allies and partners that America will eventually return to its core strengths through the democratic process rooted in the just cause provides a measure of resilience.

Democracy sadly also allows for more extreme cases where elected leaders may be ethically bankrupt. They abuse the powers of their office for personal gain or actively seek to weaken democratic institutions from within to consolidate and hold on to power.[53] An infinite-based approach weathers these dangerous periods through a shared national understanding of the just cause. The North Star attracts and self-organizes unity of effort across civil society and government to expose authoritarian schemes and preserve American democracy by upholding and defending the Constitution.

MANAGING RISKS OF AN INFINITE-BASED APPROACH

While the war in Ukraine provides insights into the benefits of an infinite approach, it also highlights risks. The risks of an infinite approach include inadvertent escalation, democratic system exploitation, and finite will and resources.

The courage to uphold the just cause on the world stage comes with actual costs; the second- and third-order effects internationally and domestically for nations standing up to Russian aggression remain to be seen. Also unknown is how well the risk of escalation will be managed. Infinite-based strategy leans into, not away from, ideological differences. It incorporates ideals into ways and means and seeks balance through strategic deterrence and patience, as NSC-68 did well. Maintaining that patience in the face of endless streaming of atrocities in the modern information age and the resultant pressure on leaders to act is difficult. The strain of the largest refugee crisis since World War II, inflation, and soaring energy costs will provide ripe fodder for populists seeking to exploit the current situation in personal quests for power. It remains to be seen whether Ukraine and the Western Alliance can sustain unity and resolve in the face of these costs.

Intense focus on China as a worthy rival may heighten the risk of inadvertent escalation. Concern over China's rapid economic rise and militarization can send U.S. threat perceptions into overdrive without due consideration of weaknesses. This risks what professor of government and founding dean of the Harvard Kennedy School Graham Allison termed the "Thucydides trap."[54] Additionally, as in the case of NSC-68, the contest with China is deeply ideological, which could push opponents into hardened stances rooted in dogmatism. Allison sees value in understanding that the struggle between the United States and China is ideological at its core but does not believe that this makes conflict a *fait accompli*, writing that in the Cold War, "harsh depictions did not freeze meaningful contact, candid conversation, and even constructive compromise."[55]

Further, concern over the threat from China has become a rare area of bipartisanship on Capitol Hill. Expressions of solidarity with Taiwan are growing from members of Congress, including signs of shifting from the long-standing policy of "strategic ambiguity" to "strategic clarity" on American commitment to defend Taiwan if attacked.[56] This bipartisanship reflects in defense spending of nearly $800 billion a year, heavily focused on preparing for peer fights. Additionally, China-bashing has become politically advantageous on both sides of the aisle.[57] Amid politically heated rhetoric, the nuance of the infinite approach may be lost; the worthy rival exposes blind spots while illuminating comparative advantages. The result should not be hostility; the rival makes the infinite player better. China represents a "best in class" authoritarian threat, but the struggle is between freedom and authoritarianism, not between China and the United States. Failure to adapt approaches to the broader threat at the grand strategic level makes America vulnerable to exploitation through other threat vectors. American grand strategy must be holistic in threat perception and develop associated ways and means to counter those threats.

A second potential risk for escalation is overly aggressive attempts to advance or enforce just cause principles abroad. An American lesson of the last twenty years is the limits of military power when given unclear, shifting, and unrealistic political objectives in war.[58] While rooted in an aspirational just cause, infinite-based strategies must also be pragmatic. In Ukraine, this includes robust economic sanctions against Russia and lethal military aid while managing concerns about escalation to avoid broader war in Europe, including the risks of nuclear engagement. Escalation management resulted in the rejection of Ukrainian requests for a NATO-enforced no-fly zone despite gut-wrenching Russian attacks on the civilian Ukrainian population. In the case of China, a military intervention deep in mainland China in support of Uyghurs facing the risk of genocide is unrealistic. The United States can call attention to their suffering and atrocities. It can contrast for the world the difference between democracies and brutal autocracies. Afghanistan taught a hard lesson that democracy and cultural change could not be gifted or imposed. Yet, homegrown democratic revolutions can arise and are what China and Russia most fear. The fall of the Berlin Wall is a reminder of how quickly tides can turn when flourishing democracies are a stone's throw or channel crossing away.[59] Only a disillusioned Chinese populace can decide for themselves to break the shackles of the CCP. The United States must compete indefinitely regardless; paradoxically, the most fundamental American advantage increases when rivals hold fast to their authoritarian systems.

The second risk is the exploitation of the democratic system when the costs of advancing the just cause impact the nation domestically. As tough as the sanctions on Russia have been on the U.S. economy, they are dwarfed by the prospect of a significant sanctions regime imposed against China, our largest trading partner. How much pain are the American people willing to endure to elevate values over interests? An austere environment that hits pocketbooks and employment creates an atmosphere ripe for what Carnegie Endowment for International Peace scholar Moisés Naím terms the "3 Ps: Populism, Polarization, and Post-truth." The 3 P's can become pathways autocrats exploit to gain power legitimately within a democratic system.[60] America is no stranger to this phenomenon and is probably better suited than other nations in the Western Alliance to manage this risk. First, the size of the U.S. economy allows the United States to absorb and manage shocks from sanctions easier than can allies and partners, and the United States is not dependent on Russian energy. Second, a free and independent press, wiser social media companies, and battered yet robust civil institutions make America better positioned to safeguard and advance the just cause even during periods of international and domestic backsliding.

The third risk is loss of will and resources—the most dangerous of the three because it is the only way to lose in the infinite game. The contest is multidimensional and multidomain, across every element of the DIME-FIL. Government organizations do not have infinite resources. The infinite approach provides three ways to manage this risk.

First, finite games can exist within infinite ones. Sinek notes, "The Just Cause provides the context for all the finite games we must play along the way."[61] The infinite game is best played at the grand strategic level. Delivering a Virginia-class submarine on time and on budget is a finite win linked to national objectives that any program manager should celebrate. An eternal journey without wins along the way would be deeply demoralizing to the players. These strategic wins within fiscal and other constraints need to occur in government and nongovernment organizations across the DIME-FIL.

Second, shared commitment to a truly just cause closes trust deficits between government and private industry. Much of the best technological innovation occurs outside the government. Many companies are joining the "B Corp" movement away from finite-minded leadership styles rooted in shareholder primacy to new approaches that balance profits while limiting the negative global impacts of ethical fading.[62] Grounding American grand strategy in safeguarding and advancing the just cause of democracy, freedom, and human equality makes collaboration with innovative private-sector leaders more likely.

Third, infinite-minded policy attracts allies and partners, expanding will and resources. The United States must increase security and economic engagement with nations committed to democracy or those that stymie our most determined rivals. This increases American comparative advantage far beyond resources the nation can muster and translate into global influence alone. The just cause should not preclude deepening relations with nondemocratic countries like Vietnam.

WINNING THE INFINITE GAME

Former Secretary of State Dean Acheson, who directed the development of NSC-68, understood the infinite nature of the geopolitical contest with the Soviet Union long before Simon Sinek benefited from his wisdom. Acheson called for eternal vigilance, advising fellow national security professionals, "We have got to understand that all our lives the danger, the uncertainty, the need for alertness, for effort, for discipline will be upon us . . . the only real question is whether we shall know it soon enough."[63] The geostrategic game is infinite, and while there will be no permanent winners, there can be permanent losers. American grand strategy must be cognizant of this fact. Despite the infinite nature of the work, Acheson imparts a sense of urgency. Beyond ongoing efforts to counter Russian aggression, there are other pressing areas where an infinite-based grand strategy can improve national security and prosperity now.

The first is in strategic competition with China. NSC-68 focused on building up U.S. military capability to exhaust Soviet resources. It worked, but the same approach with China is likely to fail; they learned from the folly of the Soviet command economy.[64] Instead, the United States must emphasize strengthening the U.S. economy. This includes revitalizing the

U.S. industrial base for high-tech manufacturing, protecting intellectual property, investing in and safeguarding critical infrastructure, and reducing reliance on China for rare materials.[65] This approach provides the means for continued U.S. military dominance through continued technological innovation and the ability to maintain a forward presence essential to deterrence. Economic prowess also makes the United States a more attractive security, trade, and development partner to rival China's Belt and Road initiatives. America must also eschew willful economic self-harm that risks U.S. credibility and national security through partisan brinksmanship on U.S. debt ceiling renewals.[66] Forced default strategies should be employed on malign rivals like Russia, not America.

Hyperpolarization weakens American power in other ways, from gridlock that hinders sound governance to partisan attempts to alter free and fair election processes.[67] Geopolitical forecaster George Friedman sees extreme polarization growing to new and harmful levels unless there are "steps taken by particular sectors of American life to rectify the situation and lessen the pain for people on both sides of the divide."[68] The just cause provides common ground that can draw Americans from all political leanings together to promote values that impart the benefits of freedom to all. Actions to hinder those freedoms invite civil strife that weakens the nation and erodes our ability to compete with opportunistic foreign rivals. The nature of the just cause resists political extremes that erode democratic values.

Whether the political pendulum swings hard left past democratic socialism all the way into communism or hard right past ethnic or religious nationalism into neofascism is largely irrelevant; each swing of the pendulum further away from shared democratic values is like kryptonite to American strength needed for the infinite game. Madeleine Albright, who escaped the Soviets with her family in 1948, offered a sage warning during her life-long commitment to the just cause:

It would be unforgivable if America's commitment to democratic principles were now to wane because there is no superpower rival to spur us, because we lack patience, or because democracy's imperfections have caused us to forget the far greater flaws of every other form of governance.[69]

Finally, the threat environment increasingly employs disinformation to undermine democracy and will require all measures of American power to fight this insidious threat while respecting constitutional protections for freedom of speech and expression. Recent events demonstrated the power of malign actors to employ disinformation to harness political violence by sowing public distrust in democracy and the electoral process that underpins it. From Moscow to Beijing, autocratic hearts warmed as they watched Americans seduced into attacking their own Capitol. The just cause raises national awareness of the existential nature of malign influence operations.

Greater emphasis should be on identifying the originators of disinformation and witting accomplices—not simply prosecuting their victims.[70] The current National Security Strategy acknowledges that defeating the cooperation between domestic and foreign malign actors requires mutually supporting foreign and domestic policies. Nongovernment advocates of the just cause will become increasingly important in countering domestic malign activity where the government is rightly restricted to preserve individual freedoms and liberties.

The only way America loses in competition or conflict is by abandoning the just cause, the idea that makes America exceptional and the most powerful nation on earth. Our democratic principles are our greatest strength. Ethical fading at home and abroad erodes U.S. competitive advantage. Abraham Lincoln understood the only way America would cease to exist was by our choice: "If destruction be our lot, we must ourselves be its author and finisher. As a nation of freemen, we must live through all time, or die by suicide."[71]

A grand strategy rooted in infinite principles ensures we live free for all time.

ACKNOWLEDGMENT

This chapter is adapted from Thomas P. Newman's Original Research Requirement at the U.S. Army War College in April 2022.

NOTES

1. Matt Pottinger, "Beijing's American Hustle," *Foreign Affairs* 100, 5 (September/October 2021), 114. https://www.foreignaffairs.com/articles/asia/2021-08-23/beijings-american-hustle.

2. H. R. McMaster, *Battlegrounds: The Fight to Defend the Free World*, Reprint edition (New York: Harper Paperbacks, 2021), 440.

3. Anonymous in accordance with USAWC nonattribution policy, "Senior Army Leader Address to USAWC Students."

4. Simon Sinek, *The Infinite Game* (New York: Portfolio, 2019), 1–4.

5. D. Robert Worley, *Orchestrating the Instruments of Power: A Critical Examination of the U.S. National Security System* (Lincoln, NE: Potomac Books, 2015), 275–277.

6. The White House, "National Security Strategy," October 12, 2022, 23. https://www.whitehouse.gov/wp-content/uploads/2022/10/Biden-Harris-Administrations-National-Security-Strategy-10.2022.pdf.

7. Office of the Secretary of Defense, "Military and Security Developments Involving the People's Republic of China 2021 Annual Report to Congress," 2021, 1–4. https://media.defense.gov/2021/Nov/03/2002885874/-1/-1/0/2021-CMPR-FINAL.PDF.

8. Rush Doshi, *The Long Game: China's Grand Strategy to Displace American Order* (New York: Oxford University Press, 2021), 4.

9. Henry Kissinger, *On China* (New York: Penguin Books, 2012), 501.

10. Kissinger, *On China*, 503.

11. Cited in *ibid.*, 506.

12. Jonathan Ward, *China's Vision of Victory* (Fayetteville, NC: The Atlas Publishing and Media Company, 2019), 37.

13. "NSC-68 United States Objectives and Programs for National Security," April 14, 1950, sec. IV.B. https://irp.fas.org/offdocs/nsc-hst/nsc-68.htm.

14. "NSC-68 United States Objectives and Programs for National Security," sec. IV.C.

15. Sinek, *The Infinite Game*, 174–176.

16. Sinek, *The Infinite Game*, 18.

17. *Ibid.*, 19.

18. *Ibid.*, 20.

19. *Ibid.*, 37.

20. *Ibid.*, 46.

21. *Ibid.*, 38–40.

22. "U.S. Declaration of Independence," 1776, Paragraph 2.

23. "NSC-68 United States Objectives and Programs for National Security," sec. II.

24. Michael J. Mazarr et al., *Hostile Social Manipulation: Present Realities and Emerging Trends* (Washington: RAND Corporation, September 4, 2019), https://www.rand.org/pubs/research_reports/RR2713.html, 17–20.

25. Sinek, *The Infinite Game*, 20.

26. Alexandra Ulmer, "Pence Says Trump Was Wrong That He Could Have Overturned 2020 Election," *Reuters*, February 7, 2022, sec. United States, https://www.reuters.com/world/us/pence-says-trump-was-wrong-that-he-could-have-overturned-2020-election-result-2022-02-04/.

27. Sinek, *The Infinite Game*, 161.

28. *Ibid.*, 161.

29. "National Security Strategy," October 12, 2022, 8–9.

30. The White House, "Joint Statement by President Biden and President von Der Leyen," March 24, 2022, https://www.whitehouse.gov/briefing-room/statements-releases/2022/03/24/joint-statement-by-president-biden-and-president-von-der-leyen/.

31. Clay Chandler, "Putin and Xi's 'No Limits' Friendship Is Put to the Test in Ukraine," *Fortune* (online article), February 25, 2022, https://fortune.com/2022/02/25/ukraine-invasion-china-xi-jinping-russia-vladimir-putin-relationship-foreign-policy/.

32. Shelby Grossman et al., "Full-Spectrum Pro-Kremlin Online Propaganda about Ukraine," Stanford University Cyber Policy Center (blog), February 22, 2022. https://cyber.fsi.stanford.edu/io/news/full-spectrum-propaganda-ukraine.

33. Davey Alba and Stuart A. Thompson, "'I'll Stand on the Side of Russia': Pro-Putin Sentiment Spreads Online," *The New York Times*, February 25, 2022, sec. Technology. https://www.nytimes.com/2022/02/25/technology/russia-supporters.html.

34. Qiao Liang and Wang Xiangsui, *Unrestricted Warfare: China's Master Plan to Destroy America*, reprint edition (Brattleboro: Echo Point Books & Media, 2015), 144–145.

35. Pete Williams, "FBI Director Wray Says Scale of Chinese Spying in the U.S. 'Blew Me Away,'" NBC News (online article), February 1, 2022, https://www.nbcnews.com/politics/politics-news/fbi-director-wray-says-scale-chinese-spying-us-blew-away-rcna14369.

36. Ryan Hass, *Stronger: Adapting America's China Strategy in an Age of Competitive Interdependence* (New Haven: Yale University Press, 2021), 23–32.

37. Hass, *Stronger: Adapting America's China Strategy in an Age of Competitive Interdependence*, 39.

38. Graham T. Allison, *Destined for War: Can America and China Escape Thucydides's Trap?* (New York: Houghton Mifflin Harcourt, Mariner Books, 2018), 235; Linda So and Jason Szep, "Threatened U.S. Election Workers Get Little Help from Law Enforcement," *Reuters* (online article), September 8, 2021, https://www.reuters.com/investigates/special-report/usa-election-threats-law-enforcement/.

39. Hass, *Stronger: Adapting America's China Strategy in an Age of Competitive Interdependence*, 40.

40. Brendan Nicholson, "U.S. Army Pacific Commander: Next War Will Be Violent, Very Human, Unpredictable and Long," *Real Clear Defense* (online article), February 21, 2022. https://www.realcleardefense.com/articles/2022/02/21/us_army_pacific_commander_next_war_will_be_violent_very_human_unpredictable_and_long_817777.html?fbclid=IwAR0fYU53UVUGfO6LyNawQZu-QsyKfZ1Si69z_Xi1kec5K5odrC7Bl368NXI.

41. Noelani Kirschner, "U.S. Companies Mobilize to Support Ukraine," *ShareAmerica* (blog), March 14, 2022, https://share.america.gov/us-companies-mobilize-support-ukraine/.

42. H. R McMaster, *Battlegrounds: The Fight to Defend the Free World*, 439.

43. Christian Brose, *The Kill Chain: Defending America in the Future of High-Tech Warfare* (New York: Hachette Books, 2020), 95–96.

44. Jude Blanchette, "Xi Jinping's Faltering Foreign Policy," *Foreign Affairs* (online article), March 28, 2022, https://www.foreignaffairs.com/articles/china/2022-03-16/xi-jinpings-faltering-foreign-policy.

45. Doshi, *The Long Game*, 333.

46. Michael Beckley, "Enemies of My Enemy," *Foreign Affairs* 101, 2 (March/April 2022), 74–75, https://www.foreignaffairs.com/articles/2021-02-14/china-new-world-order-enemies-my-enemy.

47. Madeleine K. Albright, "The Coming Democratic Revival," *Foreign Affairs* 100, 6 (November/December 2021), 101, https://www.foreignaffairs.com/articles/world/2021-10-19/madeleine-albright-coming-democratic-revival.

48. Sinek, *The Infinite Game*, 132.

49. H. R McMaster, *Battlegrounds: The Fight to Defend the Free World*, 443.

50. Joshua Wong and Ai Weiwei, *Unfree Speech: The Threat to Global Democracy and Why We Must Act, Now* (New York: Penguin Books, 2020), 239.

51. Sinek, *The Infinite Game*, 132–133.

52. Ewan Palmer, "Tulsi Gabbard Labeled a 'Russian Asset' for Pushing U.S. Biolabs in Ukraine Claim," *Newsweek* (online article), March 14, 2022, https://www.newsweek.com/tulsi-gabbard-bio-labs-ukraine-russia-conspiracy-1687594.

53. Moisés Naím, "The Dictator's New Playbook," *Foreign Affairs* 101, 2 (March/April 2022), 145, https://www.foreignaffairs.com/articles/world/2022-02-22/dictators-new-playbook.

54. Allison, *Destined for War,* xiv–xvi.

55. *Ibid.,* 236.

56. Patrick Hulme, "Taiwan, 'Strategic Clarity' and the War Powers: A U.S. Commitment to Taiwan Requires Congressional Buy-In," *Lawfare* (blog), December 4, 2020, https://www.lawfareblog.com/taiwan-strategic-clarity-and-war-powers-us-commitment-taiwan-requires-congressional-buy.

57. Benjy Sarlin and Sahil Kapur, "Why China May Be the Last Bipartisan Issue Left in Washington," *NBC News* (online article), March 21, 2022, https://www.nbcnews.com/politics/congress/why-china-may-be-last-bipartisan-issue-left-washington-n1261407.

58. Donald J. Stoker, *Why America Loses Wars: Limited War and US Strategy from the Korean War to the Present* (New York: Cambridge University Press, 2019), 6–9.

59. Robert Person and Michael McFaul, "What Putin Fears Most," *Journal of Democracy* (online article), February 22, 2022, https://www.journalofdemocracy.org/what-putin-fears-most/.

60. Moisés Naím, *The Revenge of Power: How Autocrats Are Reinventing Politics for the 21st Century* (New York: St. Martin's Press, 2022), xv.

61. Sinek, *The Infinite Game,* 34.

62. Sinek, 156–157.

63. James Chace, *Acheson: The Secretary of State Who Created the American World* (New York: Simon & Schuster, 2007), 150.

64. Ward, *China's Vision of Victory,* 229.

65. Hass, *Stronger,* 35–38.

66. Joe Walsh, "Pentagon Warns Failing to Raise Debt Ceiling Could Harm National Security," *Forbes* (online article), October 6, 2021, https://www.forbes.com/sites/joewalsh/2021/10/06/pentagon-warns-failing-to-raise-debt-ceiling-could-harm-national-security/.

67. Alex Samuels, "The States Where Efforts to Restrict Voting Are Escalating," *FiveThirtyEight* (blog), March 29, 2021, https://fivethirtyeight.com/features/the-states-where-efforts-to-restrict-voting-are-escalating/.

68. George Friedman, *The Storm before the Calm: America's Discord, the Coming Crisis of the 2020s, and the Triumph Beyond* (New York: Doubleday, 2020), 182.

69. Madeleine K. Albright, "The Testing of American Foreign Policy," *Foreign Affairs* 77, 6 (November/December 1998), 64, https://www.foreignaffairs.com/articles/china/1998-11-01/testing-american-foreign-policy.

70. P. W. Singer and Emerson T. Brooking, *Likewar: The Weaponization of Social Media* (New York: Mariner Books, 2019), 267.

71. "Abraham Lincoln's Lyceum Address," accessed March 28, 2022, http://www.abrahamlincolnonline.org/lincoln/speeches/lyceum.htm.

CHAPTER 4

Responding to China's Economic Statecraft: Maintaining America's Edge in Relationships, Global Governance, and Innovation

James A. Frick

China likely poses the greatest hindrance to America sustaining its comparative advantage in the world today. Since entering the World Trade Organization (WTO) in 2001, the People's Republic of China (PRC) has steadily expanded its economic power, becoming a global powerhouse. It surpasses all other countries in trade, has increasingly expanded investments and aid abroad, has become a leader in international finance institutions (IFIs), and its Belt and Road Initiative (BRI) is becoming a leading effort in Asian infrastructure development. Consequently, the Chinese government has applied economic means both as carrots and sticks as part of a strategy to achieve Xi's China Dream of national rejuvenation. Those countries that support China reap the benefits of interaction, while those that resist experience forms of economic retaliation.

Still, the United States currently retains its leverage within the liberal international order (LIO) that it has helped create. Its global relationships realized through both the economic and military means of national power, along with structural mechanisms of control written into this global system, combine to provide this edge. Just as these qualities offer the United States tremendous clout in its competition with China today, a retrenchment from our connections abroad exacerbated by an atrophy of U.S. leadership within this LIO can lead to a formula for decline in America's global position vis-à-vis China in the future. Furthermore, Xi's national rejuvenation potentially threatens U.S. interests and stance within the LIO. The PRC's economic statecraft has been a calculated strategy in which tools of the state have been applied to influence investment, trade, and development, while attempting to erode the order as we know it. The United States cannot allow this to happen if it is to sustain its comparative advantage. This will require a prioritization back onto America's relationships, while strengthening its structural edge in global governance and establishing a more holistic approach to innovation.

Concerns go beyond China's international connectivity or its ascendency as a global economic leader. By comparison, the literature devoted to Western countries, and for that matter emerging economies such as India, South Africa, and Brazil, does not seem to prioritize the same concerns over each of their economic statecraft. The underlying worry stems from China's regime type and just how compatible it is within the current global order. What is alarming to the United States about China's increasing influence is in how the PRC has been willing to apply its power aggressively to achieve its interests, which has, at times, contrasted with the liberal foundations of the international order. Consequently, political leaders and academics alike have been puzzled over whether an illiberal regime rising to great power status can coexist within a liberal order.

What brings China to center stage are the size of its markets, its economic strength, and rising power. These factors have emboldened the regime to apply its influence more and more in ways deemed perilous to the liberal order, and its sway has led some states to look the other way. China has proven adept at using both liberal and illiberal methods to achieve its interests, favoring the LIO when it supports PRC interests while contesting it when liberalization measures impact crucial domestic strategic interests.[1] At times, such resistance has sought to fix failures in the system in addition to furthering Chinese interests.[2] Following the 2008 global financial crisis, evidence suggests that China has not attempted to subvert the LIO[3] while its active participation within it has helped to bind it to better macroeconomic practices,[4] even if it has skirted the actual intent with underlying microlevel controls. Out of this discourse rises the apprehension that the more effective China is at economic statecraft, given its evolving global status, the greater its ability and desire to influence the order as we know it. Is Xi's China Dream compatible with this LIO? Are there any indications in China's economic statecraft to create concern? If so, what are the implications and how should the United States respond?

Why China's growing influence poses a concern to the United States is underscored by the 2022 scenario in which Russia invaded Ukraine for the second time. Here we saw much of the global community attempting to punish Russia for its violations of state sovereignty. Although somewhat a disconnected country, Russia has still been able to sway countries to minimize any retaliation, or even separate themselves from condemnation, lessoning the effectiveness of the U.S.-led efforts to influence Russia. In a similar and ever-growing likely scenario of the PRC compelling Taiwan's reunification with the mainland through force, how much more so would an extremely connected country like China be able to resist pressures from the international community and coerce other countries into inaction? Not only does China's growing influence potentially diminish the effects of an international response in such a scenario, but it also possibly reduces the effectiveness of deterrence.

To better understand what the United States must do to sustain its comparative advantage vis-à-vis China, it is important to also understand what the PRC wants and the context in which Chinese leadership frames its strategies to achieve its aims. The next segment examines existing literature exploring the PRC's grand strategy of the China Dream, situating what national rejuvenation means and how it poses a concern for U.S. interests in the LIO. The chapter then explores China's economic statecraft, finding that the PRC has increasingly strengthened its control over the means of economic statecraft and noting that China is indeed a force to be taken seriously. Next, the chapter theorizes on how the PRC's conceptualization of itself drives its implementation of economic statecraft. In so doing, it reviews Chinese economic policy development and discusses key facets of China's economic strategy and what they mean for the United States and the LIO at large. Finally, the chapter prescribes what the United States needs to do to sustain its competitive advantage in response to China's desire and actions to overtake it.

THE CHINA DREAM

Since 2009, China has identified three core interests: (1) preserving their political system and national security; (2) sovereignty and territorial integrity; and (3) stable development of their economy and society.[5] Obviously, the PRC's increased assertiveness in the South China Sea and other security-related issues fit into the first two core interests. Deng Xiaoping's reforms addressed the third core interest and were focused on economic progress and opening China to the international economy. It is through this economic development, not the military, that China has captured global attention. China has proven more adept at using economics as a tool for policy than the military element, yet the two are inexplicably entwined. Both tools of national power help to realize all three of their core interests.

Deng's "hide and abide" approach has been a guidepost for Chinese leaders as they sought "wealth and power (*fuqiang*)." But with China's evolving position in the world today, Xi Jinping has declared "that China must play a proactive international role."[6] Since assuming leadership and consolidating his authority by taking over all three key leadership positions in China (2012–2013)—head of the Communist Party, president and head of state, and head of the Central Military Commission—Xi Jinping has declared his "China Dream" of state prosperity, collective pride and happiness, and national rejuvenation.[7] This dream seeks to change the current unipolar world into a multipolar one while pushing the United States out of East Asia.[8] Specifically, national rejuvenation conceptualizes China's reestablishment as the "Middle Kingdom," an aim that if realized would establish the PRC as the de facto regional hegemon in Asia, if not globally. This gain would come at a loss in influence by the United States, its allies, and key partners.

Many states in the Indo-Pacific region already fear China's increasingly assertive behavior, wondering if it is further attempting to establish itself at the center of power within the global environment, just as it did for two thousand years in East Asia when it had previously been the Middle Kingdom.[9] Although many of these concerns derive from the PRC's implementation of its hard power as it strives toward its aims, Chinese academics are already exploring the moral facets of national rejuvenation. Informed by his research of China's strategic culture and the idea of moral realism, Yan Xuetong, for example, argues that China's future as a new hegemon should emphasize international responsibility and leadership while making alliances and moral norms for the rest of the world.[10] Such views among China's academia elite highlight that the Party may already be looking ahead to what comes next. Although the military element of China's national power may eventually play a key role in achieving this dream, still, it is through the economic element that China is progressing the most rapidly toward its goals at the finish line.

The Effectiveness of PRC Economic Statecraft

There are competing views on just how effective China can be in employing its economic statecraft to achieve the China Dream of national rejuvenation. The statist side of the argument is ripe with evidence that the PRC government does employ economics as a tool of statecraft. Frick and Hsueh contend that in the case of an illiberal regime, such as the PRC, the state may directly influence outward foreign direct investment (OFDI) through its state-owned enterprises (SOEs) and private enterprises to achieve strategic economic and political interests.[11] A 2018 OECD Investment Policy Review argues that the PRC "government's policy and strategy have been always among the most significant determinants in explaining the development of China's OFDI."[12] In this context, SOEs are necessary tools within the government's strategy for investment.[13] Other literature has found evidence to suggest that the Chinese state also manipulates private firms to achieve state interests.[14] Consequently, whether public or private, the central government has been able to manipulate firms as part of its economic statecraft.

Chinese OFDI, however, is not the only tool of economic statecraft the PRC has successfully employed. Norris' study of China's State Administration of Foreign Exchange (SAFE) found evidence that the Chinese government directed the agency to buy $300 million of Costa Rican bonds at below market rates to persuade the country to reverse diplomatic recognition from the Republic of China (ROC) to the PRC.[15] This case aligns with China's practices intended to isolate Taiwan.[16] Similarly, Chinese loans appear to be a geopolitical means for creating trade opportunities and alleviating overcapacity,[17] both of which are important to China's economic development. Correspondingly, the PRC's BRI had initially sought to alleviate overcapacity[18] and is considered a tool of China's economic statecraft for expanding its diplomatic, informational, military, and economic influence into Asia and the western Pacific.[19] The central government further emphasized this initiative by issuing the *Guidance on Further Directing and Regulating Overseas Investment Direction* in 2017. This regulation divided investments into encouraged, restricted, and forbidden categories, creating a 40% cut in China's overseas investments as the CCP attempted to reign in investment and refocus it onto advanced technology, high-standard global brands, and BRI-related activities.[20]

Frick's (2021) research into China's behavior within international governmental organizations (IGOs) further finds that Chinese authorities have become increasing active in ensuring that IGOs adequately represented their global economic position and pursued policies closely aligned to PRC interests. China has used its global economic status and support (or in some situations threatening to withhold support) to gain concessions in these organizations while also promoting its own economic pursuits.[21] Although the CCP has more broadly employed economic statecraft as a "carrot" for pursuing its interests, it has likewise been willing to use its power as a "stick" to protect and promote China's interests.[22] An Australian Strategic Policy Institute found 152 cases between 2010 to 2020 in which the PRC demonstrated coercive diplomacy.[23] Yet, China's statecraft has varied over time.

During the Hu Jintao and Wen Jiaboa era, reforms gave Chinese firms greater autonomy; commercial actors were thus increasingly able to make decisions outside of state oversight. Opposing perspectives—conservatists versus reformist—within the CCP were aligned within different governmental agencies, creating disunity. Under the Hu/Wen administration, four of the nine-member Standing Committee were considered conservative.[24] These dynamics pitted groups against one another as the Peoples Bank of China (PBOC), private banks, small and medium enterprises, and the China Securities Regulatory Commission represented liberal views and the more influential Ministry of Finance (MoF), the National Development Reform Commission, state commercial banks, and SOEs aligned with conservatives.[25] This friction was more than sufficient to allow commercial actors enough autonomy to make many of their own economic decisions.

Despite a variance in the extent of state control during this period, the Chinese government was mostly able to manage economic activities, especially within strategic sectors, while firms tended to have greater say on seeking markets and efficiency.[26] Since China's Open Door Policy in 1978, it has employed a bifurcated strategy of macro-liberalization and sectoral-level reregulation to enable economic development and achieve political goals.[27] This has also created variance over time in state control and commercial agent autonomy. All these factors appear to contribute to the alternative perspective that the PRC has, during certain periods, not been able to manipulate commercial agents as a credible demonstration of economic statecraft.

Importantly, if the Hu/Wen era allowed greater autonomy for commercial agents, then the Xi/Li administration has attempted to reign in firm-level decisions. Xi has strengthened the CCP's role in SOEs through a recentralization of policy-making, personnel, and authority over SOEs[28] while also announcing plans to extend the CCP's purview into private media and technology companies.[29] The State Council published guidelines in 2017 regulating overseas investment, further asserting its control.[30] Commercial agents may have greater autonomy in decisions in nonstrategic sectors since the introduction of the "Go Global" policy, but the Chinese government has proven itself adept at asserting its influence when it comes to strategic interests. As the CCP has repackaged "Go Global" into BRI, Xi has tightened the state's control.[31]

Despite these variations in government control, China has been very successful in its economic development policies. The PRC has not only transformed from a low-income country (LIC) to a middle-income country (MIC) over the last 30 years, but it has also jumped to the world's second largest economy with a GDP per capital of over $10,000 in 2020.[32] It has the second largest research and development (R&D) budget, ranking fourteenth on the Global Innovation Index with 69,000 international patents submitted through the Patent Cooperation Treaty in 2020.[33] It has steadily diversified its energy production to include coal, oil, natural gas, nuclear power, and renewable energy. Ultimately, all these activities have helped China to expand quality of life for its people through income, housing, education, and health care.[34] In addition, China has become more active in global governance, having joined over seventy-five IGOs by 2017, and in many cases, having expanded its influence through staffing, finance, and interaction.[35] The literature suggests that the PRC has been effective at its economic statecraft, while some of its methods seeking to insulate itself have sought to shift its markets away from the United States and reduce China's dependence on the dollar. Although the results[36] suggest that these attempts are nowhere near removing the United States from its overall position of advantage, the sheer size of China's economy warrants giving attention to the PRC's economic strategies given the potential collision course they may have with U.S. interests.

REFRAMING THE MIDDLE KINGDOM

It is the nexus of ideational and rational political thinking that influences the Chinese viewpoint today. When considering PRC economic statecraft, at the center of the analysis is the interconnection the regime has with its people, and vice versa. The two sides feed into one another. In a country with 1.4 billion people, it would be extremely foolhardy to ignore the population's wants and desires. From the position of the communist engine itself, is it not the working classes that make up this system from which the revolution originated? Mao recognized the need to integrate the constructs of communism within Chinese culture when he envisioned it taking on its own form assimilating the "democratic essence" of China's cultural traditions while discarding the residue associated with the "feudal ruling class."[37] Still today, Xi Jinping promotes the narrative of the people, noting that "the people are the true heroes, for it is they who create history" as he promoted the PRC's need to "use Marxism to observe, understand and steer the trends of our times, and continue to develop the Marxism of contemporary China and in the 21st century."[38] Xi has attempted to legitimize the regime by conflating Maoist and Marxist principles with Chinese Confucian traditions.[39]

This corresponding narrative has therefore sought to right the perceived wrongs from the last century where China lost its prestige as the center of East Asia while shedding any debt to Western imperialistic powers that might have helped them to gain advantage.[40] This narrative strives to create a nationalistic voice among the people to rebuild the prestige of antiquity. The regime could therefore not abandon Chinese culture if they seek to capture and keep the hearts and minds of the Chinese people. After all, these ideals had the opportunity to evolve over five thousand years. In fact, Communist Party members often quote from Confucius, Mencius, and other Chinese philosophers to bolster communist arguments.[41] It is, however, this narrative that will impact how the regime will act in the future and may be what constrains them to one approach over another.

For the PRC to overcome the "century of shame" imposed by the West and exacerbated by the Japanese, it had to restore the prestige it had prior to Western intervention. The CCP reached back into China's rich history to cherry-pick those qualities that would help restore national pride and unify the country. The strength of the Han dynasty conceptualized as the Middle Kingdom and the territorial gains of the Qing Dynasty became artifacts of Chinese history the PRC could build nationalistic fervor around. The term "Middle Kingdom" is a literal translation of Zhongguo (中國/中国), which is one of the names for China originating from when China was considered the center of civilization.[42] Thus began the enterprise of consolidating sovereign territories as China sought to regain power (权力) and wealth (财富). Recognizing that the PRC could fully achieve neither without an international connection, Deng Xiaoping began the process of opening China up to the international stage to enable economic development in the PRC's first step toward wealth. Xi's China Dream of national rejuvenation is the realization of the restored status that China felt it had lost under the century of shame.

A Great Modern Socialist Country

The China Dream's end state is a national rejuvenation in which "China takes its place at the center of the world stage."[43] If you align this idea with the preceding theorized reframing of the Middle Kingdom perspective, then China aims to establish itself at the center of global decisions and relationships, with expectations of deference regarding its interests and priorities. Originally, China's international connectivity was a means to achieve the power and wealth necessary to build back China; now this strategy appears to have flipped with power and wealth now becoming the means for situating itself at the center of not just Asia, but the world. Many China watchers tend to give credit to PRC strategic planning as being long-term, pragmatic, and a well-thought-out process. However, the PRC's development plan has had many struggles along the way and can be more characterized as one where there may have been a long-term vision, but the way to achieve this vision was not well understood.[44] The strategy of ways was more a trial-and-error path stitched together by multiple five-year plans.[45] As China's human capital and economic experience grew, compounded by the recognition of international opportunities, these plans have become more connected and oriented toward the Middle Kingdom goal.

Over the last decade China's ever-increasing economic power has allowed it to formulate a more coherent vision focused on building up China into a global hegemon. Xi's China Dream of national rejuvenation built upon previous party goals with the 18th Party Congress conceptualizing the first milestone of a "moderately prosperous society in all respects" with a suspense of the CCP's centenary in 2021. The Party's 12th and 13th Five-Year Plans provided the direction for achieving this suspense. China's 14th Five-Year Plan, the China 2030 economic development plan (developed in collaboration with the World Bank), and the recent Chinese Academy of Sciences (CAS) carbon peak and neutrality strategic action plan highlight the next steps in the PRC's path toward realizing the second goal, a "great modern socialist country," for achieving national rejuvenation by 2049. Once China obtains its goal of national rejuvenation it will likely not be satisfied with the status quo hierarchy within the LIO and will attempt to shift power relationships in its favor.

Xi's Vision for National Rejuvenation

During his speech to commemorate the centennial of the Chinese Communist Party in 2021, Xi acknowledged that the PRC had achieved its first centenary goal in the realization of national rejuvenation with the achievement of a "moderately prosperous society in all respects."[46] However, the origin of this idea precedes Xi. Deng Xiaoping originally conceptualized a goal of achieving a *xiaokang* (relatively well-off) life of moderate prosperity and established a per capita GNP of $800 as a benchmark.[47] Although this goal has been modified several times since Deng provided a measurement for it, Xi Jinping sought to bring it to fruition by prioritizing its implementation with the 18th National Congress, identifying the Party's centenary as a suspense for achieving the metrics of "sustained and sound economic development, greater people's democracy, a significant improvement in cultural soft power, higher living standards, and major progress in building a resource-conserving and environment-friendly society."[48] Crucial to this plan was China's continued advancement and modernization and a change to the Western-led global landscape—one more democratically led by developed and developing countries alike.[49]

China's path to national rejuvenation has followed three phases of development: standing up, getting rich, and getting strong, with the last phase requiring more than economic growth to serve as a source of legitimacy.[50] In 2017, during the 19th National Congress, Xi Jinping combined his China Dream with the concept of socialism with Chinese characteristics, codifying it into the CCP constitution. Xi also argued for a shift from rapid growth to high-quality development, improvements in the Chinese innovation system, integration of rural–urban development, improvement of the socialist market economy, consolidating the Chinese national identity by aligning the Confucian imperial past with China's adherence to Marxism–Leninism with Chinese characteristics, development of an ecofriendly system, and modernization of the military.[51] This emphasis recognizes that China would require a prioritization of quality over quantity, both economically and within its military. Attempts to maintain a harmony between domestic and external audiences have further driven alignment of ideology with culture. Consequently, the CCP's form of economic management diverges from the developed world and has recently been characterized, both within and external to China, as the "Beijing Consensus."[52] Administrative characteristics orbit around policies of "incremental reform, innovation and experimentation, export-led growth, and state capitalism" that rely on SOEs and resources based on strategic interests and authoritarianism.[53]

The next stage in the CCP's pursuit of national rejuvenation aspires to create "a modern socialist country that is prosperous, strong, democratic, culturally advanced, harmonious and beautiful."[54] This entails a modernized governmental capacity that allows the PRC to be a global leader in national strength and influence and an active member in the international community.[55] Such modernization comes with the expectations that the gap between China and advanced countries will gradually narrow, allowing China to build as a world power closer to the center of the world stage.[56] To achieve this goal, the PRC uses the developed world as its benchmark. In its attempts toward prosperity, it desires a GDP per capita equal to 70–89% of the U.S. rate (the level of developed countries); health, education, and social development to be at the top of the world rankings; and a world-class military that is one of the best in the world.[57] Although these documents say little about territorial sovereignty, Xi's centenary speech characterized "resolving the Taiwan question and realizing China's complete reunification is a historic mission and an unshakable commitment of the Communist Party of China," and that the PRC "must take resolute action to utterly defeat any attempt toward 'Taiwan independence,' and work together to create a bright future for national rejuvenation."[58]

Understanding China's Strategic Pathway

China wants to be a producer of high-end products, a green technology leader, and an innovation leader. Xi Jinping believes that these priorities will help to insulate China from future threats by solidifying its position in the global economy.[59] The pathway central to these ends emphasizes four lines of effort; the first focuses on innovation along key strategic sectors, the second highlights the use of state-owned enterprises and capital, the third is BRI, and the fourth focuses on reform in economic global governance.

Innovation. China is taking steps to elevate its production to innovative, high-end products while adjusting its domestic consumption to create both external and internal markets for these advances in industry. The 12th Five-Year Plan focused on three areas: restructuring the economy, promoting social equality, and protecting the environment.[60] Economically, the Party sought to maintain a 7% GDP growth rate while it prioritized "strategic emerging industries (SEIs)"[61] for government spending, expanded value-added industrial outputs, restructured to expand the service sector, shifted to consumption-driven growth, promoted private sector participation, and desired to develop capital markets.[62] A common theme across these objectives focused on innovation-driven growth. The 13th Five-Year Plan, likewise, made innovation the "primary driving force for development," but also emphasized synchronizing development across China's provinces while continuing to advance its ecofriendly focus as it continued to deepen its integration within the global economy.[63] Crucial to innovation was the development of China's science and technology capacity.[64] The recently published 14th Five-Year Plan picks up with the PRC achieving its first centenary milestone, noting that China will take action to promote high-quality development through developing economic strength, science and technology strength, and overall national strength by 2035. It will do so through breakthroughs in key and core technologies, a modernized economy, and in establishing itself as a "cultural powerhouse, an educational powerhouse, a talent powerhouse, a sports powerhouse, and as a Healthy China."[65] This new plan continues to focus on innovation with a heightened focus on R&D (increase of more than 7% annually) desiring to create an edge in quantum information, photonics and micro and nano electronics, network communications, artificial intelligence (AI), biotech and pharmaceuticals, and modern energy systems.[66] Bottom-line, the Chinese believe that the secret to overtaking the West is through innovation, an area that the United States has long maintained a comparative advantage in.

SOEs and Capital. The PRC continues to employ SOEs and state capital to achieve its strategic ends and uses multiple methods of state control to focus effort and capital flows into and out of China. While the 13th Five-Year Plan desired to strengthen government operational oversight and financial reforms,[67] the 14th Five-Year Plan is more explicit in the penetration of the Party's leadership into the firms' boards of directors.[68] When the Chinese government can manipulate both state owned and private firms to achieve its economic bidding, it potentially achieves a competitive advantage over the United States in the pursuit of foreign policies. However, the verdict is split as to the degree that state sponsored champions truly provide an edge in innovation.[69]

Belt and Road Initiative (One Belt, One Road). BRI is more than just a way to create internal and external infrastructure connectivity to expand access to resources and trade. BRI is a strategy that employs political, economic, informational, and cultural means for expanding PRC soft power. BRI is central to the Party's strategy for what is calls "all-around opening up."[70] The plan takes a two-pronged approach focused both internally and internationally. China intends to enable BRI thoroughfares in its buildup of its domestic infrastructure and creation of economic belts along these thoroughfares, the coastal regions, and the Yangtze River region.[71] At the international level, the PRC intends to focus on policy communication, infrastructure connectivity, trade facilitation, capital flow, and people to people exchange.[72] Problematic for the United States is that BRI methods create political and economic dependencies across the developing world that potentially pull these countries away from the United States and closer to China. The cultural and informational messages employed tend to promote qualities that are less liberal, contrasting with American values focused on human rights and individual freedoms. In addition, China seeks to use the BRI to further internationalize the renminbi (RMB). Over 30% of the countries receiving FDI from BRI projects also have currency swap agreements with China.[73]

Global Governance. China is blatant in its desire to reform economic global governance. It sees its role within IFIs as necessary to rebalance the decision-making authorities within them, along with reforms to both the international monetary system and financial regulation. Its attempts to internationalize the RMB may be part of a strategy to eventually knock the dollar out of its position as the global currency.[74] The 13th Five-Year Plan described China's role in global economic governance as one where it intended to reform and improve the system while guiding the economic agenda.[75] The 14th Five-Year Plan is more explicit in a reference to the construction of global economic governance systems and a new type of international relations.[76] Since the rest of the discussion on global economic governance speaks on the maintenance and improvement of current systems, this assertion is puzzling. Such language is possibly gesturing the CCP's intent to go beyond just reforms in the LIO. Doing so presents a direct threat to sustaining America's competitive advantage as it would reduce the mechanisms the United States has in leveraging its influence abroad.

Each of these plans demonstrates a systematic progression in the details toward modernization, which signifies that China's understanding of what it needs to do for its national rejuvenation has become even clearer. Still, accomplishment of its strategies must overcome both foreign and internal resistance.

PRC Economic Statecraft—What Are the Threats?

China has not always fulfilled international expectations as it pursued economic development. Even as China began to open to outside influence in the late 1970s, Deng Xiaoping emphasized *guochanhua*, which sought to reverse engineer others' technology to catch up.[77] Regardless of institutional improvements and statistics illustrating China's progress in international property rights, its adherence at the international level has been a concern. Repeatedly during the 1990s, the United States threatened to impose trade sanctions and block China's accession into the WTO because of concerns over intellectual property rights violations.[78] The PRC has also created animosity abroad as it reneged on investment agreements as it sought to pressure foreign firms to share their technology.[79] Additionally, China has manipulated its currency to make products more attractive abroad in attempts to meet export goals.[80] Such practices are not uncommon among developing countries. Apart from creating an unfair playing field for global economics, such "rule-breaking" behavior does not pose a threat to the LIO itself, as the international system has proven adept at influencing conformity through principles of reputation, retaliation, and reciprocity.[81] Still, such behavior undermined the Trade Related Aspects of Intellectual Property Rights (TRIPS) agreement in the 1994 Uruguay round of General Agreement on Tariffs and Trade (GATT), thus weakening trust in the WTO, which subsumed the GATT agreements.

Consequently, none of this evidence presents a true danger to the LIO other than in how China's newfound power and wealth might be leveraged to undermine U.S. global leadership or cause erosion in international standards. Primarily, tensions reside most in international monetary finance and global governance since these two areas are crucial to the United States's ability to uphold the LIO as we know it. Currently, the United States still retains a veto over IFI key decisions[82] both through its position within these organizations and in its influence over global finance, considering the dollar remains the primary global currency. It is in these two areas that China's economic statecraft exhibits a potential threat to U.S. authority.

The PRC first began attempts to undermine the dollar following the 2008 global financial crisis with President Hu Jintao and Vice Premier Wan Qishan pressing for the reconstruction of the international monetary system.[83] This emphasis was argued each year at the World Bank/International Monetary Fund (WB/IMF) Boards of Governors meetings from 2007 to 2010.[84] PBOC governor Zhou Xiaochuan offered that as part of these restructures that the Fund should consider replacing the dollar with the special drawing rights (SDR), while adding more currencies to the SDR basket.[85] There were some fundamental weaknesses with this suggestion combined with a stiff resistance from the U.S.-led G7, who effectively impeded any realistic movement toward this aim.[86] This proved to be the precursor pressing for internationalization of the renminbi in years to follow.

To achieve its Five-Year Plan goals for international monetary reform, China has followed four approaches either working collectively with Brazil, Russia, India, China, and South Africa (BRICS) or separately that helped to internationalize the RMB. First, the BRICS have begun to move away from the dollar by transferring reserves into alternate assets. Second, they have engaged in currency swaps. In addition to doing bilateral trade in national currencies, these countries have bought into one another's foreign exchange reserves. China has also promoted the renminbi in bilateral trade outside the group reaching a total of $1.7 trillion (roughly 25 percent of its annual trade) and has concluded RMB swaps with more than thirty countries.[87] A potential RMB-based oil deal between China and Saudi Arabia is the most recent example of China's attempts to expand its own currency in place of the dollar. Third, China had separately pushed for the RMB's inclusion into the IMF SDR basket with the other BRICS countries support and worked with the French during the 2011 G20 to promulgate the idea.[88] Finally, China has moved to liberalize its domestic financial markets and open its capital accounts. The RMB still has a long way to go before replacing the United States, but there are rising concerns with China's attempts to advance its currency globally. A most recent concern focuses on the $90 billion equivalent of Russian reserves held in RMB. With Western-led sanctions freezing roughly $315 billion in Russia's foreign reserves, Moscow could sidestep these measures by converting RMB into dollars or euros.[89]

Also of concern is China's behavior within IFIs that demonstrates its desires to keep the structure while modifying the balance of power in each along with its development of new organizations. Unhappy with the World Bank's power structure and policies, China has created similar organizations such as the BRICS New Development Bank (NDB) and the Asian Infrastructure Investment Bank (AIIB).[90] Even when creating organizations, China borrowed much of the same structure from the IFIs it complained about, but with a twist; it was now in charge. This suggests that the structure was not the real issue, but who held the power.[91] The PRC has been clear in its desire to reform global governance, giving itself and other developing countries greater say in global decision-making. China has pressed for these organizations to match its rising economic status. With its rise in power and wealth China has been quite effective in positioning itself accordingly within these organizations. For example, the PRC has been successful in growing its shares within the Bretton Woods organizations. In so doing, it has achieved a third ranking status (6.08 and 5.36 voting percentage in IMF and WB, respectively) compared to the first ranked United States (16.50 and 15.87 in IMF and WB, respectively) while similarly expanding its staff representation and funding in both. It has also been successful at influencing policy changes within the IMF.[92] Likewise, China has expanded its control in other developmental organizations such as the International Civil Aviation Organization (ICAO), the International Telecommunication Union (ITU), the United Nation's Department of Economic and Social Affairs (DESA), and the Food and Agriculture Organization by assuming the top management positions in each. In 2021, China also competed for but lost the top staff position at World Intellectual Property Organization (WIPO).

Implications

Reestablishing itself as the Middle Kingdom means China desires regional, if not global, leadership. Within international finance institutions the PRC has steadily pressed for reforms favoring Chinese authority, and its bilateral relations have sought to gather the necessary support to cement its position internationally. Although concerned over the dependencies created by globalization, the PRC will likely continue to expand its global outreach as it perceives this as necessary for closing the gap with developed economies. Pu and Wang offer that China has sought "continuity through change" and it no longer emphasizes a low profile with Xi Jinping focusing on global governance and Chinese solutions as part of his China Dream.[93] Still, Pu argues in his book that China has also displayed an ambivalent and even reluctant attitude toward global leadership; that central government messaging touting China as a rising power is more directed on domestic audiences while it attempts to downplay its status with international audiences.[94] Yet, Chinese scholars are in an ongoing debate about whether China is strategically overstretching in its goals and timing.[95]

Although China's behavior during its first twenty years in international organizations may downplay China's scope of leadership, its behavior over the last twenty years suggests that China is maneuvering into a position of global prominence. The PRC's rule-changing behavior in LIOs emphasizes that China will use whatever leverage it has to better position itself on the international stage. Recently, during the 19th CCP Congress, Xi Jinping lauded China's move to global leadership describing China as either a "great power" or "strong power" 26 times during the report of the event.[96] China's willingness to establish new organizations, its push to rival the U.S. dollar with the RMB, its efforts to become a technological frontrunner, combined with a myriad of bilateral economic strategies, all suggest a posturing for global leadership. China's behavior in IFIs further suggests that it desires to emulate the U.S. position within them as a rising power. In fact, China's deliberate and systematic approach to increasing its influence suggests that it wants to retain the current system while slowly transitioning into a leadership role.

MAINTAINING AMERICA'S COMPETITIVE ADVANTAGE THROUGH RELATIONSHIPS, INSTITUTIONS, AND INNOVATION

While serving as an Army planner in 2006 just north of Baghdad, I distinctly recall a fellow officer pontificating that the U.S. should pull back within the forward operating bases (FOBs) in response to the ever-increasing attacks. It was his supposition that if the U.S. military were not outside the gates intermixing with the populace and operating right at the doorstep of the insurgency there would be much less violence, at least against U.S. forces. In truth, the very opposite happened. When U.S. forces withdrew, the insurgency overtook the population and pushed the fight to right outside the compound's walls, and since we shared the base with Iraqi soldiers whose families also lived outside the FOB, we sometimes found the fight within our own walls. Instead, we found greater success pushing outside our fortifications to secure and enable the population. Counterinsurgency doctrine has taught us that the only way to successfully fight an insurgency is to win over the population, and to do that, you must be with, protect, and demonstrate that there is a better way than what the other side is offering.[97]

I offer this experience as an analogy for where the United States finds itself today at the global level in its great power competition with China. Drawing upon these operational considerations offers lessons that can be applied at the strategic level in dealing with China. We are seeing the PRC infiltrating the developing world, offering its illiberal methods as a counter to the more liberal West. Today we see Chinese, their capital, and their companies embedded within much of the global south. Deng's "hide and abide" policies and Xi's now more assertive stance with Mao's revolutionary tactics of consolidating power until the conditions are set are like the insurgency faced in Iraq. We are also seeing China's increased presence within global governance and a pressure to decrease the U.S. status. If so, the U.S. response should not be to retreat from the international system it created and led for almost 70 years. Instead, the United States must become even more engaged.

The findings of this paper demonstrate how the PRC has employed an economic statecraft that used its influence from within international organizations, its trade connections, its capital, and its understanding of the developing world to steadily build its power vis-à-vis the United States. Chinese leadership have targeted those crucial areas in which the United States has held a comparative advantage for so many years—relationships, global governance, and innovation. Withdrawing is not the answer, neither is any type of war (military or economic) with the PRC. Instead, the United States should become even more involved and proactive within the LIO it helped build. Through this connection, the Unites States should take back on the mantle of encouraging what Robert Zoellick described as "being a responsible stakeholder,"[98] and the United States should work with its allies and partners becoming more involved in LIC development to ensure the lessons these countries learn orient them toward a liberal system. Finally, the United States must maintain its own innovative edge, which has allowed it to achieve the forefront in technological knowhow and development but has also allowed the United States to come up with solutions to often complex problems. To do so, the United States should focus on three lines of effort centered on maintaining relationships, leveraging its structural edge in global governance, and renewing its edge in innovative technologies.

Maintain U.S. Relationships

During the 2022 Association of the United States Army (AUSA)–hosted LANPAC Symposium and Exposition, the Chief of the Australian Army, Lt-Gen. Richard M. Burr, argued during his keynote presentation that "it doesn't matter what the question is, the answer is always partnerships." For the United States, these words have held true throughout much of its history. In Europe, the United States alliance with the NATO proved crucial to deterring communist expansion during the Cold War era. Likewise, U.S. alliances with Korea, Japan, the Philippines, Thailand, and Australia as part of the "hub and spoke" approach, proved necessary for adverting similar threats in the Asia–Pacific region. Just as the Foster and Hillison chapter argued, "the best of all possible worlds . . . is one in which NATO and the European Union work closely with each other and the United States," this chapter likewise argues that NATO and treaty alliances like it remain crucial to maintaining an edge against China. The binding qualities of a treaty can help to assure a lasting relationship, while promoting shared values and shared costs. Such enduring relationships can provide the United States with the necessary coalition to constrain China and encourage them to behave more responsibly.

Still, as a military planner in the Asia–Pacific region, I remember having discussions on country prioritization, since we did not have the forces or capital to share equally across the theater. I can remember feeling proud of our product that prioritized allies and critical core partners. However, the recent agreement between China and the Solomon Islands[99] has caused me to now question this conceptualization, wondering if we needed to have broadened our thinking. While the United States has often focused its priorities, and capital, onto its allies, China has deliberately and systematically applied its economic prowess onto developing countries to siphon their support away, and they did so at a fraction of the cost. Any suggestion to change U.S. methods may seem ludicrous from the American perspective where "might makes right" centering around the idea that capability and capacity have often carried the day. We are more direct in our actions, and games like chess and American football provide good metaphors for how the United States likes to approach problems. We care little about the "pawns" and focus on the "queen" or the 240-pound linebacker who can run a sub-47 second 40-meter dash. Even the U.S. military touts the idea of mass and firepower as helping to achieve victory in past conflicts. So, of course, the United States desires arrangements with partners such as Japan, South Korea, Thailand, and the Philippines over island countries that may not even have a military. But, with the Chinese, the United States is playing two games—one is chess, and the other is Go. This latter game is all about the numbers and prioritizes influence and posture in the developing world. This developing world provides votes, business, and geographic positioning. Where the United States isn't involved, China is. Therefore, the United States must reconcile our penchant for focusing on the center of gravity and acknowledge that it is also participating in a game of give and take where an indirect approach might be called for. It must maintain its alliances and core partners who have the capabilities and capacity to share in stability concerns, but also do what is necessary to limit the PRC's hold over other countries within the U.S. sphere of influence.

Still, the United States cannot be everywhere. For America to maintain its competitive advantage over China, we may need to rethink our previous strategies that often discounted lesser developed states. The United States should continue with a bifurcated strategy of maintaining its alliances as foundational to its foreign policy while considering ways to buttress developing nations against Chinese manipulation. This does not mean the need to stop contributing to alliances, but instead to shift to an approach where the United States diversifies its support. We do it in stocks and bonds, why not in countries where not doing so cedes influence to the Chinese? Instead of spending $10 billion on an ally, spend $9 billion on the ally and $1 billion spread out among multiple developing countries. In some cases, the smaller amount spent onto a smaller state may make the difference in losing a partner to China. Yet, this goes beyond what the U.S. government can, or should do on its own. As in the past, the United States must leverage international organizations like the IMF and WB, and other international economic institutions to promote liberal economic values through economic and technical assistance. This also means the United States needs to reinvigorate its leadership within these organizations. Likewise, the United States should encourage its allies to share the burden in supporting developing countries.

Under this theme, the United States should not concede developing country modernization to the Chinese. Partnerships also help to expose countries to U.S. values while providing them alternatives to China. It appears the United States has, at times, forgotten that for developing countries to bridge the gap from LIC to MIC they must create infrastructure as well as domestic institutions. The irony that we see today is that the CCP claims that it has been Marxist ideology conjoined with Chinese characteristics that has allowed modernization, when in actuality, the conservative adherence to these traditions has presented the greatest roadblocks to China's economic reforms.[100] Regardless, the PRC has aligned its own interests with the infrastructure needs of other developing countries by diversifying its access to resources and trade, thus creating a "win–win" strategy within the global south. To counter this, the United States should work closely with IFIs, its allies, and core partners to determine how to offer developing countries better alternatives than BRI that also do not leave them easily leveraged by China. The current Biden administration's Build Back Better World initiative is one such example that focuses on creating choices.

The U.S. management of relationships does not stop with its allies and partners. It must also carefully manage its relationship with China. The U.S.–Sino relationship, by its very nature, will not be another Cold War.

The United States and the Soviet Union were completely decoupled, so a strategy of containment with literal and ideological walls between them made sense. Contrastingly, the United States and China are very much connected, with their economies intertwined. Although there may be similar strategies that can be learned from the Cold War era and applied today, both countries must be cognizant that malign approaches against one another can have unfavorable effects on both parties. The 71st U.S. Secretary of State Antony Blinken has characterized this relationship as one of constant adjustment where the United States "is competitive where it needs to be, cooperative when it can be, and adversarial when it must be."[101] With ever increasing PRC assertiveness and a belief by party leadership that in a game of "chicken" the United States will likely back down, the United States must counter China through a coalition of allies and partners while clearly communicating its interests and red-lines. Where ambiguity may have afforded previous administrations a degree of flexibility in their foreign policy, the circumstances today are different. The U.S. competitive advantage in Asia is through its allies and partners. Compound this assumption by the fact that one core partner (Taiwan) and one crucial ally (South Korea) dominate the semiconductor industry. Allowing the PRC to forcefully occupy Taiwan would not only undermine the United States' reputation as a credible partner but it also cedes a vast portion of an industry vital to the United States into Chinese hands.

Maintain U.S. Structural Edge in Global Governance

Early in its opening up period, the PRC recognized that to learn from the international system and to make changes within this system, that China had to be a part of it. Nowhere is this more evident than from the CCP's internal discourse leading up to its decision to join the WTO. The party determined that regardless of its reservations over the concessions necessary for WTO membership, the only way it could make changes was through membership. The United States is, however, already a leading member in most international organizations today, and in most cases, still exhibits active leadership. Through its influence in these organizations, the United States must continue to press for liberal practices. The United States must not succumb to China's pressure for more so-called democratic solutions. As the United States is a proponent for spreading democracy worldwide, this may at first glance seem appealing. However, China's desire is for developing countries to have an increased say on the running of global governance, though not necessarily in following established rules such as the protection of intellectual property and free and open markets. China offers its system as the example for emerging economies to pursue for modernization. By asserting ever increasing leverage through its trade and investments, and promoting an illiberal path to modernization, there is a potential that giving developing countries greater voice in the LIO could erode their liberal foundations. Therefore, it is important for the United States to remain engaged to ensure that illiberal ideology does not take these organizations onto a less optimal path. In keeping with this theme, the United States should join institutions and organizations where China and the developing nations are members. For example, although the United States adheres to United National Convention on the Law of the Sea (UNCLOS), it is not a member. Joining such an institution would provide greater weight to its argument that UNCLOS members, such as China, should follow its prescriptions. Likewise, instead of boycotting China-led international organizations, the United States should instead seek membership. The AIIB is one such example. To date, the organization has not exhibited any undue influence by the PRC.[102] U.S. membership combined with the influence of other like-minded members could help persuade China to allow this organization to maintain its autonomy. Additionally, greater U.S. presence working with allies and partners within more IGOs can help ensure liberal standards that some countries attempt to evade through forum shopping.

Second, an active U.S. participation helps to ensure a quality control over liberal principles within IFIs. The United States should therefore continue to use its position as a global power to encourage the PRC to act responsibly within the LIO. There are many examples where China's membership has caused it to conform to liberal economic practices. China's membership in the WB and IMF has led to much of its domestic institutional advancements that have allowed it to achieve economic modernization so quickly; yet, its participation has also required China to incorporate more liberal institutions into its domestic structures and international interactions.[103] China's membership in the WTO helped influence its commitment to the IMF's Article VIII[104] (which obligates a state to promote currency convertibility) and persuaded adjustments to property rights laws and an improved relationship with WIPO.[105] Likewise, the IMF's decision to include the RMB into the SDR pool has made it more difficult for China to maintain control over its exchange rate policies adding some merit to the "binding" contention.[106] Additionally, regulatory institutions that have international sway like the U.S. Federal Aviation Administration (FAA) can create a degree of leverage over China in ensuring quality control.[107] Under the context of this argument, the United States should join the Comprehensive and Progressive Agreement for Trans-Pacific Partnership (CP-TPP) and work to include the PRC if it adheres to the high expectations of membership.[108] This agreement has provisions to limit the role of SOEs, and with the combined oversight by the United States and Japan, could help constrain China to follow even more liberal trade practices. Clearly, the PRC's behavior will need monitoring as China's participation in other IFIs has exhibited questionable behavior at times.[109] From within, the United States could use the binding qualities of these institutions to hold PRC behavior to that of a responsible stakeholder.

Third, the strength of the dollar and leadership over international finance mechanisms provide the United States structural capacity that China does not currently have a response to. Therefore, China has attacked the strength of the dollar, desiring for a replacement. With Russia's recent occupation of Ukraine, the United States was able to lead a Western response in retaliation. Key to the severity of this reaction was the status of the dollar and the use of organizations like the Society for Worldwide Interbank Financial Telecommunication (SWIFT) in international finance allowing for sanctions to have greater impact on Moscow. Given the PRC's stance on the invasion, imagine how this would have played out in an international order under less democratic leadership with the RMB as the global currency. However, the threat to the dollar's prominence is less an external problem than it is an internal one. The U.S. national debt levels and its persistent balance of trade deficits create an even bigger risk, as does the extensive use of sanctions by the United States.[110] Therefore the United States must maintain its leadership within the LIO, and from this position, establish a liberal standard of economics for all developing countries, while also getting its own house in order. Maintenance of the dollar as the international currency remains one of the United States' key tools for global leadership.

Maintain U.S. Edge in Innovation

"Over the past century the United States has earned and maintained its status as the world's leader in technological innovation," but "losing this edge translates into a loss of hard power, which in turn translates into a loss of soft power, which in turn translates into an inability to influence matters that affect U.S. national security and interests."[111] Still, the United States continues to lead global competitors, to include China, when considering its ranking by the Global Innovation Index (GII).[112] Throughout much of this time the U.S. government has driven this innovation, but the paradigm is changing, with private and academic sectors taking the lead and small start-up firms making much of the significant advances.[113] Consequently, research organizations like the National Research Council argue for the United States to adjust its policy approach to innovation.[114] In the semiconductor sector, for example, an area that China aspired to take a lead in, U.S. private companies dominate "high-value research and design" while Chinese firms account for "lower-value assembly, testing and packaging."[115] The preceding review of PRC strategies highlights that China has sought to gain an innovation edge by greater government oversight and control. Conversely, innovation derives from greater intellectual freedom and competition. Therefore, if the United States is to maintain a competitive edge in innovation, its government must adjust its long-held practices by working together with private industry and academia[116] by focusing more on open technological strategies.[117] But, in an era of globalization in which the PRC has ever increasing leverage, the United States should not restrict its open strategies to its borders; its response must also include working closely with its allies if it is to adequately limit malign Chinese behavior.[118] The Committee on Foreign Investments in the United States (CFIUS) work with the European Union in the U.S.–EU Trade and Technology Council (TTC) to control Chinese investments in key industries and minimize intellectual property theft is one such example.

As argued in the section under relationships, maintaining the United States' edge in innovation also does not mean completely isolating China. There are areas where Chinese technological advancements have been beneficial to U.S. innovation.[119] Instead, the U.S. government should also be vigilant in differentiating constructive from malign Chinese innovative policies applying its own tools of economic statecraft combined with relationships and structural mechanisms of global governance to constrain China to more responsible behaviors.

CONCLUSION

A review of Chinese strategic documents suggests the CCP has instituted a comprehensive planning effort for achieving the China Dream of national rejuvenation. These combined with party rhetoric and analysis of China's economic statecraft suggest that the state is very much in control of its economic path forward. Although there were periods where the ways and means of economic statecraft were not as coherent in achieving desired ends, the state has readily adapted. Its planning efforts have been fluid with adjustments made to build upon success or modify as required. Under Xi, greater centralization of economic statecraft is evident, which is allowing the CCP to better manage the state's path along its economic development. Consequently, China has been somewhat effective in overcoming the friction between liberal market practices and its illiberal ideologies and continues to exhibit progress toward achieving its short- and long-term goals.

Most concerning to the America is the CCP's blatant desire to reform the LIO while presenting itself as the model vis-à-vis the United States. China has used economic statecraft to gain wealth, but now it is doing so to gain power hoping to eventually dethrone the United States. Its methods have been directed at those areas where the United States has long been strong. China has attempted to siphon away from U.S. relationships, make reforms in the current system of global governance, and become a leading global innovator. To deflect China's attempts at undermining America's competitive edge, the United States must prioritize back onto its relationships abroad, while retaining and even expanding its structural edge in global governance and establishing a more holistic approach to innovation.

NOTES

1. For literature discussing China's use of liberal and illiberal methods see Roselyn Hsueh, *China's Regulatory State: A New Strategy for Globalization* (Ithaca, NY: Cornell University Press, 2011); Clara Weinhardt, and Tobias ten Brink, "Varieties of Contestation: China's Rise and the Liberal Trade Order," *Review of International Political Economy* 27, 2 (2002): 258–280, https://doi.org/10.1080/09692290.2019.16 99145.

2. Catherine Wearver, "The Rise of China: Continuity or Change in the Global Governance of Development?," *Ethics & International Affairs*, 29, 4 (2015), 424–425.

3. Daniel W. Drezner, *The System Worked: How the World Stopped Another Great Depression* (New York, NY: Oxford University Press, 2014).

4. Alastair Iain Johnston, *Social States: China in International Institutions, 1980–2000* (Princeton, NJ: Princeton University Press, 2008), and James Frick, "China in International Organizations: National Interest, Rules and Strategies." Ph.D. dissertation, Temple University, ProQuest Dissertations and Theses.

5. Christopher Herrick, Zheya Gai, and Surain Subramaniam, *China's Peaceful Rise: Perceptions, Policy and Misperceptions* (Manchester, UK: Manchester University Press, 2016), 57–58.

6. Tom Miller, "Great Leap Outward: Chinese ODI and the Belt & Road Initiative," in *To Get Rich Is Glorious: Challenges Facing China's Economic Reform and Opening at Forty* (Baltimore, MD: John Hopkins University Press, 2019), 233–260.

7. Stig Stenslie and Chen Gang, "Xi Jinping's Grand Strategy," in *China in the Era of Xi Jinping: Domestic and Foreign Policy Challenges*, eds. Robert S. Ross and Jo Inge Bekkevold (Washington, DC: Georgetown University Press, 2014).

8. Aaron L. Friedberg, *A Contest for Supremacy: China, America, and the Struggle for Mastery in Asia* (New York, NY: W.W. Norton, 2011); Avery Goldstein, *Rising to the Challenge: China's Grand Strategy and International Security* (Stanford, CA: Stanford University Press, 2001).

9. Steven W. Mosher, *Hegemon: China's Plan to Dominate Asia and the World* (San Francisco, CA: Encounter, 2000).

10. Yan Xuetong, *Ancient Chinese Thought, Modern Chinese Power* (Princeton, NJ: Princeton University Press, 2011). Dr. Xuetong is a member of the CCP who serves as a distinguished professor and dean at the Institute of International Relations at Tsinghua University. He is the founder of "moral realism" and in 2008, *Foreign Policy* named him one of the world's "Top 100 Global Thinkers."

11. James Frick and Roselyn Hsueh, "OFDI: A Chinese Foreign Policy Tool." Unpublished work presented at the annual conference of the 2021 American Political Science Association.

12. OECD Investment Policy Review (2018), 66.

13. Randall Morck, Bernard Yeung, and Minyuan Zhao, "Perspectives on China's Outward Foreign Direct Investment," *Journal of International Business Studies* 39 (2008), 337–350; M. Sanfilippo, "Chinese FDI to Africa: What Is the Nexus with Foreign Economic Cooperation?" *African Development Review* 22, 1 (2010), 599–614; Shun-Chiao Chang, "The Determinants and Motivations of China's Outward Foreign Direct Investment: A Spatial Gravity Model Approach," *Global Economic Review* 43, 3 (2014), 244–268; Roselyn Hsueh and Michael B. Nelson, "China and the Telecommunications Revolution in Africa: The Politics of Infrastructure Development for Resources and Markets" (2018), https://ssrn.com/abstract=3069863; and Milan Babic, Javier Garcia-Bernardo, and Eelke Heemskerk, "The Rise of Transnational State Capital: State-Led Foreign Investment in the 21 Century," *Review of International Political Economy* 27, 3 (2020), 433–475.

14. Roselyn Hsueh, *China's Regulatory State: A New Strategy for Globalization* (Ithaca, NY: Cornell University Press, 2011); Roselyn Hsueh and Michael B. Nelson, "China and the Telecommunications Revolution in Africa: The Politics of Infrastructure Development for Resources and Markets" (2018), https://ssrn.com/abstract=3069863; Elizabeth C. Economy, "China's New Revolution: The Reign of Xi Jinping," *Foreign Affairs*, 97, 3 (2018), 60–74; Min Ye, *The Belt Road and Beyond: State-Mobilized Globalization in China: 1998–2018* (New York, NY: Cambridge University Press, 2020).

15. William Norris, *Chinese Economic Statecraft: Commercial Actors, Grand Strategy, and State Control* (Ithaca, NY: Cornell University Press, 2016), 170. Wise also found evidence that China's economic relationship with Costa Rica was explicitly based on enticing the country to change diplomatic relations. Carol Wise, *Dragonomics: How Latin America Is Maximizing (or Missing Out on) China's International Development Strategy* (New Haven, CT: Yale University Press, 2020), 231–233.

16. OFDI research covering the period from 2001 to 2017 found that China invested OFDI into only 4 of the 19 countries that recognized Taiwan. Two of these four countries switched their diplomatic relations to the PRC during this period. Frick and Hsueh, "OFDI: A Chines Foreign Policy Tool," 6.

17. Stephen B. Kaplan, "Banking Unconditionally: The Political Economy of Chinese Finance in Latin America," *Review of International Political Economy* 23, 4 (2016), 644–651; and Stephen B. Kaplan, *Globalizing Patient Capital: The Political Economy of Chinese Finance in the Americas* (New York, NY: Cambridge University Press, 2021).

18. Mingjiang Li, "The Belt and Road Initiative: Geo-economics and Indo-Pacific Security Competition," *International Affairs* 96, 1 (2020), 169–187.

19. Xiaoyu Pu and Chengli Wang, "Rethinking China's Rise: Chinese Scholars Debate Strategic Overstretch," *International Affairs*, 94, 5 (2018), 1032.

20. Francesca Congiu and Christian Rossi, "China 2017: Searching for Internal and International Consent," *Asia Maior*, 28 (2017), 76–77.

21. Following the 2008 global financial crisis, China aligned itself with BRIC countries to press for a greater voice through the G20 while increasing pressure on the WB and IMF to increase its leadership role in the organizations and achieve some concessions particularly in the IMF. Specifically, it influenced the IMF to add the RMB to the SDR basket. James Frick, "China in International Organizations: National Interest, Rules and Strategies." Ph.D. dissertation, Temple University (2021), ProQuest Dissertations and Theses, 215.

22. Elizabeth C. Economy, "China's New Revolution: The Reign of Xi Jinping," *Foreign Affairs* 97, 3 (2018), 187.

23. The report finds that although the CCP would use trade restrictions to punish or change the behavior in target states while providing unrelated official reasons to disguise its coercion and allow deniability. In so doing, the CCP would balance between punishing a country just enough to change the behavior without driving the country away. The report provided examples of where China would block one sector while increasing another to illustrate its position. The report also found that China used tourism as an economic lever. Given that China accounts for more than 20% of global tourism, it has been able to punish states by restricting its outbound tourism market. F. Hanson, E. Currey, & T. Beattie, *The Chinese Communist Party's Coercive Diplomacy* (Australian Strategic Policy Institute, 2020), 11–17. http://www .jstor.org/stable/resrep26121.

24. This retrenchment continued with the conservative ratio increasing under the Xi-Li leadership as four of seven Standing Committee members were conservatives. Jiang, "The Limits of China's Monetary Diplomacy," 160.

25. Yang Jiang, "The Limits of China's Monetary Diplomacy," in *The Great Wall of Money*, Eds. Eric Helleiner and Jonathan Kirshner (Ithaca, NY: Cornell University Press, 2014), 159–160.

26. OECD Investment Policy Review, 2018.

27. Roselyn Hsueh, *China's Regulatory State: A New Strategy for Globalization* (Ithaca, NY: Cornell University Press, 2011).

28. Wendy Leutert, "Firm Control: Governing the State-Owned Economy Under Xi Jinping," *China Perspectives* (2018), 1–2, 27–36.

29. Elizabeth C. Economy, "China's New Revolution: The Reign of Xi Jinping," *Foreign Affairs*, 97, 3 (2018), 60–74.

30. Additionally, privately owned companies seeking to invest overseas must be filed and approved by the National Development and Reform Commission (NDRC) and the Ministry of Commerce (MOFCOM). SOEs may also need to seek additional approval through the State-owned Assets Supervision and Administration Commission (SASAC).

31. Party direction such as the 12th and 13th five-year plans and the Made in China 2025 emphasize the Chinese government's attempt to establish state unity. For example, these documents prioritize overseas transfer of high-end equipment, advanced technology, access to resources, and market-seeking investments in high-tech and medium-high-tech intensity areas. Chinese Communist Party, *The 13th Five-Year Plan for Economic and Social Development of the People's Republic of China: 2016–2020* (Beijing, China: Central Compilation & Translation Press), https://en.ndrc.gov.cn/policies/202105/P020210527785800103339.pdf and China State Council (Made in China 2025), 31–37, http://www.cittadellascienza .it/cina/wp-content/uploads/2017/02/IoT-ONE-Made-in-China-2025.pdf.

32. State Council Information Office of the PRC (SCIO), *China's Epic Journey from Poverty to Prosperity* (Beijing, China, 2021), 9.

33. *Ibid.*, 10.

34. The quality-of-life Engel coefficient has improved for urban residents from 57.5% in 1978 to 29.2% in 2020; rural residents' coefficient has improved from 67.7% to 32.7%. It has established the world's largest education system, ranking in

the upper-middle category worldwide in modern education. Average life expectancy is 77.3 years and infant mortality is down to 5.4 per 1000. *Ibid.*, 23.

35. Frick (2021) argues that within the context of relative shifts in power, China has exhibited varying behaviors in different organizations based upon the extent to which IGOs conformed to China's national interests. China has followed a rational behavior approach consisting of strategies of rule-taking, rule-breaking, rule-changing, and rulemaking. As China's power has increased so has its bargaining power, "leading to assertive rule-changing attempts to adapt IOs to allow its ascendancy as a rule-maker." "China in International Organizations: National Interest, Rules and Strategies." Ph.D. dissertation, Temple University, ProQuest Dissertations and Theses.

36. The amount of China's GDP spent on electronic components has hardly changed since 2012 (2.7% in 2012 and 2.6% of GDP in 2020), and it still relies heavily on Western and Western-leaning markets for its high-tech exports. There is also no evidence that the renminbi has begun to replace the dollar in Asia or elsewhere, with cross-border payments averaging around 2% (as recorded by SWIFT). Additionally, the Chinese banking system plays a small role in global finance when compared to the United States with Chinese markets more heavily influenced by G7 economies. "Fortified but Not Enriched," *The Economist* (May 28th, 2022), 18–20. https://www.economist.com/briefing/2022/05/26/china-is-trying-to-protect-its-economy-from-Western-pressure.

37. Creel, H. G., *Chinese Thought from Confucius to Mao Tse-tung* (Chicago, Illinois: University of Chicago Press, 1953), 255–256.

38. Xi Jingping, Speech at the Ceremony Marking the Centenary of the Communist Party of China (July 1, 2021), 5 and 7, http://www.xinhuanet.com/english/special/2021-07/01/c_1310038244.htm.

39. Francesca Congiu and Christian Rossi, "China 2017: Searching for Internal and International Consent," *Asia Maior*, 28 (2017), 62.

40. H. G. Creel, *Chinese Thought from Confucius to Mao Tse-tung* (Chicago, Illinois: University of Chicago Press, 1953), 247.

41. *Ibid.*, 256.

42. Joseph Esherick, "How the Qing Became China," in *Empire to Nation: Historical Perspectives on the Making of the Modern World* (Rowman & Littlefield, 2006), 232–233.

43. Angang Hu, Yilong Yan, Xiao Tang, and Shenglong Liu, *2050 China: Becoming a Great Modern Socialist Country* (Springer Press, 2021), 17. https://doi.org/10.1007/978-981-15-9833-3.

44. Frick's research exploring China's relationship within the World Bank and IMF contends that China underwent periods of turmoil as it attempted to negotiate change between reformists and conservatists in the party. This period was characterized by experimentation and a slowly evolving adjustment to traditional mindsets as capitalist ideas required reconciliation with Marxist constructs. James Frick, "China in International Organizations: National Interest, Rules and Strategies," (Ph.D. dissertation, Temple University, 2021), ProQuest Dissertations and Theses.

45. In 1956, the 8th Party Congress in support of the first Five-Year Plan proposed the "four modernizations" of industry, agriculture, transportation, and national defense. Since then, each Party Congress has made modifications based upon experimentation and an evolving environment. In 2017, the 19th Party

Congress further modified the second goal to become a "great modern socialist country that is prosperous, strong, democratic, culturally advanced, harmonious, and beautiful." Angang Hu, Yilong Yan, Xiao Tang, and Shenglong Liu, *2050 China: Becoming a Great Modern Socialist Country* (Springer Press, 2021), https://doi.org/10.1007/978-981-15-9833-3, 2–5.

46. Xi Jingping, Speech at the Ceremony marking the Centenary of the Communist Party of China (July 1, 2021), 1–4, http://www.xinhuanet.com/english/special/2021-07/01/c_1310038244.htm.

47. State Council Information Office of the PRC, *China's Epic Journey from Poverty to Prosperity* (September 2021), 4.

48. *Ibid.*, 6.

49. Xi has highlighted China's role in reforming the rules of the global order in response to the changing international landscape during multiple venues. In 2017, he pointed this out during a speech at the World Economic Forum (WEF). Francesca Congiu and Christian Rossi, "China 2017: Searching for Internal and International Consent," *Asia Maior*, 28 (2017), 65. In 2019, a whitepaper published by the State Council Information Office of the PRC, specifically addressing reforming the global governance system to meet the changes in the international system. State Council Information Office of the PRC Whitepaper, *China and the World in the New Era*.

50. Congiu and Rossi contend that the "getting rich" phase (1978–2012) was achieved through Deng's economic reforms giving a larger role to market forces, but with this new wealth a new ideology was necessary to address the imbalance through inadequate development. This would require responding to social conflicts through nationalism and restoring China as a world leader in science, technology, economics, and business. Francesca Congiu and Christian Rossi, "China 2017: Searching for Internal and International Consent," *Asia Maior*, 28 (2017), 67–68.

51. Xi Jinping, "Secure a Decisive Victory in Building a Moderately Prosperous Society in All Respects and Strive for the Great Success of Socialism with Chinese Characteristics for a New Era. Delivered at the 19th National Congress of the Communist Party of China" (October 18, 2017), http://www.chinadaily.com.cn/china/19thcpcnationalcongress/2017-11/04/content_34115212.htm.

52. Peter Ferdinand and Jue Wang, "China and the IMF: From Mimicry Towards Pragmatic International Institutional Pluralism," *International Affairs* 89, 4 (2013), 905.

53. John Williamson, "Is the 'Beijing Consensus' Now Dominant?" *Asia Policy* 13 (2012), 6–7.

54. SCIO, *China's Epic Journey from Poverty to Prosperity*, 67.

55. *Ibid.*, 67.

56. Angang Hu, Yilong Yan, Xiao Tang, and Shenglong Liu, *2050 China: Becoming a Great Modern Socialist Country* (Springer Press, 2021), 31, https://doi.org/10.1007/978-981-15-9833-3.

57. *Ibid.*, 57–59.

58. Xi Jingping, Speech at the Ceremony marking the Centenary of the Communist Party of China (July 1, 2021), 11, http://www.xinhuanet.com/english/special/2021-07/01/c_1310038244.htm.

59. "Fortified but Not Enriched," *The Economist* (May 28th, 2022), 18–20. https://www.economist.com/briefing/2022/05/26/china-is-trying-to-protect-its-economy-from-western-pressure.

60. APCO, 2010, "China's 12th Five-Year Plan: How It Actually Works and What's in Store for the Next Five Years," 3–5, https://sustainabledevelopment .un.org/index.php?page=view&type=400&nr=700&menu=1515.

61. These seven industries are biotechnology, new energy, high-end equipment manufacturing, energy conservation and environmental protection, clean-energy vehicles, new materials, and next-generation IT. The party views these industries as the backbone of China's economy and desires these firms to succeed on a global scale. *Ibid.*, 3.

62. Asian Development Bank, The 12th Five-Year Plan: Overview and Policy Recommendations, No. 2011-3 (2011), 5–6.

63. Central Committee of the CPC, *The 13th Five-Year Plan for Economic and Social Development of the PRC: 2016–2020*, 20–21.

64. Central Committee of the CPC, *The 13th Five-Year Plan for Economic and Social Development of the PRC: 2016–2020*, 22–31.

65. Central Committee of the CPC, *The 14th Five-Year Plan for Economic and Social Development of the PRC: 2021–2025*, 7.

66. Central Committee of the CPC, *The 14th Five-Year Plan for Economic and Social Development of the PRC: 2021–2025*, 8 & 11.

67. Central Committee of the CPC, *The 13th Five-Year Plan for Economic and Social Development of the PRC: 2016–2020*, 40–46 and 65.

68. Central Committee of the CPC, *The 14th Five-Year Plan for Economic and Social Development of the PRC: 2021–2025*, 47. As previously discussed in the literature review, this desire to increase party presence within firm boards corresponds to findings on board restructuring.

69. "Return to Picking Winners," *The Economist* 442, 9279 (January 15, 2022), S4–S6, https://www.proquest.com/magazines/return-picking-winners/docview /2619675116/se-2?accountid=4444.

70. Central Committee of the CPC, *The 13th Five-Year Plan for Economic and Social Development of the PRC: 2016–2020*, 141.

71. Central Committee of the CPC, *The 13th Five-Year Plan for Economic and Social Development of the PRC: 2016–2020*, 79–83, and 103.

72. Interestingly, the party points out IFIs are important to BRI, and specifically calls out AIIB and the New Development Bank. Like its internal plan, the PRC desires to build infrastructure networks to connect subregions and create economic corridors along these routes. The plan also highlights cultural exchanges as a tool for BRI, noting tourism as one of the mechanisms under this category. Central Committee of the CPC, *The 13th Five-Year Plan for Economic and Social Development of the PRC: 2016–2020*, 146–148.

73. Theodore H. Cohn and Anil Hira, *Global Political Economy: Theory and Practice* (New York, NY: Routledge, 2021), 163.

74. Hyo-Sung Park posits that China's motives for internationalizing the RMB are for trade facilitation, management of inflation, getting out of the "dollar trap," reforming the international monetary system, and for geopolitical considerations. The latter two reasons would be attempts to replace the dollar. Park, Hyo-Sung Park, "China's RMB Internationalization Strategy: Its Rationales, State of Play, Prospects and Implications," M-RCBG Associate Working Paper Series, No. 63. Harvard Kennedy School (2016), 7–18.

75. Central Committee of the CPC, *The 13th Five-Year Plan for Economic and Social Development of the PRC: 2016–2020*, 149.

76. Central Committee of the CPC, *The 14th Five-Year Plan for Economic and Social Development of the PRC: 2021–2025*, 102–103.

77. Lihong Li, "Localizing WIPO's Legislative Assistance: Lessons from China's Experience with the TRIPs Agreement," in *Implementing the WIPO's Development Agenda*, Ed. Jeremy De Beer (Ottawa, Canada: The Center for International Governance Innovating and Wilfrid Laurier University Press, 2009), 122.

78. *Ibid.*, 126.

79. U.S. Department of State Investment Climate Statements: China—2017 Executive Summary. https://2017-2021.state.gov/reports/2017-investment-climate-statements/china/index.html; and 2021 Executive Summary, https://www.state.gov/reports/2021-investment-climate-statements/china/.

80. James Frick, "China in International Organizations: National Interest, Rules and Strategies," Ph.D. dissertation, Temple University (2021), 197, ProQuest Dissertations and Theses.

81. *Ibid.*, 275.

82. In both the World Bank and the International Monetary Fund, the United States holds a de facto veto over structural changes within these organizations. Its shares provide the United States the threshold necessary for preventing key rule changes.

83. Peter Ferdinand and Jue Wang, "China and the IMF: From Mimicry towards Pragmatic International Institutional Pluralism," *International Affairs* 89, 4 (2013), 910.

84. Based upon statements given by both Xie Xuren (MoF) and Zhou Xiachaun (PBOC Governor). IMF, 2009 statement given by Xie Xuren at Annual Governors Meeting; and 2010 statement given by Zhou Xiaochuan at Annual Governors Meeting.

85. J. Wang, "China–IMF Collaboration: Toward the Leadership in Global Monetary Governance," *Chinese Political Science Review*, 3 (2018), 62–80, https://doi.org/10.1007/s41111-017-0085-8.

86. The SDR, unlike the U.S. dollar, lacked broad circulation, combined with the IMF not having the resources to effectively manage "a super-sovereign reserve currency." *Ibid.*, 74.

87. Cynthia Roberts, Leslie Armijo, and Saori Katada, *The BRICS and Collective Financial Statecraft* (New York: Oxford University Press, 2017), 105.

88. *Ibid.*, 102.

89. Laura He, "4 Ways China Is Quietly Making Life Harder for Russia," *CNN Business* (March 18, 2022), https://www.cnn.com/2022/03/17/business/china-russia-sanctions-friction-intl-hnk/index.html.

90. Klaus Rohland, the WB's China director, believe the AIIB was in response to the WB not increasing China's voting capital commensurate with its economic development. Klaus Rohland (WB China Director, 2010–2015), Personal Communichon, November 18, 2019. A consultant for the development of the AIIB, Dr. David Dollar, understood the act as an attempt by China to get the WB to refocus back onto infrastructure. David Dollar (WB China Director, 2004–2009), Personal Communication, October 24, 2019.

91. James Frick, "China in International Organizations: National Interest, Rules and Strategies." Ph.D. dissertation, Temple University (2021), 257, ProQuest Dissertations and Theses.

92. *Ibid.*, 257.

93. Xiaoyu Pu and Chengli Wang, "Rethinking China's Rise: Chinese Scholars Debate Strategic Overstretch," *International Affairs*, 94, 5 (2018), 1024–1025.

94. Xiaoyu Pu, *Rebranding China: Contested Status Signaling in the Changing Global Order* (Stanford, CA: Stanford University Press, 2018), 100.

95. Xiaoyu Pu and Chengli Wang, "Rethinking China's Rise: Chinese Scholars Debate Strategic Overstretch," *International Affairs*, 94, 5 (2018), 1019–1035.

96. *Ibid.*, 1019.

97. U.S. Joint Chiefs of Staff, Counterinsurgency, Joint Publication 3–24. Washington, D.C.: US Joint Chiefs of Staff, April 25, 2018, xiii.

98. Robert Zoellick, "Can America and China Be Stakeholders?" Speech given December 4, 2019, to U.S.–China Business Council.

99. On March 24, 2022, the Solomon Islands announced a policing agreement and work on a broader security agreement that could allow China to send police or military forces to the Solomon Islands. After the Solomon Islands switched their recognition from Taiwan to the PRC in 2019, China gave US$730 million to the country. Christopher Cairns and April Herlevi, "China and the Solomon Islands: Driver of Security Cooperation," *CNA* (2022), https://www.cna.org/our-media/indepth/2022/04/china-and-the-solomon-islands.

100. James Frick, "China in International Organizations: National Interest, Rules and Strategies," Ph.D. dissertation, Temple University (2021), ProQuest Dissertations and Theses.

101. Camille P. Dawson, Deputy Assistant Secretary, Bureau of East Asian and Pacific Affairs, U.S. Department of State, outlined Secretary Blinken's priorities on China during a panel discussion on "Strategic Environment: Preparedness to Meet Future Challenges" at the AUSA LANPAC Conference, May 17, 2022.

102. James Frick, "China in International Organizations: National Interest, Rules and Strategies." Ph.D. dissertation, Temple University (2021), 257, ProQuest Dissertations and Theses, 257.

103. *Ibid.*, 261–281.

104. IMF China Director Steven Dunaway believed that the committal to the IMF's Article VIII proved to be a necessary step toward entry into the WTO. Interview on April 13, 2020. Linkage of Article VIII with WTO ascension was corroborated by interviews with other IMF staff.

105. Paolo Davide Farah and Elena Cima, "China's Participation in the World Trade Organization; Trade in Goods, Services, Intellectual Property Rights and Transparency Issues," in *Trade with China: Business Opportunities, Legal Uncertainties*, Ed. Aurelio Lopez-Tarruella Martinez (Valencia: Roca Junyent, 2010), 102, 106–107.

106. Andrew Walter, "China's Engagement with International Macroeconomic Policy Surveillance," in *The Great Wall of Money*, Eds. Eric Helleiner and Jonathan Kirshner (Ithaca, NY: Cornell University Press, 2014), 154.

107. Scott Kennedy, "Touching the Elephant: Explaining Patterns of China's Innovation," in *China's Uneven High-Tech Drive: Implications for the United States*, Ed. Scott Kennedy (CSIS, 2020), 5.

108. Allen argues that joining an organization like CP-TPP, with such high standards, would serve as a "level-set" measuring tool for resolving technological and economic differences. Craig Allen, "For Cooperative Innovation, China Must Lead the Way," in *China's Uneven High-Tech Drive: Implications for the United States*, Ed. Scott Kennedy (CSIS, 2020), 53.

109. James Frick, "China in International Organizations: National Interest, Rules and Strategies," Ph.D. dissertation, Temple University (2021), ProQuest Dissertations and Theses.

110. Daniel Dresner, "The United States of Sanctions," *Foreign Affairs*, 100, 5 (September/October 2021).

111. Tina Srivastava, *Innovating in a Secret World: The Future of National Security and Global Leadership* (Potomac Books, University of Nebraska Press, 2019), 133, 140.

112. Qiu Mingda, "A Larger but Not Leaner Fat Tech Dragon," in *China's Uneven High-Tech Drive: Implications for the United States*, Ed. Scott Kennedy (CSIS, February 2020), 9–10.

113. Darren E. Tromblay and Robert G. Spelbrink, *Securing U.S. Innovation: The Challenge of Preserving a Competitive Advantage in the Creation of Knowledge* (Lanham, MD: Rowman & Littlefield, 2016), 243.

114. Tina Srivastava, *Innovating in a Secret World: The Future of National Security and Global Leadership* (Potomac Books, University of Nebraska Press, 2019), 4–5.

115. Alexander Hammer, "Exporting U.S. Innovative Capacity to China? A Case Study of Semiconductor Manufacturing Equipment," in *China's Uneven High-Tech Drive: Implications for the United States*, Ed. Scott Kennedy (CSIS, February 2020), 41.

116. Darren E. Tromblay and Robert G. Spelbrink, *Securing U.S. Innovation: The Challenge of Preserving a Competitive Advantage in the Creation of Knowledge* (Lanham, MD: Rowman & Littlefield, 2016), 244.

117. Srivastava defines open innovation as a paradigm that assumes that firms "can and should use external ideals as well as internal ideas," *Innovating in a Secret World: The Future of National Security and Global Leadership* (Potomac Books, University of Nebraska Press, 2019), 7.

118. Kevin G. Nealer, "Trading Iron Curtains for Chinese Walls: Is It Different This Time?" in *China's Uneven High-Tech Drive: Implications for the United States*, Ed. Scott Kennedy (CSIS, 2020), 49.

119. Kennedy argues that market-driven innovations have pushed firms beyond China to adapt and improve. Scott Kennedy, "Touching the Elephant: Explaining Patterns of China's Innovation," 3. Hammer argues that the U.S. semiconductor industry is "still so far ahead of China's and that migration of certain components of the industry abroad is helping the U.S. strengthen its overall place in the industry." Alexander Hammer, "Exporting U.S. Innovative Capacity to China? A Case Study of Semiconductor Manufacturing Equipment," 41.

CHAPTER 5

Reconceptualizing Small State Alignment in a Multipolar World

John A. Mowchan

INTRODUCTION

Following the demise of the Soviet Union, the United States became the world's sole great power owing to its majority control of global resources and preeminent military capabilities. Over the last several decades, however, a rising China and resurgent Russia have eroded America's material superiority and ability to exert influence around the globe. As America's relative competitive advantage continues to diminish, China and Russia now have an enhanced ability to assert regional dominance and exert influence over small states.[1] This is particularly true for small states with no formal membership in a military alliance or economic bloc located in strategically important locations around the world. As a result, small states have the potential to acquire a higher degree of strategic freedom to align with one or more of America's great power adversaries. Such actions could in turn further diminish America's global competitive advantage. This strategic reality illuminates the central role alliance politics will play in the twenty-first century and the necessity of discerning the alignment tendencies of small states with one or more great powers.

In the post–Cold War era, scholars have developed a wide array of typologies to capture the nature of alliance formation ranging from hard balancing to bandwagoning.[2] These typologies have some merit; however, the locus of analysis predominately focuses on the most powerful states in the international system. When scholars have considered small states, they have often employed realist approaches that focus on anarchy, the balance of military power, and other material-based considerations.[3]

Certainly, realist explanations carry a high degree of parsimony, but such an approach does not adequately account for the full range of state alignment options in the post–Cold War era. Furthermore, academics and practitioners often erroneously define "alliance" and "alignment" in the same way when they convey markedly different meanings. By conflating the term "alliance," which is based on a formal written agreement between states, with the term "alignment," which conveys an informal or less-binding set of expectations, this prevents a more holistic understanding of alliance politics. Taken together, a fresh approach is needed to understand the alignment tendencies of small states in a multipolar world.

Given the preceding, this chapter provides military strategists and policy analysts an ideational-based reconceptualization of small state alignment. This new analytical framework is grounded in several key assumptions. First, alliance politics is best understood as a continuum where small states will first seek to align with one or more great powers by establishing mutually beneficial foreign policy relations. A small state's alignment with a great power may or may not eventually lead to formal membership in a military alliance or economic bloc with a great power. Second, a constructivist approach to alliance politics focused on the ideas and beliefs of leading political elites can often better explain small state alignment than realist-based models. The validity of a leader-centric analysis takes its lead from Bruce Bueno de Mesquita, who notes "leaders, not states, choose actions. Leaders and their subjects enjoy the fruits and suffer the ills that follow from their decisions."[4] These assumptions set the foundation by which small states, with no alliance affiliation, will pursue one of two alignment strategies with the great powers: positive bandwagoning or great power bridging.

Positive bandwagoning occurs when ruling elites in a small state believe cooperation with only one great power is in the best, long-term good of the state and their individual political interests. In this case, the state adopts more cooperative foreign policies with one great power over another even when there are no guarantees such actions will lead to the country's accession into a military alliance with which the target great power is associated. Great power bridging occurs when political leaders in a small state almost equally cooperate with two or more great powers. It is through bifurcated cooperation that these ruling elites seek to maximize the economic, political, or military benefits to the state and their interests, while also seeking to avoid a perception of favoritism, which could elicit a backlash from any great power. These new alignment strategies can subsequently frame Washington's bilateral cooperation with West-leaning states seeking an alignment with the United States.

The alignment of small states located in strategically important regions of the world can help bolster America's competitive advantage in relation to its adversaries in several ways. First, small states can provide geographical access that allows U.S. military forces to respond to national security threats in more timely manner. Second, small states rich in natural resources or rare earths can allow America to be less dependent on adversarial great powers currently supplying the global market. Third, small state alignment with the United State and its allies provide Washington an opportunity to promote democracy and a liberal world order. These benefits illuminate the importance of being able to identify and then subsequently foster mutually beneficial bilateral relations with small states outside the formal U.S. alliance network.

This chapter is divided into five sections. The first section provides a brief literature review of the post–Cold War alliance strategies. Using key concepts developed by Randall Schweller, the second section presents a new model on small state alignment. This model theorizes that in a multipolar world, small states with no formal membership in a military alliance will pursue one of two alignment strategies: positive bandwagoning and great power bridging. The third section details the research design. In the fourth section, Ukraine is used as a case study to demonstrate how both strategies were employed from 1995 to 2015. The last section provides a conclusion.

ALLIANCE STRATEGIES IN THE POST–COLD WAR ERA: A REVIEW

In an era where America's relative material advantage vis-à-vis other great powers continues to narrow, the United States must bolster its global network of friends and partners. By establishing new bilateral relationships with small states outside the current alliance network, Washington will be better positioned to advance U.S. core national interests in key geographical regions around the world. More importantly, these new foreign policy relations may be able to improve America's competitive advantage over its adversaries. Because other regional great powers, such as China and Russia, are also vying for influence with these nonaligned states, it is imperative U.S. policymakers and strategists be able to identify early on ruling governments and political elites with a proclivity to align with the United States.

In reviewing the literature on alliance politics, scholars have developed various typologies to capture the nature of alliance formation since the end of the Cold War. A critique of the leading approaches, however, finds many of the major alliance strategies focus almost exclusively on the great powers, giving little attention to the strategic options available to small states. For example, following the collapse of the Soviet Union, realists argued hard balancing would remain the predominate tendency of states in the international system. This was certainly true with China and Russia where both countries implemented large-scale defense modernization programs and created new alliances such as the Collective Security Treaty Organization and the Shanghai Cooperation Organization (SCO) to balance against the United States.[5] Additionally, since 2005, Russia and China have conducted numerous combined military exercises, thereby improving Moscow's and Beijing's ability to counter Washington's moves on the global stage.[6] Although hard balancing can help us to understand the actions of Russia and China, this strategy predominately focuses on those countries residing near the top of the international hierarchy. Small states are not considered because they often do not have the material resources or military capabilities to effectively implement this type strategy. The underlying logic here is that only a great power has the capacity to hard balance against another great power.

Some scholars have attempted to discern when balancing may apply to small states. Remaining true to structuralist assumptions, a small state residing in the shadow of one great power may attempt to balance against this threat by siding with a great power from a different region of the world. Although on the surface this may appear as bandwagoning, in reality a small state will seek to enter into an alliance with an extraregional power as a way to balance against the closer, more powerful state.[7] In this strategy, geography is the key driver moving small states to join a military alliance with one great power over another. Stephen Walt applies this balance-of-threat approach to explain why some former Soviet states including Estonia, Latvia, Lithuania, Poland, and Hungary sought out NATO membership as a way to increase their material power and guarantee their security against a historically aggressive Russia. The advantage with this approach is a small state can rely on the superior material capabilities of an extraregional great power to help guarantee its security against another more powerful or threatening state. The risk with this strategy, however, is that military assistance may arrive too late to help a small state survive an armed attack. Additionally, a small state's balancing actions could, in itself, elicit a military or economic backlash from the neighboring great power.

On one level, a strategy of regional balancing against a more powerful (or more threatening) great power appears to be a valid strategy in a multipolar world. On another level, if one proceeds on assumption that Russia is the most powerful and/or most threatening country in Europe, what then explains a decision by Belarus and Armenia to enter into an alliance with its former imperial center? Certainly, one can argue close historical and/or cultural ties explain the alliance patterns of these former Soviet states; however, these explanations fall short in Ukraine where at times the country has sought closer ties with the West under Presidents Viktor Yushchenko, Petro Poroshenko, and Volodymyr Zelensky, but under Viktor Yanukovych, the ruling government aggressively pursued closer economic and security relations with Russia. This puzzle illuminates the shortcomings of alliance politics as we understand them today.

One of the most frequently employed alliance strategies by small states is bandwagoning, which can occur in several different ways. One way takes its cue from Walt, who focuses his analysis on what can be described as the "shadow of the future threat."[8] In this case, a small state may bandwagon with a great power to stave off the potential for a future armed attack.[9] By proactively engaging a great power, a small state can increase its diplomatic and economic ties to create a mutually beneficial relationship that makes a future conflict more unlikely. The negative here, however, is a great power may be better positioned to meddle in the small state's internal affairs. Such actions could then create political or social instability that could undermine the ruling government's hold on power. Second, small states with no other options with alternative great powers may enter into an alliance with a threatening great power. In Mearsheimer's words, "bandwagoning is employed mainly by minor powers that stand alone against hostile great powers. They have no choice but to give in to the enemy because they are weak and isolated."[10] While this approach saves the small state from incurring the death and destruction of a losing war, the ruling government often cedes much of its global autonomy and in some cases sovereignty.

Randall Schweller is critical of Walt and Mearsheimer, noting how a balance of threat approach assumes alliance tendencies are based exclusively on security. For Schweller, profit, rewards, and gains are the primary factors driving alliance decisions to bandwagon with a more powerful state. As Schweller notes, "when profit rather than security drives alliance choices, there is no reason to expect that states will be threatened or cajoled to climb aboard the bandwagon; they do so willingly."[11] In advancing his balance of interest theory, Schweller concludes that when the status quo between established and rising great powers is in flux, states that care little about costs may join with a revisionist state in anticipation of sharing in the gains.[12] The disadvantage, however, is that if a revisionist state fails to change the status quo, those that joined in may incur great costs, or even an attack by the status quo power.

There are several notable weaknesses and strengths with Schweller's bandwagoning for profit theory. First, like others, Schweller suffers from a great power bias, which he explicitly admits, noting how his "bandwagoning argument and . . . cases focused on great powers, not weak ones."[13] Second, Schweller often uses the terms "alliance" and "alignment" interchangeably to advance his theory. This tendency was most prevalent in his critique of other alliance strategies. For example, in critiquing Walt's balance of threat theory, Schweller notes how, "Balance-of-threat theory is designed to consider only cases in which the goal of *alignment* is security, and so it systematically excludes *alliances* driven by profit."[14] Conflating the terms "alignment" and "alliance" prevents a more holistic understanding of how these are two different activities with the former best understood as a subset of the latter.

In summary, balancing and bandwagoning as alliance strategies remain valid approaches to understanding the actions of the great powers. Less developed, however, is how our theoretical conceptions of alignment remain underspecified because the terms "alliance" and "alignment" are often used to mean the same political behavior. Such a fallacy erodes our ability to understand how political elites may pursue broader alignment strategies designed to realize their national security goals. Given the preceding, it becomes necessary to clearly define the terms "alliance" and "alignment." To achieve a higher degree of precision, alignment is best understood as *an informal or less-binding formal set of expectations between a small state and a great power.* An alliance is then defined as a *formal written agreement or treaty between two or more sovereign states to take some form of military or economic-related action based on established protocols and provisions.* By giving the terms "alignment" and "alliance" distinct meanings, one can more readily understand alliance politics for small states as a continuum where the former is a potential interim step for some small states on the way to the endstate condition of the latter.

Next, the scholarly tendency to view bandwagoning as a negative state-based endeavor is myopic. In nearly all cases, scholars define bandwagoning as a form of appeasement, kowtowing, or ceding autonomy in exchange for long-term survival. Additionally, empirical support for state bandwagoning is often derived from cases that occur immediately prior to, during, or just after the international system experienced a great power conflict. One notable exception is Schweller's conception of wave-of-the-future bandwagoning, which views this form of political behavior in a positive light.[15] Schweller also does well to support this approach using examples from the Cold War when the system was best characterized by great power competition, not conflict. While Schweller grounded his argument in ideological considerations, his approach represents an opportunity from which this study seeks to build upon in the next section.

TOWARD A NEW THEORY OF SMALL STATE ALIGNMENT

This study builds on select elements of Randall Schweller's balance of interest theory as a way to articulate two new forms of state alignment: positive bandwagoning and great power bridging. The primary determinant of these two alignment strategies is a political leader's ideas and beliefs about the identity of the state in world politics. The validity of this leader-centric analysis draws support from Daniel Byman and Kenneth Pollack, who conclude "scholars fail to acknowledge that common international behavior—balancing against a threat, choosing a grand strategy, or marching off to war—results from *decisions made by individuals.*"[16] It is through a leader-centric approach one can better understand the nature of alliance politics for small states in an international system dominated by multiple great powers.

Positive Bandwagoning. Positive bandwagoning is based on two important aspects of Schweller's balance of interest theory. First, small states are best viewed as actors residing in the shadows of two or more powerful status quo and/or revisionist states that are in strategic competition with one another. When this occurs, political leaders in small states have time to implement strategies and policies designed to improve the country's security and economic standing. However, small states often cannot achieve these goals on their own owing to material deficiencies (e.g., lack of natural resources, technical know-how, large army) or internal constraints (ineffective institutions, social instability, etc.). These limitations subsequently energize political leaders in small states to pursue proactive foreign policies with a target great power that has the capacity to provide the necessary political, economic, and military support.

Second, bandwagoning is not viewed exclusively as a negative state action, where political leaders appease, give in, kowtow, or give up political autonomy to achieve a greater good. On the contrary, bandwagoning can be a positive endeavor, where small states may engage in cooperative military, diplomatic, and economic activities with a great power to achieve what ruling political leaders believe to be in the best, long-term good of the state and their political interests.[17] Of significance, positive bandwagoning can occur even when there are no guarantees a small state's alignment with a great power will lead to the country's accession into a military alliance or regional economic grouping in the short or long term. When a small state follows a positive bandwagoning alignment, this study posits ruling elites will tend to exhibit higher levels of verbal and material cooperation with one great power over another.

Great Power Bridging. A second alignment strategy available to small states is great power bridging. In this approach, political leaders attempt to align with two or more great powers to the extent that such efforts do not cross a tipping point of membership in a formal military alliance or economic union. When a small state follows a great power bridging alignment, ruling elites exhibit higher levels of verbal and material cooperation with two or more great powers. Cooperation can occur across numerous lines of operation including political/diplomatic, economic, and military activities designed to align the small state with the target great powers. It should be understood that a small state does not necessarily have to cooperate with a great power along all three lines of engagement. For example, a small state may eschew military cooperation, focusing more on economic relations with different great powers, and vice versa. Furthermore, the levels of verbal and material cooperation can oscillate between both great powers over time.

Unlike positive bandwagoning, which has a higher potential for leading to membership in a military alliance or economic bloc, foreign policy elites pursing this alignment strategy perceive binding economic or military commitments may produce negative second- or third-order effects. For example, accession into a regional economic union could undermine lucrative trade relations with non-member countries. Separately, when two great powers are engaged in a strategic competition, signing a defense pact with one regional great power may elicit a political, economic, or military backlash from the other.

RESEARCH DESIGN

To test the viability of this ideational-based model, this study uses a structured, focused comparison of the alignment tendencies of Ukrainian presidents between 1995 and 2014. Ukraine represents a viable test case for three reasons. First, Ukraine is fertile ground for realists and their theoretical propositions on small state alignment. Whether it is Kenneth Waltz's balance of power theory or Stephen Walt's balance of threat theory, realists expect former Soviet states to bandwagon with their former imperial center owing to power asymmetries or Moscow's historical aggressive tendencies with countries on its geographical periphery. While Belarus may fit realist expectations with Minsk opting to align and subsequently enter into a military alliance and economic bloc with Moscow, Ukraine has often exhibited a zig-zag alignment trajectory between the United States (and its European allies) and Russia since the end of the Cold War. Second, Ukraine has become an epicenter for a strategic competition between the United States and the Russia Federation. Of particular relevance for this study's ideational model, in January 2022 British intelligence revealed Moscow intended to install a pro-Russian president in Ukraine as one way to prevent the country's accession into NATO.[18] This revelation illuminates the critical importance of understanding the alignment tendencies of Ukrainian political elites and how this information can allow Washington to formulate viable policies that protect our national security interests while avoiding a direct military confrontation with Russia. Third, with Russia's decision to invade Ukraine in February 2022, the preceding factors are playing out in real time today.

The three Ukrainian presidents considered in this study are Leonid Kuchma, Viktor Yushchenko, and Viktor Yanukovych, who were the lead actors in Ukraine's foreign policy, and their ideas about the state's identity.[19] As Jeffrey Mankoff reminds us, "[a state's] international behavior is above all a function of [its] self-perception and the *identity constructed by their elites*."[20] State identity is best understood as the public articulation of the ruling group's ideas of the country's purpose in international affairs.[21] These ideas about the purpose of the state represent a quintessential touchstone that leads members of the foreign policy establishment utilize to identify and prioritize the national interests of the country. It is from these national interests that political elites develop and implement foreign policies that define the state's actions in world affairs. The key takeaway here is that the actions of the state are first determined by the *decisions* of those occupying the highest levels of a ruling government. Given the preceding, this study posits positive bandwagoning will occur when the state identities between a great power and small state converge. When state identities diverge or oscillate over time, a small state will pursue a great power bridging alignment strategy.

To measure each president's ideas about the state's identity, this study used the MaxQDA analytical software program.[22] A conceptual analysis using available speeches, interviews, and addresses for each president while in office is employed to understand the level of significance each president gave to a specific state identity concept. The methods employed in this conceptual analysis follows the steps outlined by Kathleen Carley in her 1993 article "Coding Choices for Textual Analysis: A Comparison of Content Analysis and Map Analysis."[23]

The level of analysis is the single word, which takes an accounting of eighteen state identity concepts. These concepts are then placed under five general categories: **political** (independent, former Soviet, freedom, liberty), **cultural** (European, Slavic, Western, Eastern, Eurasian), **economic** (market and modern), **ideological** (democracy, liberal, and communism), and **religious** (Orthodox, Christian, Islamic, secular).[24] The scale of measurement employed for state identity is an ordinal measurement to discern the order of magnitude for each of the state identity concepts by tabulating the frequency with which these terms appear in the presidential speeches on an annual basis. Because the number of speeches will vary, the by-year order of magnitude of each concept will be divided by the total number of words in all the speeches for that year.[25] Additionally, interviews with area specialists and online surveys help establish a baseline of the state identity narratives associated with the United States and Russia. In the United States, political leaders have historically described the state as democratic, modern, independent, and Western. Russian state identity narratives are more ambiguous but in this study are defined as Russian, Slavic, Eastern, Eurasian, and independent. To measure alignment, this study uses Harvard's Integrated Crisis Early Warning System (ICEWS) event dataset and qualitative analyses.[26]

UKRAINE'S ALIGNMENT TRAJECTORY: 1995–2014

A content analysis of presidential speeches, interviews with area specialists, and the use of online surveys and event data highlights several key findings. First, each Ukrainian president under examination espoused multiple state identities. Leonid Kuchma often gave emphasis to Ukraine's state identity as being independent and for his time in office tended to prefer a great power bridging alignment strategy. Viktor Yushchenko viewed Ukraine as Western European and based on democracy and free-dom. As a result, Yushchenko tended to favor a positive bandwagoning alignment with the United States. Frequently referring to Ukraine as an eastern country, Viktor Yanukovych preferred a positive bandwagoning alignment with Russia. The following sections provide a summary of each president's alignment strategy.

Leonid Kuchma's Alignment Strategy: Great Power Bridging

From 1995 to 2004, Leonid Kuchma routinely used great power bridg-ing to align Ukraine with Russia and the United States. An analysis of Kuchma's foreign policy speeches, interviews, and press statements indi-cates he primarily viewed the state as independent (political). Support-ing narratives included European (cultural) and democratic (ideological). As depicted in Figure 5.1, President Kuchma's level of verbal coopera-tion with either Washington or Moscow varied during the ten years under examination.

According to one regional expert, Kuchma's state identity narratives were somewhat unpredictable at times. On one hand, Kuchma would emphasize Ukraine's identity as one and the same with Russia based on their shared histories. On the other hand, Kuchma would seemingly turn his back on Russia by embracing Western and European ideals centered on democracy, freedom, and civil liberties.[27] Kuchma's vision of Ukraine with one foot in the West and the other in the East contributed to his preference for great power bridging with Washington and Moscow for his time in office.

A primary state identity narrative for Kuchma was his tendency to view Ukraine as an independent actor in world politics. In 1996, Kuchma

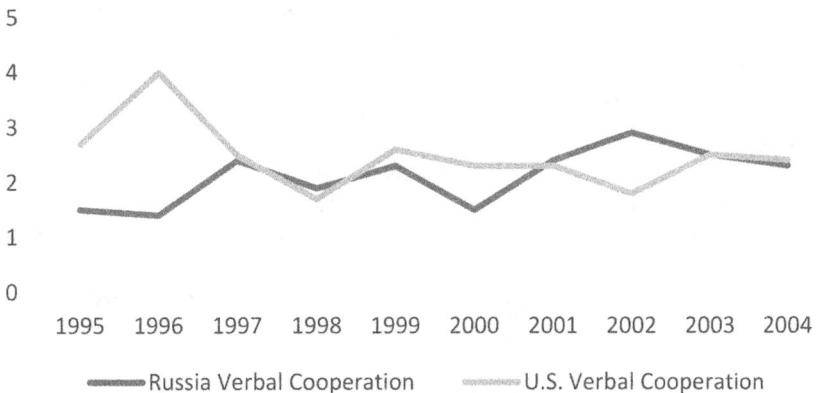

Figure 5.1 Leonid Kuchma's Alignment Preferences with the United States and Russia, 1995–2004

heavily emphasized the country's independent status, asserting, "as far as Ukraine's independence is concerned, let us proceed from the fact that we have an independent state and that it depends primarily on us [and an] . . . independent and sovereign Ukraine . . . is a signpost on our road to Europe."[28] Kuchma's strict adherence to the idea that Ukraine was an independent state was a point not missed by academics and practitioners alike. For example, John Edwin Mroz and Oleksandr Pavliuk find Kuchma was best viewed as an ardent defender of Ukraine's independence and his nationalist fervor represents a linchpin in the development of the state's foreign policies.[29]

Kuchma's ideas and beliefs about the state have their origins, in part, from his time as a leading figure in the Soviet Union's rocket and space technologies industry. From these experiences, Kuchma's expert management skills served him well as a pragmatic leader who would pursue a strategy of great power bridging, dealing with Russia and the United States in ways beneficial to not only his ruling government but Ukraine as well. Ukraine's sense of independence under Kuchma was critical as he made every effort to establish a precedent for not making the country a part of the new Russian Federation in any way. In his book, *Ukraine Is Not Russia*, Kuchma makes reference to the many divisions in the country, singling out the differences between Ukrainians in the western part of the country, who wanted to align with the West, with those in the east, who preferred close foreign relations with Russia.

Additionally, Kuchma's ideas and beliefs about the state's identity and external threats were based on his keen awareness of Russian and Ukrainian history. In this case, the centuries of domination by the Russian empire and Soviet Union instilled a sense of urgency for Kuchma to craft a set of foreign policies that made Ukraine an independent actor in international affairs. In his 1995 Independence Day speech, Kuchma stated, "the emergence of the Ukrainian independent state on the political map of the world became one of the most outstanding events in the world history of the second half of the 20th century and is a natural result of our people's centuries-long aspiration to be masters in their own home."[30]

As a secondary state identity narrative, President Kuchma also tended to view Ukraine in European terms. During his first several years in office, Kuchma frequently highlighted Ukraine's placement in the world order as being a member of the group of European countries. For example, in a 1996 interview with reporters from Radio Liberty, Kuchma noted how "the future of Ukraine lies in Europe, for our country is located in the very heart of Europe."[31] Kuchma's European discourse continued at a steady rate throughout much of his presidency. In April 2002, Kuchma stated, "Ukraine is a European country. Naturally, integration in the European structure is one of our foreign policy priorities.[32]

Although President Kuchma may have given a secondary emphasis on categorizing Ukraine as European, his discourse must be placed into proper context. First, Kuchma believed that by defining Ukraine as European, such actions would create a gateway for a political and economic alignment with the United States. However, President Kuchma's European state identity narrative must be placed in its proper context where the Ukrainian president's conceptualization of this state identity is much more inclusive than one might suspect. In this case, Kuchma also viewed Russia as part of the European community along with other CIS countries and the United States. Speaking in 1998 after returning from a visit to Russia, Kuchma added clarity to his ideas about the state's identity and his alignment with multiple great powers, noting, "the European choice of Ukraine—nonetheless does not mean one-way orientation towards Europe. We are talking about a system of Ukraine's relations in the following area: Russia, Europe, North America, the Black Sea region and the CIS."[33]

Finally, in addition to Kuchma's attentiveness to articulating an independent and European state identity narrative for Ukraine, he also cultivated a democratic discourse in his foreign policy speeches. While in office, Kuchma routinely touted the state's democratic leanings, noting in 1997 how "Ukraine's relevant norms and tenets are among the most democratic in the world."[34] Kuchma's narrative on the centrality of democracy in the Ukrainian state continued through 2001. Following a series of meetings with the EU, Kuchma stated, "I am nevertheless convinced that Ukraine, as virtually the whole world, and first and foremost Europe, will follow the democratic path."[35]

President Kuchma's ideas about democracy represented a conduit for his alignment preferences with the United States; however, his ideas about the Ukrainian state as democratic must be qualified. While a content analysis may identify democracy as a secondary state identity narrative for Kuchma, area experts and secondary sources do not completely validate this reality. For example, in 1994 U.S. President William Clinton commended President Kuchma and his efforts to advance democratic reforms in Ukraine.

We congratulate you, Mr. President, and all Ukrainians on your remarkable achievements in the almost 3 years since regaining your freedom. You held a historic referendum and began the hard work of reform and building democratic institutions.[36]

But beginning in 1999 and continuing through to the end of his presidency, democracy took a back seat to Kuchma's desire to stay in power. During this time, the Ukrainian president's authoritarian tendencies became increasingly visible in the state bureaucracy and throughout society. In

1999, Freedom House issued a negative report on Ukraine, observing a democratic retrenchment owing to, "increased government pressure on the independent media, presidential elections in October–November which were not free and fair, and attempts by the executive after the elections to increase presidential powers at the expense of parliament."[37] The following year, Ukraine's Polity Score (a measure of the degree of democratization) also registered a setback in 2000. According to a 2010 Polity IV Report, Kiev began to feel "the strains of post-communist transition and [the state] began to unravel during the 1999 election campaign as Kuchma began to assert his independence and attempted to consolidate power."[38]

President Kuchma's proclivity for great power bridging was also born out in his foreign policies. Beginning with Russia, the first, and perhaps one of the most important, events occurred in 1997 when President Kuchma and his Russian counterpart, Boris Yeltsin, signed a Treaty of Friendship and Cooperation. Taking nearly two years to sign, this treaty became the backbone of Ukraine's alignment with Russia, laying a foundation for Kiev's strategic partnership with Russia (and vice versa) in the realm of politics, economic development, energy supplies, military–technical cooperation, and culture. The 1997 Friendship Treaty with Russia was immediately followed in 1998 with a ten-year comprehensive economic cooperation agreement. However, the collapse of Russia's economy a few months later stymied this bilateral effort with economic progress first taking effect in 2000 when Ukraine's exports and imports to and from Russia went from $2.4 billion and $5.6 billion to $3.5 billion and $5.8 billion respectively.[39]

Kuchma's alignment with Russia was also made evident in his decision to replace his foreign minister, Boris Tarasyuk, who advocated for Ukraine's exclusive alignment with the West. In his place, Kuchma installed Anatolii Zlenko, who viewed Russia as the state's most important strategic partner. Although Zlenko was a strident supporter of Kuchma's beliefs about aligning Ukraine with Russia, he also believed the state's foreign policies must be multi-vector in nature where an alignment with other great powers such as the United States was in the state's best interests. According to Zlenko, developing a strategic partnership with Russia and the United States would be the "key guarantee of our [Ukraine's] security."[40]

In 1996, Kuchma entered Ukraine into a strategic partnership with Washington through the establishment of the Kuchma–Gore Commission. This bilateral arrangement served as a foundational mechanism by which Ukraine's alignment with the United States would develop across various lines of operation including trade, investment, and national security. In the military realm, Ukraine signed NATO's Charter on a Distinctive Partnership in 1997, which subsequently facilitated the opening of a NATO Liaison Office in Kiev in 1999. In 2002, Kuchma then announced his intention for Ukraine to join NATO, leading to the establishment of the NATO–Ukraine

Action Plan. By 2004, Kiev had signed the Ukraine–NATO Memorandums on Host Nation Support and Strategic Airlift. The memorandum on Host Nation Support established protocols for the conduct of joint military exercises in Ukraine, while the Strategic Airlift memo detailed how NATO would be able to use the Ukrainian Air Force's transport planes for operations in Afghanistan. In 2003, Kuchma then supported U.S.-led military operations in Iraq by approving the deployment of approximately 2,000 soldiers for various missions in the theater of operations.

Viktor Yushchenko's Alignment Strategy: Positive Bandwagoning with the United States

A content analysis of President Yushchenko's foreign policy speeches reveals he gave emphasis to the following state identity narratives: European, Western, democracy, and market. Figure 5.2 provides a visual illustration of President Yushchenko's levels of verbal cooperation with Russia and the United States from 2005 until 2009. During this time Yushchenko consistently maintained a higher level of verbal cooperation with Washington over Moscow. With Yushchenko's ideas about the state defined as European and Western as well as democratic and market-oriented, this pattern of verbal cooperation is best understood as a preference for positive bandwagoning with the United States. Although Yushchenko sought out a strong alignment with the West, he still believed a cooperative approach with the Kremlin was necessary. According to one area specialist, Yushchenko was an ardent pro-West president, who viewed the country's integration into Euro-Atlantic security and economic structures as essential. By becoming part of NATO and the EU, Yushchenko believed these organizations could help ensure the country's long-term security and economic prosperity.[41]

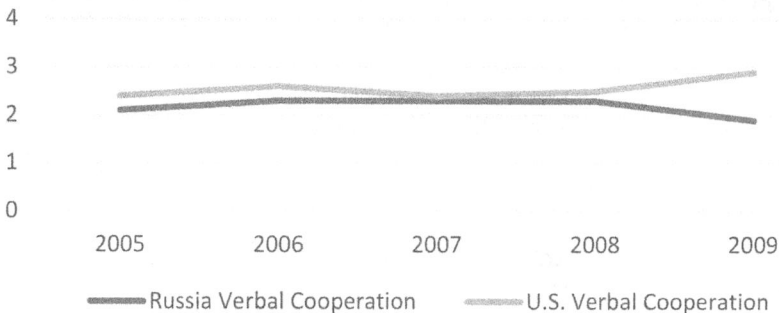

Figure 5.2 Viktor Yushchenko's Alignment with the United States and Russia, 2005–2009

President Yushchenko's ideas about Ukraine as a European state on the global chessboard were brought to the fore during a 2006 visit to the College of Europe in Warsaw, Poland. In his speech, Yushchenko averred how he had always believed Ukraine was part of Europe. In his words, "I was born at Europe's center. I am convinced that Ukraine is a European country. Our values were and continue to be European values."[42] President Yushchenko's efforts to frame Ukraine as European was driven by a belief that the country's path to stability resided in the West with Europe and the United States. From this stability, Ukraine would achieve economic success and a newfound sense of security from external threats.

President Yushchenko's ideas about Ukraine as Western manifests itself in several narratives, which highlight the state's role as a productive member in the Euro-Atlantic economic and security communities. For example, during his visit to Washington, DC in 2005, President Yushchenko delivered a speech to the U.S. Congress detailing his ideas of how Ukraine is a country that belongs in the West.

The Orange Revolution gave evidence that Ukraine is an advanced European nation, sharing the great values of the Euro-Atlantic civilization. I'm convinced that the European and Euro-Atlantic aspirations of Ukraine may not be viewed as an additional hindrance. Ukraine's integration is not a problem, but rather great new opportunities opening before our civilization.[43]

Given Ukraine had become a less important foreign policy agenda item in Washington at the time, Yushchenko hoped to convey the idea that ruling leaders in Kiev shared the same ideas and values as political elites in Washington and in other European capitals.

The "Orange Revolution" and Yushchenko's rise to the presidency was from the very beginning equated with democracy's return to Ukrainian politics. From his first days in office, Yushchenko consistently espoused a state identity narrative that cast the state (and country) as democratic. For example, in a 2005 Council of Europe speech, Yushchenko framed his ideas about the state and democracy in the following way:

I will take every effort to secure the irreversibility of democratic changes in my country. The Ukrainian public authorities will do their utmost to secure the fundamental principles of the Council of Europe—human rights protection, pluralist democracy and the rule of law.[44]

Yushchenko's consistent focus on the central importance of democratic ideals were also reflected in a December 2009 address on human rights, where he noted, "we have today made our place as a democratic state, we have eliminated any forms of political persecutions. Freedom of speech, conscience and choice is already a standard . . . [in Ukraine]."[45]

Polity analysts recognized Yushchenko's commitment to democracy and the follow-on results where Ukraine's Polity score rose (indicating a more democratic government) between 2004 and 2009.[46]

Finally, Yushchenko espoused an economic discourse viewing Ukraine as a market economy. In 2005, Yushchenko's ideas about this state identity narrative were made clear in a speech to the U.S. Chamber of Commerce. According to Yushchenko, "our government and our power is going to turn [Ukraine's] market into a transparent market and a competitive market without providing preferences to anybody . . ."[47] The following year, Yushchenko raved about how major foreign powers had validated his ideas. Specifically, in 2005 he proclaimed, "over the past eight months we have succeeded in having Ukraine and the Ukrainian economy, recognized by the European Union and the United States as a market economy."[48] Yushchenko's discourse viewing Ukraine as a market economy reflects his beliefs about an alignment preference with the United States over Russia. By fostering an economic alignment with the United States, economic growth would follow, thus making Ukraine a key member in the global economy.

During his time in office, President Yushchenko's foreign policies reflected his positive bandwagoning with the United States and the West. On one hand, he took action to align Ukraine with the United States by developing strong political, economic, and military ties with Washington. Concurrently, the Ukrainian president took on a more confrontational orientation with Russia owing to Moscow's continued meddling in the country's internal affairs. Beginning in 2005, Yushchenko was granted a rare privilege to address a joint session of the U.S. Congress. Prior to this time, the only other former Soviet state leader to address Congress was Boris Yeltsin in 1992. During this singular event, President Yushchenko also signed a joint statement with President Bush on establishing a new strategic partnership with the United States. Formalized in 2008 as the U.S.–Ukraine Charter on Strategic Partnership, this agreement helped expand security cooperation between both countries.

In the economic arena, Yushchenko's domestic economic reforms translated into foreign policy victories with the United States and Europe. By focusing on open market principles and ways to integrate Ukraine into the global economy, Yushchenko was able to garner American support. From 2005 through 2007, Washington provided Ukraine technical assistance for its accession to the WTO, $50 million for economic reform programs, and the repeal of 1974 Jackson-Vanik amendment. Enhancing these activities, President Yushchenko opened six public information centers across Ukraine to help inform the public on his aspirations to join the WTO as well as other Euro-Atlantic organizations such as the EU and NATO.[49] By 2008, Ukraine signed the Trade and Investment Cooperation Agreement with the United States. This agreement created a bilateral Trade and

Investment Council, which authorized economic officials to discuss ways of bolstering economic ties between both countries.[50] Taken together, these efforts significantly helped Yushchenko to realize his goal of WTO membership in 2008.

Yushchenko was also able to strengthen Ukraine's alignment with the West when in April 2005 when during the NATO–Ukraine Commission meeting in Vilnius, alliance leaders invited Kiev to begin an Intensified Dialogue on Membership. The Intensive Dialogue program was a pivotal precursor step prior to the issuance of a Membership Action Plan (MAP). In February 2006, during a meeting in Brussels, the Ukrainian president told NATO officials his country was ready for a MAP, proceeding to establish an interagency commission, which would oversee the country's accession into the alliance.[51] President Yushchenko's continued commitment to positive bandwagoning with the United States produced positive results when President Bush signed into law the 2007 NATO Freedom Consolidation Act, which provided Ukraine with $30 million of U.S. military assistance from 2008 to 2012.[52]

From the beginning of his presidency, Yushchenko regarded Russia as an economic and military threat to Ukraine. On the economic front, in 2006 and 2009 Moscow cut gas supplies to Ukraine and other countries in Western Europe. Beyond Moscow's use of the energy lever against Kiev, Russia often issued military threats against Ukraine. For example, in February 2008, Russian President Vladimir Putin threatened to target Ukraine with nuclear weapons if the country joined NATO or allowed the United States to deploy assets as part a European missile defense program. Taken together, these events reinforced Yushchenko's threat perceptions of Russia, resulting in routine diplomatic or economic activities with Moscow. Of significance, Yushchenko's threat perceptions of Russia reinforced his desire to follow a positive bandwagoning alignment with the United States, even if there were no guarantees Ukraine would be admitted to NATO or the EU.

Viktor Yanukovych's State Identity Narratives and Ukraine's Alignment

From 2010 to 2013, Viktor Yanukovych used a positive bandwagoning strategy to align Ukraine with the Russian Federation. During his abbreviated time in office, President Yanukovych articulated two primary state identity narratives: Eastern and Eurasian. As a supporting conception, Yanukovych also viewed Ukraine as European. As reflected in Figure 5.3, President Yanukovych's level of verbal cooperation was consistently higher with the Russian Federation than the United States.

President Yanukovych's ideas about the state are best defined as an amalgamation of Eastern, Eurasian, and to a lesser degree European. This

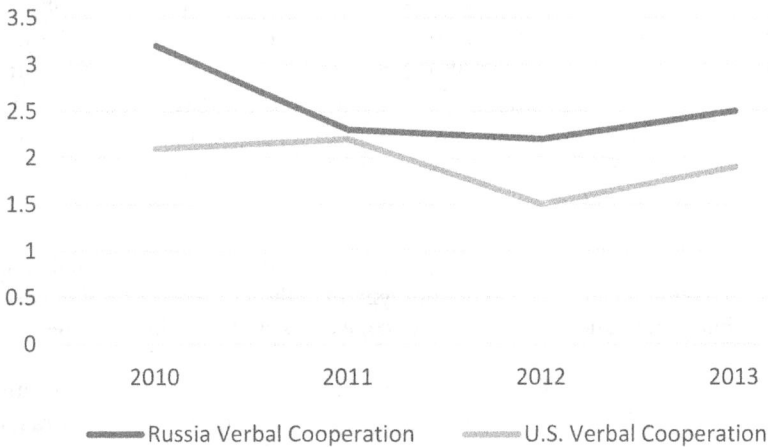

Figure 5.3 Viktor Yanukovych's Alignment with the United States and Russia, 2010–2013

pattern of verbal cooperation is best understood as a preference for positive bandwagoning with Russia. Although Yanukovych gravitated toward an alignment with Russia, he still viewed cooperation with the West, particularly Europe, as important but not necessarily a vital component of his foreign policies. Interviews with various regional experts and an analysis of the secondary literature validate this correlation. According to one area expert, Yanukovych strived to reposition Ukraine into a closer geopolitical orbit with Russia; however, he also sought to maintain economic relations with countries in the West. From one perspective, Yanukovych was pro-EU, not pro-NATO.[53]

Shortly after entering office, President Yanukovych's articulation of a European narrative also conveyed a positive view on economic cooperation with Europe. For example, during a meeting with the president of the European Commission, Yanukovych noted how "European integration is the key priority of our foreign policy, and also a strategy for carrying out systemic social and economic reforms."[54] However, while Yanukovych may have opted to first visit Brussels on March 1, 2010, by the time he got to Russia, he was espousing the merits of Ukrainian–Russian brotherhood. Specifically, in his meetings with President Medvedev on March 5, Yanukovych proclaimed he was in total agreement with the Kremlin's patriotic beliefs about the importance of Soviet history in modern times. Yanukovych then suggested veterans from Ukraine and Russia should hold a joint celebration to commemorate the 65th anniversary of their victory over Adolf Hitler and fascism.[55]

An Eastern and Eurasian discourse was also an integral part of President Yanukovych's repertoire of state identity narratives. As a former governor in the Donetsk region, Yanukovych frequently linked the identity of the Ukrainian state and the country to Russia. For example, in 2010 Yanukovych noted how "the citizens of our countries . . . are linked by kinship and by the history of our nations."[56] By 2013, there was little change in his ideas about the state's Eastern and Eurasian identities. This was particularly made clear in his annual address to the Ukrainian parliament in 2013 when Yanukovych stated,

[The] European integration vector along with the consistent realization of national interests on the Eurasian track, active employment of the CIS platform and intensification of bilateral relations constitute the scope of Ukrainian national interests.[57]

Clearly, Europe may have retained some degree of importance for Yanukovych, but by 2013 his ideas about Ukraine and its relevancy in Eurasia had become a central component in his public addresses.

Yanukovych's tendency to align with Russia over the United States and Europe was clearly visible in his policies on military and economic issues. One of his first actions to align with Russia occurred in April 2010 when Yanukovych signed an agreement to extend Moscow's lease for the Russian Navy's Black Sea Fleet in Sevastopol until 2042 and authorize the country's armed forces to partake in the Russian-led 'Union Shield' exercises albeit at a low-level of participation. In the field of economics, Viktor Yanukovych was able to foster an alignment with Russia by quickly concluding a series of new security and economic agreements with Moscow. For example, in May 2010 during a visit by Russian President Dimitry Medvedev to Ukraine, Yanukovych signed agreements on intelligence sharing, border demarcation, as well as inter-bank and science and technical cooperation. The clearest indicator of Yanukovych's ideas about the state's Eastern identity came in 2013 when he decided to take a $15 billion Russian economic package in lieu of signing the EU Association agreement. According to one area specialist, Yanukovych's decision amounted to a "civilizational choice."[58] Russia's invasion of Crimea and support to rebels in 2014 would later bifurcate that civilizational choice between a Russia-leaning Donbass and a European-leaning west with aspirations for alignment and eventual alliance with NATO and membership with the EU, thus burning the great power bridging strategy.

With the United States, however, the Ukrainian president's orientation was best understood as cordial. For Yanukovych, good relations with America were important; however, he firmly believed Washington's meddling in Ukraine's internal affairs was the primary catalyst for the 2004 Orange Revolution. As a result, Yanukovych sought to maintain routine interstate activities with the United States in order to stave off any social and political instability that could threaten his hold on power. This freedom of action subsequently allowed him to devote all of his attention to a positive bandwagoning strategy with Russia.

CONCLUSION

The dawn of the twenty-first century has witnessed a return of great power competition with the United States, Russia, and China vying for influence and position in all corners of the world. In reflecting on the events over the twenty years covered in this case study, Russia and the West were slowly moving into a new Cold War, with NATO and the EU continuing its eastward expansion and Russia responding with hybrid warfare that included a mix of coercive diplomacy, nefarious cyber operations, and military force in places such as Georgia and Ukraine. This conclusion is not missed by the highly Robert Legvold, who noted back in 2015 that

the new Cold War began the moment we [the West] went over the cliff, and that happened with the Ukraine crisis. I trace this qualitative shift to be a result of what happened in Ukraine, with the Russian annexation of Crimea and Russia's direct support for separatism in the Donbas. Going over the cliff did not happen suddenly though, it happened as a result of a series of steps that cover most of the post-Cold War period. I see it as a phased process, in which no one, including the leadership of Washington, Moscow and Berlin recognized the phases we were in.[59]

Given this new strategic reality, the United States must expand its bilateral partnerships with small states outside our current alliance network as one way to regain a competitive advantage with its adversaries such as Russia. To accomplish this task, this chapter offered a fresh approach to understanding alliance politics in the twenty-first century. First, it was argued the scholarly tendency to use "alliance" and "alignment" interchangeably has prevented a more nuanced understanding of alliance politics for small states in a multipolar world. Although small states will seek out alliances as a way to ensure their security, they will first attempt to establish an alignment with a great power as an interim step to formal membership in a military or economic bloc. Second, in marked contrast to realist explanations that often only consider the actions of the most powerful states in the system, this study advanced a constructivist-grounded model of alliance politics focused on the ideas and beliefs of leading political elites. It is these ideas and beliefs that give content to the state's identity, which in turn helps us to understand how small states may pursue one of two alignment strategies with a great power: positive bandwagoning or great power bridging.

To test this model, this study examined the alignment tendencies of three Ukrainian presidents (Presidents Leonid Kuchma, Viktor Yushchenko, and Viktor Yanukovych) between 1995 and 2014. During the period

under examination, all three conceptualized the state's identity in different political, cultural, and ideological terms. President Kuchma's primary
state identity narrative centered on independence, which correlated with
a preference for great power bridging with the United States and Russia.
Presidents Yushchenko focused on defining the state as democratic, free,
and Western European, which correlated with a tendency for positive
bandwagoning with the United States. Finally, President Yanukovych
viewed the state in terms of Eastern, Eurasian, and European, which was
correlated with a positive bandwagoning preference with Russia.

The preceding analysis suggest several conclusions and recommendations for academics and practitioners as well as opportunities for future
research. First, this study concludes a president's ideas about the state's
identity are an important variable to better understand the nature of alliance politics for small states in an international system comprised of multiple great powers. By examining a president's level of verbal cooperation
with two or more great powers allows one to infer a non-statistical correlation between a president's ideas about the state's identity and their alignment preferences. Second, as was seen with President Kuchma, a political
leader's ideas about the state have the potential to vary based on key sys-
temic- and/or domestic-level events. The point here is that while a political leader's ideas and beliefs may be a critical factor affecting alignment
preferences, systemic and/or domestic events can also carry impact and
subsequently shape a president's ideational lens of world politics.

This ideational-based model of alliance politics is, however, not perfect.
While an analysis of a president's level of verbal cooperation may serve as
an indicator of their inclination to align with one or multiple great powers,
it is plausible a head of state may advance a less than genuine state identity
narrative in hopes of eliciting short-term support from a target great power.
For example, a president may emphasize a commitment to democracy and
human rights to garner political or economic support from the United
States or the EU, but then take no substantive internal action to make those
propositions a reality. This might be especially true when states are trying to meet the conditions for membership in organizations like the EU or
NATO. Alternatively, a president's beliefs and alignment preference may be
authentic; however, domestic politics or public opinion may impede their
ability to transform words into action. This shortfall may require expanding the model in this study to consider other independent variables such
as a political leader's ideas about external threats, public opinion, and the
nature of domestic political opposition. This shortfall represents an opportunity for future research where these additional variables may be able to
provide increased specificity on a political leader's alignment preferences
and the state's foreign policies vis-à-vis a great power.

This study's focus on state identity can assist military strategists and
policymakers to develop more viable strategies and policies to protect or

advance U.S. national security interests in key regions where Russia and China are attempting to assert their political or economic dominance.

This particularly true in Ukraine where U.S. policymakers and strategists can use select instruments of national power to influence political elite decision-making. For example, U.S. leaders can use economic and military incentives such as foreign aid, trade promotion, arms sales, and security assistance to reinforce President Volodymyr Zelensky's and his successors' alignment with the West. If in the future a pro-Russian president rises to power in Kiev, Washington could suspend its military and economic support while using diplomacy and public policy statements that tout the merits of an alignment with the United States and its allies. It is through the information element of power that practitioners can either promote or counter the strategic narrative advanced by political leaders in Ukraine or other small states.

Furthermore, by considering how ideological factors may be equally as important, if not more important than, material-based factors, strategists and policy analysts can develop a more holistic understanding of alliance politics in small states. With an increased emphasis on a political leader's ideas and beliefs, national security professionals can begin to see how simple balance of power or threat calculations do not necessarily translate into predictable small state behavior. A focus on state identity can also allow policy analysts an opportunity to acquire new insights into the creation and longevity of modern-day security alliances and regional economic unions. This opens the door to future studies that investigate the relationship between identity politics and national security matters. In Ukraine, an analysis of President Volodymyr Zelensky's ideas and beliefs prior to, during, and after the Russia–Ukraine war would help discern his alignment trajectory going forward. Currently, a cursory examination of Zelensky's ideas about the state's identity points toward positive bandwagoning with the United States and its allies; however, the evolving strategic conditions in the region may cause him to follow a great power bridging strategy with the West and Russia in the future.

This study may also be helpful reminder for U.S. intelligence analysts, who are often required to provide a personality assessment on foreign political and military leaders. Beyond those foreign leaders currently in power, intelligence professionals would do well to expand their database to include political elites, who have the short-term potential to become members of a small state's foreign policy establishment. By analyzing the foreign policy speeches of rising political figures, analysts would be better positioned to warn policymakers about the potential alignment of small states with America's adversaries and the second- and third-order effects that could negatively impact U.S. national interests. At the individual level, personality assessments could enhance a policymaker's ability to negotiate or influence the outcome of bilateral engagements with foreign elites, who are or who will one day become president or prime minister. At the policy level, personality assessments can help U.S. political leaders develop and implement proactive foreign policies with a target small state to improve America's competitive advantage with an adversarial great power.

Finally, this research endeavor elucidates the complexity of twenty-first century interstate relations and how even the smallest of states can still have a voice in world politics. Donald Milsten captures this point succinctly when he wrote, "small powers can and do initiate policies which evoke all manner of reactions from other nations. Small powers are sought after, and they seek; they may act to enhance peace, or they may promote conflict and chaos. They may be overawed, overpowered, or simply dominated, but they cannot be ignored."[60] Milsten's words certainly ring true today where small states with a high degree of geostrategic importance such as Ukraine, Georgia, Vietnam, or the Philippines have the capacity to shape political outcomes that can enhance or erode the influence of one or more great powers. When one factors in China and Russia's respective geographical proximity and historical ties to these countries, discerning the alignment trajectory of small states can help the United States expand its global network of friends and partners in order to improve its declining competitive advantage with its adversaries not only in the short term but the long term as well.

ACKNOWLEDGMENT

This chapter is adapted from John Mowchan, "East or West? Understanding Strategic Alignment in the Post-Soviet Space," Ph.D. dissertation, West Virginia University, 2018.

NOTES

1. This study uses Michael Handel's conceptualization of a "small state," which is defined in terms of power (military power) rather than size (e.g., geographical area). See Michael Handel, *Weak States in the International System* (London: Frank Cass (1981), 10–11.

2. Realists define hard balancing as the act of creating or entering into a formal military alliance. Hard balancing can also occur when a state devotes significant internal resources to build up their military capabilities against another more powerful adversary. As conceptualized by Christopher Layne, leash-slipping occurs when a state within an alliance led by a benevolent hegemon increases its military capabilities to independently pursue other foreign policy goals. See Christopher Layne, "The Unipolar Illusion Revisited: The Coming End of the United States' Unipolar Moment," *International Security* 31, 2, (2006), 9, 30.

3. Stephen M. Walt, *The Origins of Alliances* (Ithaca, NY: Cornell University Press, 1987), 29; Jack S. Levy, "The Causes of War: A Review of Theories and Evidence,"

in *Behavior, Society, and Nuclear War, Vol. 1,* Eds. Philip E. Tetlock, Jo L. Husbands, Robert Jervis, Paul C. Stern, and Charles Tilly (New York: Oxford University Press, 1989).

4. Bruce Bueno de Mesquita, "Domestic Politics and International Relations," *International Studies Quarterly* 46, 1 (2002), 4.

5. See Steven Pifer, "Pay Attention, America: Russia Is Upgrading Its Military," The Brookings Institution, February 5, 2016. http://www.brookings.edu /research/opinions/2016/02/05-russian-military-modernization-us-response -pifer; Yuan-Kang Wang, "China's Grand Strategy and U.S. Primacy: Is China Balancing American Power?" *The Brookings Institution-Center for Northeast Asian Policy Studies,* 2006, http://www.brookings.edu/fp/cnaps/papers/wang2006.pdf.

6. See Richard Weitz, "Assessing Chinese-Russian Military Exercises: Past Progress and Future Trends." Center for Strategic and International Studies (CSIS), July 9, 2021, https://www.csis.org/analysis/assessing-chinese-russian -military-exercises-past-progress-and-future-trends.

7. See John J. Mearsheimer, "The Future of the American Pacifier," *Foreign Affairs* 80, 5 (2001); John J. Mearsheimer, *The Tragedy of Great Power Politics* (New York and London: W. W. Norton & Company, 2014); Stephen M. Walt, *Taming American Power: The Global Response to U.S. Primacy* (New York: W. W. Norton, 2006).

8. Mearsheimer, 2014; Thomas S. Mowle and David H. Sacko, *The Unipolar World: An Unbalanced Future* (Palgrave Macmillan, 2007); Stephen M. Walt, *Taming American Power: The Global Response to U.S. Primacy* (New York: W. W. Norton, 2006).

9. Walt, 2006, 108.

10. Mearsheimer, 2014, 163.

11. Randall L. Schweller, "Bandwagoning for Profit: Bringing the Revisionist State Back In," *International Security* 19, 1 (1994), 79.

12. Schweller, 1994, 106–107.

13. Randall L. Schweller, "New Realist Research on Alliances: Refining, Not Refuting, Waltz's Balancing Proposition," *American Political Science Review* 91, 4 (1997), 928.

14. Schweller, 1994, 79; emphasis added.

15. As conceptualized by Randall Schweller, states may bandwagon with a larger regional great power because ruling elites view such a course of action as the "wave of the future." In this case, ideology is the primary driver moving states along an alignment trajectory toward a regional great power. During the Cold War, many ruling governments in the developing world willingly aligned with Moscow because of its communist ideology. One requirement outlined by Schweller is that wave-of-the-future bandwagoning requires the larger power to have a charismatic leader, as was the case with Adolf Hitler in Nazi Germany, Joseph Stalin in the Soviet Union, and Mao Tse-tung in China. See Schweller, 1994, 96–97.

16. Daniel L. Byman and Kenneth M. Pollack, "Let Us Now Praise Great Men: Bringing the Statesman Back In," *International Security* 25, 4 (2001), 145, emphasis added.

17. Schweller views small states as lambs, revisionist states as wolves, and status quo states as lions. Schweller, 1994, 74–75, 93, 100–103.

18. Foreign, Commonwealth and Development Office, "Kremlin Plan to Install Pro-Russian Leadership in Ukraine Exposed," 2022, https://www.gov.uk /government/news/kremlin-plan-to-install-pro-russian-leadership-in-ukraine -exposed.

19. The foreign policy establishment is comprised of a set of senior political elites, who have the authority to develop and implement the state's foreign policies. The lead actor in this group is assumed to have the most influence in the decision-making process to commit or withhold state resources to align with external powers.

20. Jeffrey Mankoff, *Russian Foreign Policy: The Return of Great Power Politics* (New York: Rowman and Littlefield, 2009), 41, emphasis added.

21. This definition of state identity draws upon the work of Masahiro Matsumura, Shibley Telhami, Michael Barnett, and Marc Lynch. See Masahiro, Matsumura "The Japanese State Identity as a Grand Strategic Imperative," Working Paper for the 2006–2007 Visiting Fellows Program at the Brookings Institution's Center for Northeast Asian Policy Studies (CNAPS) (Washington, DC: The Brookings Institution Center for Northeast Asian Policy Studies, 2008), 3; Shibley Telhami and Michael N. Barnett, *Identity and Foreign Policy in the Middle East* (Ithaca, NY: Cornell University Press, 2008); Marc Lynch, "Abandoning Iraq: Jordan's Alliances and the Politics of State Identity," *Security Studies* 8, 2–3 (1998), 349.

22. An overview of this analytical software can be found at https://www .maxqda.com/qualitative-analysis-software?gclid=CjwKCAiAm7OMBhAQ EiwArvGi3MYPptfU-s8be6xl__sS6-8N-JDmtsLrX4iBEBeIGtuo2lhXkb4BzRo CAyUQAvD_BwE.

23. Kathleen Carley, "Coding Choices for Textual Analysis: A Comparison of Content Analysis and Map Analysis," *Sociological Methodology* 23 (1993).

24. These eighteen concepts of a state's identity were derived from an analysis of the secondary literature, elite interviews, and a word cloud analysis of a random sample of forty foreign policy speeches (ten from each case). The word cloud analysis revealed Azerbaijani, Georgian, Moldovan, and Ukrainian presidents frequently identified the state as either independent, European, or developing, with emphasis also being given to the CIS and NATO. In the secondary literature, Giorgi Gvalia and his colleagues find modernizing, European, and Western integration are three of Georgia's state identity narratives under President Saakashvili. Ronald Suny notes how some states can view themselves as democratic, communist, and Slavic. Elite interviews revealed that political leaders in former Soviet states often viewed the state in a variety of different ways including Western, Eurasian, free, liberal, secular, and orthodox. Taking into consideration these potential narratives, this study adds Christian, Islamic, liberal, former soviet. See Giorgi Gvalia, David Siroky, Bidzina Lebanidze, and Zurab Iashvili, "Thinking Outside the Bloc: Explaining the Foreign Policies of Small States," *Security Studies* 22, 1 (2013), 110; Ronald Grigor Suny, "Provisional Stabilities: The Politics of Identities in Post-Soviet Eurasia," *International Security* 24, 3 (2000), 152–159.

25. Carley, 1993, 81–87.

26. An overview of ICEWS can be found at https://support.dataverse.harvard .edu. The coding scheme for ICEWS data follows the Conflict and Mediation Event Observations (CAMEO) ontology, which was originally developed by Deborah J. Gerner, Philip A. Schrodt, Ömür Yilmaz, and Rajaa Abu-Jabr. This study uses Schrodt's version of CAMEO, which consists of twenty top-level categories of events. Although these twenty categories are aggregated into four groups—verbal cooperation, material cooperation, verbal conflict, and material conflict—for simplicity this study only uses the verbal cooperation category. Events in this category capture the oral statements or remarks associated with a political leader's intentions

to pursue cooperative, friendly relations with a target actor. CAMEO categories for these events are as follows: Statement (1), Appeal (2), Express Intent to Cooperate (3), Consult (4), Engage in Diplomatic Cooperation (5). See Deborah J. Gerner et al., *Conflict and Mediation Event Observations (CAMEO): A New Event Data Framework for the Analysis of Foreign Policy Interactions* (New Orleans: International Studies Association, 2002); Phillip A. Schrodt and Omar Yilmaz, *CAMEO Conflict and Mediation Event Observations Codebook* (Lawrence: University of Kansas, 2007).

27. Interview on file with author.

28. Leonid Kuchma, "Kuchma Interview on State of Nation," *Krymskaya Gazeta*, August 17, 1996; Leonid Kuchma, "Kuchma Interviewed on Poland, NATO, Russia," *Zycie Warszawy*, June 25, 1996.

29. John Edwin Mroz and Oleksandr Pavliuk, "Ukraine: Europe's Linchpin," *Foreign Affairs* 75, 3 (1996), 53.

30. Leonid Kuchma, "President Kuchma Independence Day Speech," *Kiev Radio*, August 24, 1995.

31. Leonid Kuchma, "Kuchma Discusses 'Road to Europe'," *Kiyevskiye Novosti*, December 6, 1996.

32. Leonid Kuchma, "Ukrainian President on Mideast Peace, Ties with Iraq, Iran, Russia, Europe." *Beirut Al-Nahar Online*, April 23, 2002.

33. Leonid Kuchma, "Kuchma's Speech at Moscow University." *UT-1 Television Network*, February 28, 1998, emphasis added.

34. Leonid Kuchma, "Kuchma Addresses II Forum of Ukrainians," *Uryadovyy Kuryer*, August 23, 1997.

35. Leonid Kuchma, "President Kuchma Sums Up Ukraine-EU Summit," *UT-1 Television Network,* September 11, 2001.

36. William Jefferson Clinton, "Remarks Welcoming President Leonid Kuchma of Ukraine," *The American Presidency Project—William J. Clinton*, November 22, 1994, http://www.presidency.ucsb.edu/ws/index.php?pid= 49506.

37. Freedom House, "Ukraine," *Freedom in the World 1999*, (1999), https://freedomhouse.org/report/freedom-world/1999/ukraine.

38. Polity IV Country Report, "Ukraine" (2010), http://www.systemicpeace.org/polity/Ukraine2010.pdf.

39. World Bank, "World Integrated Trade Solution (WITS)," 2016, http://wits.worldbank.org/about_wits.html.

40. Anatolii Zlenko, "Ukrainian Foreign Affairs Minister Anatoliy Zlenko on Changes in Foreign Political Course," *Kiyevskiye Vedomosti*, January 29, 2001.

41. Interview on file with author.

42. Viktor Yushchenko, "Address of Ukrainian President Viktor Yushchenko to the Students of the College of Europe, 12 May 2006," College of Europe; Natolin, *Poland*, 1.

43. Viktor Yushchenko, "Ukraine's Yushchenko Addresses Joint Session of U.S. Congress," U.S. Department of State, April 6, 2005, 2, http://iipdigital.usembassy.gov/st/english/texttrans/2005/04/200504061638281cjsamoht0.3202631.html#axzz41g7IIUAg

44. Viktor Yushchenko, "Address of the President Viktor Yushchenko at the Parliamentary Assembly of the Council of Europe," Council of Europe, January 25, 2005, 3 http://assembly.coe.int/Sessions/2005/Speeches/Yuschenko_250105_E.htm.

45. Viktor Yushchenko, "Ukraine Makes Its Place as Democratic State," *Ukrinform*, December 10, 2009, https://www.ukrinform.net/rubric-other_news

/888303ukraine_makes_its_place_as_democratic_state___president_yushchenko _175322.html.

46. Polity IV, *Authority Trends 1991–2013: Ukraine*, https://www.systemicpeace .org/polity/ukr2.htm.

47. Viktor Yushchenko, "Remarks by Ukrainian President Viktor Yushchenko to the U.S. Chamber of Commerce," U.S. Chamber of Commerce, Washington, D.C., April 4, 2005, http://www.usubc.org/AUR/aur465.php.

48. Yushchenko, 2006, 3.

49. U.S. Department of State, Bureau of European and Eurasian Affairs, *Country Assessments and Performance Measures—Ukraine*, 2007, https://www.state.gov/p /eur/rls/rpt/92794.htm; U.S. Department of State, Bureau of European and Eurasian Affairs, *Country Assessments and Performance Measures—Ukraine*, 2006, https://www.state.gov/p/eur/rls/rpt/63181.htm; U.S. Department of State, Bureau of European and Eurasian Affair,. *Country Assessments and Performance Measures—Ukraine*, 2005, https://www.state.gov/p/eur/rls/rpt/55799.htm.

50. International Trade Administration, "Ukraine-Trade Agreements," U.S. Department of Commerce, 2017, https://www.export.gov/article?id=Ukraine -Trade-Agreements.

51. Interagency Commission to be set up for Ukraine's Ascendance to NATO. *Ukrainian Government Web Portal*, March 3, 2006, http://www.kmu.gov.ua/control /en/publish/article?art_id=31466745&cat_id=244315200.

52. Albania, Croatia, Georgia, and Macedonia were also included in this act. See NATO Freedom Consolidation Act, U.S. Senate Report 110–34, March 9, 2007, U.S. Government Publishing Office, https://www.gpo.gov/fdsys/pkg/CRPT -110srpt34/html/CRPT-110srpt34.htm.

53. Interview on file with author.

54. Viktor Yanukovych, "Ukraine's Yanukovich Pledges to Work for EU Integra- tion," *Euractiv*, March 2, 2010, https://www.euractiv.com/section/europe-s-east /news/ukraine-s-yanukovich-pledges-to-work-for-eu-integration/.

55. Luke Harding, "Viktor Yanukovych Promises Ukraine Will Embrace Russia." *The Guardian*, March 5, 2010, https://www.theguardian.com/world/2010 /mar/05/ukraine-russia-relations-viktor-yanukovych.

56. Viktor Yanukovych, "News Conference Following Russian–Ukrainian Sum- mit Talks," *President of Russia Website*, May 17, 2010, http://en.kremlin.ru/events /president/transcripts/7781.

57. Viktor Yanukovych, "Annual Address of the President of Ukraine to the Ukrainian Parliament," *Embassy of Ukraine to the United Kingdom of Great Britain and Northern Ireland*, June 12, 2013, http://mfa.gov.ua/en/news-feeds/foreign-offices -news/13010-shhorichne-poslannya-prezidentaukrajini-viktora-janukovicha-do -verkhovnoji-radi-ukrajini.

58. Michael Kelly, "Ukraine Just Made a 'Civilization Defining' Decision—and It Picked Russia Over the West," *Business Insider*, November 21, 2013, http://www .businessinsider.com/ukraine-wont-sign-eu-agreement-2013-11.

59. Robert Levgold, "Robert Legvold on the New Cold War," *The World Post*, September 24, 2015, https://www.huffingtonpost.com/samuel-ramani/robert -legvold-on-the-new_b_8514120.html.

60. Donald Milsten, "Small Powers—A Struggle for Survival," *Journal of Conflict Resolution* 13, 3 (1969), 388.

CHAPTER 6

Energy Security: Competition and Cooperation

Brett D. Weigle

In 2006, Pulitzer Prize–winning energy expert Daniel Yergin warned, "In a world of increasing interdependence, energy security will depend much on how countries manage their relations with one another."[1] His remarks came during a time of increasing international demand for oil amid a supply shortage of this commodity. His insight on relations highlighted the tensions at the time between major oil-consuming nations (China, European countries, Japan, the United States) and major oil-producing nations (Russia and members of the Organization of Petroleum Exporting Countries [OPEC] in Africa, Latin America, and the Middle East). This competition for energy security, viewed through the lens of oil and natural gas supply, has been a fixture in international relations for most of the twentieth and twenty-first centuries.

Energy security looms large in the relations between the United States and two nations with outsize influence on its national interests: China and Russia. President Joseph Biden's 2021 *Interim National Security Strategic Guidance* portrays them as "rivals" and "competitors."[2] In 2021, the relations between these three global actors are partially characterized by their oil-and-gas identities. The latest available data (2020) shows the United States is both the world's largest consumer and producer of oil, China is the world's largest oil importer[3] but only the fifth largest producer, and Russia is the world's third largest oil producer.[4] According to 2021 data, the United States is the world's largest gas producer with Russia in second place,[5] while China is the world's largest gas consumer.[6] These supplier/customer relationships for geographically limited, carbon-based energy sources embody *competition* by individual states to advance their own interests, a tenet of the *realism* school of international relations. Of course, similar tensions would arise among states competing for the geographically limited minerals that enable use of carbon-free energy sources (e.g., cobalt, copper, lithium, nickel, rare earth metals).

A scholar of the *liberalism* international relations school, however, would argue that these three major powers could enhance their energy security by *cooperating* within the norms of multilateral institutions. Rather than focusing on carbon-based energy, such cooperation could (1) advance their global standing in the campaign to mitigate climate change, (2) simultaneously prepare their economies for the transition to noncarbon ("green") energy sources, and (3) increase their collective soft power in less-developed nations through investments in green energy that enhance their achievement of goals (1) and (2). Cooperation in any venue with Russia at this time is difficult in light of the financial sanctions imposed by the United States and the European Union after Russia's invasion of Ukraine in February 2022. However, where green energy cooperation is feasible, the mutual benefits to China, Russia, and the United States would extend to the rest of the world community and sustain America's competitive advantage through new opportunities for its technological and financial power. This argument will advance in three stages.

First, we will examine the broad concept of energy security through the framework of an equilateral "energy security triangle" (supply, economics, and the environment) proposed by the World Economic Forum and paralleled by the International Energy Agency (IEA). Next, we will review how carbon-based energy amplifies realist tendencies in the Chinese, Russian, and American pursuit of economic power, distorting the shape of the energy security triangle. We will conclude with the benefits of cooperation by the three major powers in developing and distributing energy derived from noncarbon sources within extant programs hosted by two other multilateral institutions: the International Renewable Energy Agency (IRENA) and the International Atomic Energy Agency (IAEA). Such cooperation could restore the energy security triangle to an equilateral shape for the benefit of the planet.

SUSTAINING AMERICA'S COMPETITIVE ADVANTAGE: ENERGY SECURITY

America's competitive advantage rests on the *energy security* bestowed by its robust oil and gas production, a blessing of geography and the technical expertise of U.S. energy companies using hydraulic fracturing and directional drilling to extract oil and gas from previously uneconomical shale formations. The United States enjoys a further advantage in that nearly all oil and gas transactions in the global market are denominated in U.S. dollars; however, Russia and China are encouraging (and in some cases, mandating) use of the ruble and the yuan, respectively, in their oil and gas trades.[7]

But there is no universal definition of energy security. The U.S. Congress defines it as "assured access to reliable supplies of energy and the ability to protect and deliver sufficient energy to meet mission essential requirements."[8] However, this narrow focus on *supply* of energy ignores the complexity of the global environment.

The IEA, located in Paris, offers a more balanced view of energy security, as a system whose components are *supply*, *economics*, and the *environment*, with two different time horizons:

[Energy security is] the uninterrupted availability of energy sources at an affordable price. Energy security has many aspects: **long-term** energy security mainly deals with timely investments to supply energy in line with economic developments and environmental needs. On the other hand, **short-term** energy security focuses on the ability of the energy system to react promptly to sudden changes in the supply-demand balance.[9] [emphasis added]

In 2014, the World Economic Forum (WEF) in Geneva reinforced the three-part nature of energy security as "the ability to provide a secure, affordable and environmentally sustainable energy supply," introducing an *energy security triangle*.[10] It is important to note that the IEA and WEF definitions are focused on energy *consumers*. Conversely, governments of energy-*producing* nations that depend on energy exports to finance their rentier economies favor a view of the supply-cost relationship that maximizes their revenue.

The tensions between the three vertices of the energy security triangle are helpful in explaining the dilemma of emphasizing one system component more than the other two. Ideally, a nation balances the three components to create an energy system represented by an equilateral triangle (Figure 6.1). Refinement of the components for this discussion will be helpful.

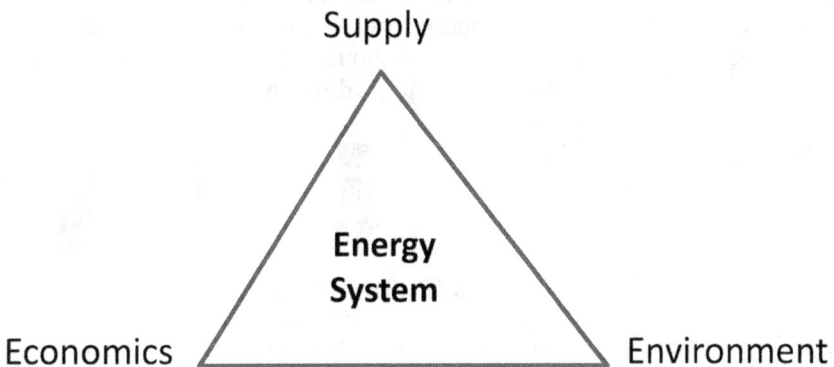

Figure 6.1 The Energy Security Triangle in Balance

In our discussion of sustaining a competitive advantage, we will replace two components to focus our understanding of the argument. First, replacing the "economics" component with "cost" affords a more fine-grained depiction of its inverse relationship with supply. Second, like many human activities, production, distribution, and consumption of energy—regardless of energy source—have potentially negative effects on the land, water, and atmosphere of the planet.[11] However, the "environment" component of the triangle could be replaced with "greenhouse gas emissions" since their reduction is necessary to "[prevent] 'dangerous' human interference with the climate system" according to the UN Framework Convention on Climate Change (UNFCCC).[12]

President Biden's *Interim National Security Strategic Guidance* declares that a vital national interest of the United States is "to protect the security of the American people . . . from threats like climate change."[13] Achieving this interest generates the potential to decrease competition by increasing cooperation with Russia and China in the energy sphere of foreign policy. Two cases illustrate this opportunity.

The first case takes a macro analysis perspective. Decreasing greenhouse gas emissions requires reducing the supply of carbon-based energy sources on a global scale. As of 2020, oil and natural gas provide the overwhelming majority of transportation fuels while coal generates most of the globe's electricity.[14] Reducing the supply of carbon-based energy without reducing demand (or replacement with a noncarbon substitute) will increase the cost of energy for governments and individual persons (Figure 6.2). The IEA warns that energy cost inflation threatens "human well-being and . . . a country's economic development."[15] This result affects the domestic and foreign policy of our three global actors, including their carbon-reduction targets under the UNFCCC.

The second case focuses on the micro level of analysis. If market forces operate without policy constraints (such as UNFCCC targets), cost to individual consumers can only decrease with a concomitant increase in energy supply. Without government market interventions, individual consumers bear the increased energy costs resulting from constricting carbon-based energy supply to reduce greenhouse gas emissions (Figure 6.3)—unless replaced by similarly priced noncarbon energy. The governments of China, Russia, and the United States, despite their global power, are not insulated from this domestic political challenge.

The impacts of energy costs on energy-consuming citizens, businesses, and governments result in a

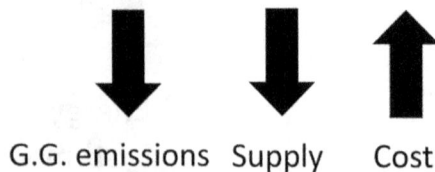

G.G. emissions   Supply   Cost

Figure 6.2 Reducing Environmental Harm Increases Carbon-based Energy Costs

tension between short-term economic impacts and long-term environmental benefits (the latter should, in time, theoretically enhance the economy). Governments in less-developed countries use consumption subsidies for carbon-based fuels for cooking and transporta-

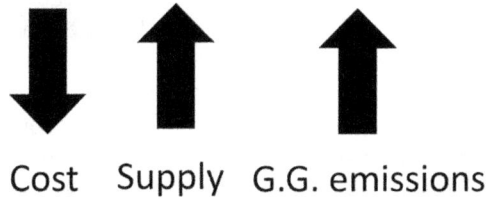

Cost   Supply   G.G. emissions

Figure 6.3 Minimizing Carbon-based Energy Costs Harms the Environment

tion to alleviate poverty; these market interventions are a favored policy tool for authoritarian and democratic regimes alike to maintain domestic stability.[16] These low, subsidized prices encourage greater consumption and thus result in disincentives for energy efficiency and cleaner energy sources. Eliminating such subsidies forms a key part of the UNFCCC strategy to reduce greenhouse gas emissions, as agreed in the November 2021 COP26 Glasgow climate summit.[17]

These tensions between energy cost and reduction of greenhouse gas emissions distort the shape of the energy security triangle from its stable equilateral form. The supply vertex seems sensitive to competition for energy security between the producing and consuming nations over the finite carbon-based energy sources (oil, gas, coal) concentrated in certain geographic locations.

CARBON-BASED ENERGY COMPETITION

To determine the aspects of this competition, we need to understand the role that carbon-based energy plays in each of the global powers' economies, affecting how they seek to achieve their national interests. There is little difference in how they consume energy in their transportation sector, while electricity generation depends primarily on coal and natural gas with varying proportions of nuclear and renewable energy. We will begin with the carbon-based fuel consumer most dependent on imports, China, followed by Russia as a major producer, and conclude with the largest consumer and producer, the United States.

China

China's determination to be recognized as a global actor is underpinned by its rapid economic growth and the concurrent expansion of its energy needs. China's energy system predominantly consists of state-owned enterprises (SOEs) producing petroleum products and electricity, with strong involvement by the ruling Communist Party.[18] The government's latest national energy strategy shifted its top priority from "sustainable development" in 2018 to "energy security" in 2020; one analyst noted this emphasis on access to energy is explicitly coupled with lower energy prices for China's "people/society" (see Figure 6.3).[19] As noted earlier, China is the world's largest oil importer and gas consumer.

Almost half of China's 2020 crude oil imports came from, in order, Russia, Saudi Arabia, Angola, and Iraq.[20] Analyst Erica Downs characterizes the energy relationship between China and Russia as "arguably more robust than it has ever been," anchored on twin advantages: one geostrategic and one financial.[21] Russia enjoys preeminence because its land border with China allows movement of oil and natural gas that bypasses maritime chokepoints, such as the Strait of Malacca, where an adversary (read: U.S. Navy) could threaten shipments by tanker vessels from Africa and the Middle East. Russian oilfields connect to Chinese refineries via a spur from the Eastern Siberia–Pacific Ocean pipeline. Russia enjoys a position as China's second largest supplier of natural gas through its Power of Siberia pipeline and is China's sixth largest supplier of liquefied natural gas (LNG) delivered via tanker vessels from its Siberian Yamal terminal on its Arctic coast.[22]

From a financial aspect, China has provided loans to Russian energy companies since 2005 to cushion the effects of low oil prices and U.S. economic sanctions against Russia. Chinese loans partially financed construction of both pipelines and the Yamal LNG facility in return for long-term oil and gas delivery contracts at prices comparable to Russia's large European customers.[23] Amid rising tensions with Europe even before Russia's 2022 invasion of Ukraine, China signed new ten-year deals to import Russian oil and gas.[24] After the invasion, China advanced its desire to make the yuan a global currency by using it to pay for Russian energy imports when Russia was cut off from much of the U.S. dollar-denominated global financial transfer system.[25]

Both nations enjoy aspects of this growing energy interdependence. Their bilateral trade rewards China's investments in the Russian energy sector and shelters its purchases of Russian gas and oil from the international financial sanctions imposed due to Russia's invasion of Ukraine. Russia benefits from its connections to the growing Chinese market through pipelines secure within its sovereign territory, unlike the risk to its pipelines to Europe, which transit Belarus, Turkey, and Ukraine and are vulnerable to risk of interference. An example of this risk is Belarus's November 2021 threat to cut Russian gas flows to Europe over a dispute about migrants policy.[26] China gains a reliable supplier of both piped natural gas and LNG that buffer it from possible U.S. sanctions.

The potential threat to China's maritime oil and gas supply chain posed by the Strait of Malacca partly motivates China's claims to contested areas in the South China Sea. Additionally, in the vicinity of the Spratly Islands, the U.S. Geological Survey estimates undiscovered reserves may average

"2.5 billion barrels of oil and . . . 25.5 trillion cubic feet of natural gas."[27] These potential gas reserves could offset China's heavy reliance on coal for electricity generation.

Since 2011, China's power generation sector primarily runs on coal, with China "consum[ing] more coal than the rest of the world combined."[28] Hydropower, wind, and nuclear generation occupy distant second, third, and fourth places in power generation within China.[29] Despite holding the third largest global coal reserves, China has been a net coal importer since 2009, today buying principally from neighbors Indonesia, Mongolia, and Russia.[30] The resumption in 2021 of China's coal imports from the United States vividly illustrates Daniel Yergin's linkage of energy security to three international relationships.[31]

First, a leading import source, North Korea, was dropped by China in 2017 in support of UN sanctions against the North Korean nuclear weapons program.[32] However, there is evidence that the Chinese government has begun ignoring private shipments of coal from that country in the face of increasing international coal prices.[33] Second, Australia was the leading supplier of coal to China until a call by the Australian prime minister for an investigation into the origin of the coronavirus causing COVID-19 sparked a Chinese ban on Australian coal in early 2020.[34] Consequently, U.S. coal producers saw their export business resume as the Chinese government slashed its tariffs on American coal as a signal to the Biden administration of its willingness to renormalize trade relations hampered by sanctions under the Donald Trump administration.[35]

The extensive air pollution resulting from China's dependence on coal motivated its adoption of nuclear power generation. Since 1991, China has commissioned 48 operating reactors that generate 45 gigawatts (GW) of electricity, "making its fleet the third-largest in the world behind France (63 GW) and the United States (98 GW)" according to CSIS scholar Jane Nakano.[36] One component of China's Belt and Road Initiative is the sale of its nuclear power generation technology, but the sole purchase by Pakistan may have resulted more from China's friendship with that nation to balance the fraught China–India relationship. Largely, China seems to favor financing nuclear projects deemed less attractive to other investors as a way to create demand for future Chinese nuclear sales, as in Argentina and the United Kingdom.[37] Unlike China, Russia has a more extensive energy supply portfolio.

Russia

Russia's status as a major producer and exporter of oil and natural gas is a source of both economic strength and vulnerability. Oil and natural gas revenues provided 30% of Russia's federal budget revenues in 2020, a decrease from just over 50% as recently as 2014.[38] Over half of the country's exports come from oil and natural gas, with 60% of its oil exports and about 75% of its natural gas exports destined for nations in the European Union (EU).[39]

Consequently, Russia enjoys leverage over nations with strong ties to the United States. This represents a vulnerability as most U.S. European allies are in the North Atlantic Treaty Organization (NATO) and the EU. As the European Commission notes, "the United States is the EU's largest trade and investment partner by far."[40] Russian attempts to use energy leverage against EU nations has not been universally successful, especially in light of economic sanctions after its Ukraine invasion. Russia's prominent position as a supplier of coal, gas, and oil to China, as previously noted, is vulnerable to China's continued willingness to finance the delivery pipelines and help Russia evade international sanctions.

Like China, Russia's largest oil and gas companies are state-owned enterprises, allowing the government to use their production as instruments of national power.[41] Despite lacking a marquee label like Belt and Road, Russian energy companies such as Gazprom nevertheless invest in energy projects around the world, from exploration and production in Algeria to infrastructure for natural-gas vehicles in Vietnam.[42] The exit of Western multinational petroleum companies from their joint projects with Russia since the 2022 Ukraine invasion, while depriving the Russian SOEs of some capital and expertise, has left the SOEs with a larger share of the global oil market.[43]

European nations, having endured temporary interruptions of Russian gas supplies transiting Ukraine in 2006 and 2009, have slowly sought diversification alternatives. They have increased LNG imports from the United States, through discussions with the Trump and Biden administrations, with several northern European nations constructing LNG terminals in the past decade. The Southern Gas Corridor pipeline, bringing gas from Azerbaijan to Italy, gives Europeans another alternative to Russian pipelines.[44]

Russia has countered this diversification with its TurkStream pipeline route, shipping its gas under the Black Sea through continental Turkey into the European market.[45] Russia seeks to maintain its energy leverage against Europe by reducing risk of interruption of flow through its Soviet Union–legacy pipelines transiting Ukraine and Belarus. The NordStream 2 pipeline (completed in 2021, but not approved to open as of 2022) was designed to deliver gas under the Baltic Sea to a terminal in Germany. It would have doubled the capacity delivered since 2011 by its sister, NordStream 1. Opposed by the Trump administration, NordStream 2 is majority-owned by a Russian SOE, and its approval to begin operations was withdrawn by the German government days before the 2022 Russian war against Ukraine.[46]

Finally, Russia buttresses its position in the global energy market through multilateral organizations and through financial investments. As a member of OPEC+, Russia sometimes collaborates with OPEC members (notably Saudi Arabia) and other oil-producing nations to influence global oil prices by adjusting production.[47] The Saudi–Russian entente's joint suppression of oil prices in 2020 was seen as an attempt to undermine U.S. power by starving its robust shale oil–production industry of operating capital. However, this cooperation is not without danger. When Saudi Arabia disagreed with Russia on production volumes, it subsequently cut its oil export price further to capture Russian market share as world oil demand slowly revived during the COVID-19 demand crisis.[48] The Gas Exporting Countries Forum provides Russia with another international venue to leverage its economic power, having held the office of secretary-general twice since 2010.[49]

While Russia mostly burns natural gas to generate its electricity, hydropower, coal, and nuclear generation are also critical sources of supply.[50] Extensive Russian coal reserves make it the third-largest exporter after Indonesia and Australia, with China and other Asian nations the prime customers.[51] In 2020, coal generated 12% of Russia's electricity while hydropower produced another 7%.[52] Its undeveloped hydropower resources are ranked second in the world, with only 20% of the potential developed; the positioning of most hydropower generation in Siberia makes China a lucrative electricity export market.[53]

The Russian nuclear power fleet consists of thirty-eight reactors supplying about 18% of the country's electricity demand.[54] The commercial success of Russia's nuclear energy SOE, Rosatom, is a testament to the resurrection of the industry after the Chernobyl accident in 1986. Rosatom boasts contracts to build thirty-five reactor units in Bangladesh, Belarus, China, Egypt, Finland, and India and for NATO members Hungary and Turkey.[55] Additionally, Rosatom was awarded a $10 billion contract by the Iranian government to build nuclear power reactors under the 2015 Joint Comprehensive Plan of Action (JCPOA). The Biden administration assured the Russian government in April 2022 that international sanctions resulting from the Ukraine invasion would be waived for the Rosatom JCPOA agreement—an attempt to alleviate fears of other JCPOA parties that Iran would use Rosatom's absence as an excuse to resume uranium enrichment.[56]

Nuclear power plays a crucial role in Russia's Arctic strategy as well, as Rosatom operates four relatively new nuclear-powered icebreakers, with three more under construction. They are part of a forty-vessel fleet of icebreakers that will enable a year-round capacity to escort merchant vessels along the Northern Sea Route skirting Russia's Arctic coast; Russia advertises this route as a fuel-saving alternative to transit between Europe and Asia that is free of risky chokepoints like the Suez and Panama Canals.[57] Tankers with ice-strengthened hulls have used this route to deliver LNG cargoes from Norway and Russia to customers in China and Korea. Russia hopes this route will make its LNG more economically attractive in Asia to compete with LNG exported from terminals on the U.S. Gulf Coast.

United States

The United States achieved a large measure of energy security when it perfected the technology to produce its shale oil and gas reserves, allowing it to become a natural gas net exporter in 2017 and a net exporter of oil and refined products in 2020.[58] It has also been a major coal exporter since at least 1950.[59] The United States' top oil and LNG export destinations are China, Europe, and northeast Asia. The international nature of energy security is evident when one considers that the United States still imports oil from several OPEC members, some Caribbean nations, Canada, and Mexico because it is cheaper to refine than U.S. crude oil, based on the location and capability of a particular U.S. refinery. Likewise, America both imports and exports electricity and natural gas to and from Canada and Mexico, based on relative cost.[60]

The United States generates most of its electricity from natural gas, followed by nearly equal contributions from coal and nuclear, and again by similar proportions of wind and hydropower.[61] Unlike its global competitors, nearly all segments of the American energy sector are operated by private companies (the U.S. government does own some hydroelectric generating facilities in western and southeastern states).[62] U.S. nuclear-power manufacturing expertise resides in three companies: Bechtel, Fluor, and Westinghouse Electric. The latter's newest sales were four reactors completed in China in 2019 while all three companies provide maintenance services to nuclear plants around the world.[63] Fluor is also a major investor in NuScale Power, whose innovative small modular reactor was the first such design to receive certification by the U.S. Nuclear Regulatory Commission in 2020.[64]

The energy industry is a major source of economic power that underpins America's diplomatic, informational, and military instruments of national power. The American Petroleum Institute estimates that the oil and gas industry contributed 8% of U.S. gross domestic product in 2020. The think tank Energy Futures Initiative reports that all energy sectors "employed more than 8.27 million workers, accounting for 5.4 percent of all jobs in the United States" during that same year.[65]

This economic contribution—and the avoided costs of being a net energy importer—generates funds for the Department of State and the U.S. Agency for International Development (USAID) to extend U.S. diplomatic influence by energy investments in less-developed nations, through programs such as *Power Africa*, *Asia EDGE*, and the *Energy Resource Governance Initiative* (ERGI).[66] These energy programs serve to strengthen U.S. bilateral relations while offsetting China's and Russia's foreign energy investments. They also advance U.S. government climate policy goals to assist partner nations in their transition to noncarbon energy sources.

The United States can leverage its energy security through two aspects of information power. First, it is no longer constrained by oil imports from the Middle East, which reduces the appearance that the United States might "go to war for oil," a possibility implied by the 1980 Carter Doctrine[67] and an accusation leveled by critics of the 1991 and 2003 U.S. invasions of Iraq.[68] Second, the United States wants to be seen as a reliable source of oil and natural gas for allies and partners to diminish Russian energy economic coercion and to counter potential Chinese interference with tanker ships transiting the Strait of Malacca. Finally, U.S. military power is enhanced through a secure domestic oil supply while decreased U.S. security demands in the Middle East allow the government to shift military forces to compete with China for influence in Asia. But competition is not the only way to enhance energy security. Russia, China, and the United States have opportunities for cooperation.

COOPERATION THROUGH MULTILATERAL INSTITUTIONS

Cooperation with Russia and China to develop and promote clean energy technologies has established precedent in U.S. foreign policy throughout the twenty-first century. President George W. Bush's 2006 *National Security Strategy* highlights the *Asia-Pacific Partnership for Clean Development and Climate* that it formed with China—and with Australia, Canada, India, Japan, and South Korea.[69]

At the G8 summit on July 15, 2006, President Bush announced that "he intended to speed up the process to conclude the necessary agreement that would permit bilateral civil nuclear cooperation between the United States and Russia."[70] While the process was slowed down by the Russian invasion of Georgia in 2008, what became known as the "123 Deal" went into effect in 2011.[71] President Barack Obama's State Department established the U.S.–Russia Energy Working Group in 2012 to cooperate on "public-private partnerships, city-to-city pairings, trade missions, and university links."[72] However, effective cooperation was later stymied by Russia's invasion of Ukraine in 2014.[73] Given Russia's second invasion of Ukraine in 2022 and repeated Russian threats of potential nuclear escalation in that conflict,[74] it is unlikely that the United States and Russia will be able to pursue any meaningful energy cooperation in the near term.

After a lapse in cooperation with China on energy issues during the Trump administration, President Biden's official position is to "welcome the Chinese government's cooperation on issues such as climate change, . . . where our national fates are intertwined."[75] The two nations issued a joint declaration pledging cooperation on climate action after COP26 in Glasgow, even though Chinese President Xi Jinping (and Russian

President Vladimir Putin) did not attend the summit.[76] Where could this cooperation occur in the future?

Within the international order embraced by liberalism, such cooperation would occur within the framework of multilateral institutions. Two such bodies already count our three global powers among their members: the IRENA, headquartered in Abu Dhabi, and the IAEA, located in Vienna. Three existing program areas suggest nonconfrontational venues: (1) IRENA promotion of renewable hydrogen; (2) IAEA collaboration on nuclear power generation; and (3) IRENA joint ventures in clean energy solutions in less-developed nations. The World Trade Organization might also play a useful role, especially in dispute resolution.

Green Hydrogen

Renewable ("green") hydrogen is produced using electricity generated from noncarbon sources, such as wind and solar power; its combustion does not add carbon to the atmosphere. China, Russia, and the United States participate in IRENA's *Collaborative Framework on Green Hydrogen*, a venue for their cooperation in the development and deployment of green hydrogen.[77] Their participation is strengthened since they enjoy robust positions in three sectors of the renewable energy domain that prime them to cooperate as green hydrogen suppliers.

First, they possess significant reserves of the minerals used to manufacture photovoltaic (PV) solar panels and wind turbines: silicon, copper, and the elements known as lanthanides (also known as *rare earth elements*).[78] The three nations are the leading global producers of silicon (the base material for PV panels) and major refiners of copper (the electrical conductor in PV panels and turbines). China and the United States are the top two global producers of lanthanides, which are used in the powerful magnets driving the generators in wind turbines.[79]

Second, the three nations enjoy geographic advantages for solar and wind power generation, providing the carbon-free electricity needed for green hydrogen.[80] Their extensive coastlines are graced with offshore winds while expansive plains provide space for both PV panels and wind turbines.

Third, China is the world's predominant manufacturer of PV panels, followed by U.S. partners Malaysia, Vietnam, and South Korea.[81] In 2020, an American company ranked second globally for new wind power installations while Chinese companies held the third and fourth rankings.[82] The robust nuclear power manufacturing industries of China, Russia, and the United States have been noted earlier. This confluence of mineral reserves, green electricity generation potential, and solar and wind generation manufacturing enables trilateral collaboration to produce green hydrogen.

This renewable fuel can ameliorate the revenue losses felt by Russia and the United States during the global transition away from oil and natural gas endorsed under COP26. Their collaboration could also reduce barriers to gains in global manufacturing and installation of PV panels. China's cooperation on green hydrogen could moderate aspects of trade competition with the United States. For example, the U.S. government had imposed tariffs in the past on imported Chinese PV panels—tariffs upheld by the World Trade Organization—and such cooperation could figure into a U.S. decision to permanently remove the tariffs.[83]

Nuclear Power Generation

The IAEA finds that "nuclear power can play an important role in responding to climate related challenges" and providing "grid stability" to offset the intermittent nature of solar and wind generation.[84] The COP26 proceedings did not explicitly endorse nuclear power, but the IAEA noted broad support for its role in reducing carbon emissions.[85] The European Commission included nuclear power (and natural gas) in its 2022 taxonomy of sustainable investments.[86] Aside from the French company EDF, the capacity to design and build nuclear power generating facilities resides in China (under the State Nuclear Power Technology Corporation), Russia (Rosatom), and the United States (Bechtel, Fluor, and Westinghouse). International support and market share suggest a rich area for trilateral cooperation.

International and bilateral venues already exist. All three nations are members of the IAEA's International Project on Innovative Nuclear Reactors and Fuel Cycles which seeks to "promote the sustainable development of nuclear energy."[87] The U.S. Energy Department's Office of International Nuclear Energy Policy and Cooperation leads a U.S. interagency collaboration with several nations in the nuclear power generating realm. Since 2007, the *U.S.–China Bilateral Civil Nuclear Energy Cooperative Action Plan* has supported joint work on, for example, fast reactor technologies, while the *U.S.–Russia Civil Nuclear Energy Cooperation Action Plan* has allowed exploration of reactor demonstration projects since 2006, to name two areas of cooperation.[88] Russia and the United States also offered assistance to Japan to clean up its Fukushima nuclear generating station damaged by a tsunami in 2011.[89] However, as mentioned earlier, the Russian invasion of Ukraine in 2022 complicates cooperation in nuclear energy.

Clean Energy Investments in Developing Nations

Investing in projects to bring clean energy to less-developed nations could tamp down competitive friction between the United States, Russia, and China, especially in the Western Hemisphere. IRENA programs offer structures for cooperation and cost-sharing in two regions: Latin America and Africa.

IRENA encourages foreign investment through its 2019 *Regional Action Plan for Latin America*, partnering with the Inter-American Development Bank to "develop a pipeline of investment mature projects" and producing analysis to "improve the bankability of projects."[90] IRENA's existing *Clean Energy Corridor for Central America* program and its International Conference on Hydropower Investment in Developing Countries planned for October 2022 offer collaborative venues where the three global powers could increase their influence without exacerbating tensions in other areas of competition.[91] USAID could formally connect its Latin America initiative *Strengthening Utilities and Promoting Energy Reform* (SUPER) with the IRENA programs and invite Russia and China to collaborate in parallel development efforts.[92] Actions to stabilize Latin American and Caribbean governments would also provide rich investment opportunities for the renewable-energy industries of the three global actors.[93]

IRENA provides another two "ready-made" cooperation venues with its *Africa Clean Energy Corridor* and *West Africa Clean Energy Corridor*. These initiatives seek to connect African regional power grids; electricity transmission infrastructure companies from the three global powers might realize construction opportunities by cooperating in these venues. China's investments in Africa through its Belt and Road Initiative and capital projects sponsored by the Russia–Africa Energy Committee now compete with American companies seeking "sustainable, long term investment in African energy" promoted by the Biden administration during the 2021 U.S.–Africa Energy Forum.[94] Instead, all three nations could avoid duplicate efforts by cooperating under the umbrella of IRENA's programs, much like USAID's *Power Africa* memorandum of understanding with IRENA.[95] This multilateral approach could provide a neutral venue for cooperation that maximizes the benefit to African customers by preventing investment gaps while building local goodwill for all three global powers.[96]

CONCLUSION

Energy security does not have to mean energy competition. Geographically limited, carbon-based energy sources engender competition due to their inherent "have/have not" nature. Their movement from supplier to customer, and the resulting flow of payments, are subject to economic, political, and even military interruptions. However, the almost universally available sources of renewable energy (sun and wind) make their generation an ideal area for cooperation, both bilaterally and within multilateral institutions. The multilateral energy organizations IEA, IAEA, and IRENA offer venues for Russia and China to participate in a rules-based international order that does not privilege a U.S. perspective (like the Bretton Woods international monetary system).[97]

All three nations have roughly proportional capabilities in different aspects of the energy system, making cooperation among de facto peers more appealing since resources could be aligned synergistically to achieve bilateral or even trilateral national interests. The collective carbon emission reductions would help the three nations meet their nationally determined contributions under the 2015 UNFCCC Paris Agreement.[98]

New revenues would be generated by increased production of raw materials and finished products for solar and wind energy generation, and by construction and servicing of new nuclear power generation stations and installation of electric transmission lines. This increased cash flow would ease the economic shock on the economies of China, Russia, and the United States as the global economy transitions away from carbon-based fuels. Russian commercial activities in renewable energy do not appear to be sanctioned by the United States or the European Union in response to its Ukraine invasion,[99] so this "acceptable" revenue stream might entice cooperation among the three nations. Most importantly for the United States, resources in the diplomatic and economic domains could be diverted to the remaining spheres of competition with China and Russia to sustain America's competitive advantage.

NOTES

1. Alexander Jung and Georg Mascolo, "The War over Resources: 'Energy Security Will Be One of the Main Challenges of Foreign Policy,'" interview with Daniel Yergin, *Spiegel International*, July 18, 2006, https://www.spiegel.de/international/spiegel/the-war-over-resources-energy-security-will-be-one-of-the-main-challenges-of-foreign-policy-a-427350.html

2. Joseph R. Biden, Jr., *Interim National Security Strategic Guidance* (Washington, DC: The White House, March 2021), 6, 8, https://www.whitehouse.gov/wp-content/uploads/2021/03/NSC-1v2.pdf. President Biden's 2022 *National Security Strategy* includes similar language.

3. "Country Analysis Executive Summary: China," U.S. Energy Information Administration (EIA), September 20, 2020, p. 5, https://www.eia.gov/international/content/analysis/countries_long/China/china.pdf

4. EIA, "Frequently Asked Questions: What Countries Are the Top Producers and Consumers of Oil?" May 10, 2022, https://www.eia.gov/tools/faqs/faq.php?id=709&t=6

5. Bill Brown, "United States Continued to Lead Global Petroleum and Natural Gas Production in 2020," EIA, July 19, 2021, https://www.eia.gov/todayinenergy/detail.php?id=48756

6. Eric Yep and Shermaine Ang, "Analysis: China Steps Up Natural Gas Supply for 2021–22 Winter-Spring Season," *S&P Global Platts*, September 24, 2021, https://www.spglobal.com/platts/en/market-insights/latest-news/lng/092421-analysis-china-steps-up-natural-gas-supply-for-2021-22-winter-spring-season

7. Phil Rosen, "China Is Buying Russian energy with Its Own Currency, Marking the First Commodities Paid For in Yuan since Western Sanctions Hit Moscow,"

Business Insider, April 7, 2022, https://markets.businessinsider.com/news/commodities/dollar-vs-yuan-china-buys-russian-oil-coal-ukraine-sanctions-2022-4?op=1

8. 10 U.S. Code § 2924 (3)(A).

9. International Energy Agency (IEA), "Energy Security," December 2, 2019, https://www.iea.org/areas-of-work/ensuring-energy-security

10. World Economic Forum/Accenture, *The Global Energy Architecture Performance Index Report 2014*, Geneva, December 2013, 11, https://www3.weforum.org/docs/WEF_EN_NEA_Report_2014.pdf

11. Energy Explained, "Energy and the Environment Explained," EIA, August 24, 2020, https://www.eia.gov/energyexplained/energy-and-the-environment/

12. United Nations Framework Convention on Climate Change, "What Is the UNFCCC?" n.d., https://unfccc.int/process-and-meetings/the-convention/what-is-the-united-nations-framework-convention-on-climate-change

13. Biden, *Interim National Security Strategic Guidance*, p. 9. President Biden's 2022 *National Security Strategy* includes similar language.

14. International Energy Outlook 2021, "Consumption," EIA, October 6, 2021, https://www.eia.gov/outlooks/ieo/consumption/sub-topic-01.php; "World Gross Electricity Production by Source, 2019," IEA, August 6, 2021, https://www.iea.org/data-and-statistics/charts/world-gross-electricity-production-by-source-2019

15. IEA, "Defining Energy Access: 2020 Methodology," October 13, 2020, https://www.iea.org/articles/defining-energy-access-2020-methodology

16. "COP26: How Much Is Spent Supporting Fossil Fuels and Green Energy?" *BBC News*, November 15, 2021, https://www.bbc.com/news/59233799; see also Harro van Asselt and Jakob Skovgaard, "Chapter 11: The Politics and Governance of Energy Subsidies," in T. Van de Graaf et al. (eds.), *The Palgrave Handbook of the International Political Economy of Energy* (London: Palgrave Macmillan, 2016), 269–284.

17. UNFCCC, "Glasgow Climate Pact," Conference of the Parties 26 (COP26), November 13, 2021, https://unfccc.int/sites/default/files/resource/cop26_auv_2f_cover_decision.pdf

18. "State-Owned Assets Supervision and Administration Commission of the State Council (SASAC)," State Council of the People's Republic of China, last modified June 29, 2021, http://en.sasac.gov.cn/index.html; Orange Wang and Zhou Xin, "China Cements Communist Party's Role at Top of Its SOEs, Should 'Execute the Will of the Party'," *South China Morning Post*, January 8, 2020, https://www.scmp.com/economy/china-economy/article/3045053/china-cements-communist-partys-role-top-its-soes-should

19. Yuki Yu, "China's Energy Strategy 2020: Shifting Focus & Future Directions," *Energy Iceberg: Chinese Clean Power Policy Intelligence & Market Insights*, July 1, 2020, https://energyiceberg.com/china-energy-policy-2020

20. China Power Team, "How Is China's Energy Footprint Changing?" Center for Strategic and International Studies, January 30, 2021, https://chinapower.csis.org/energy-footprint

21. Erica Downs, "China–Russia Energy Relations," Center for Naval Analyses, testimony before U.S.–China Economic and Security Review Commission, March 21, 2019, https://www.uscc.gov/sites/default/files/Downs_Testimony ... pdf

22. "Russian Pipeline Gas Exports to China Nearly Triple in 2021," *RT*, September 20, 2021, https://www.rt.com/business/535299-russia-china-gas-supply-triple/

23. Downs, "China–Russia Energy Relations"; Michael Ratner and Heather L. Greenley, *Power of Siberia: A Natural Gas Pipeline Brings Russia and China Closer* (Washington, DC: Congressional Research Service, April 21, 2020).

24. Rosemary Griffin, "Russia, China Sign New Energy Deals Following Ukraine Tension," *S&P Global Platts*, February 4, 2022, https://www.spglobal.com/platts /en/market-insights/latest-news/oil/020422-russia-china-sign-new-energy -deals-following-ukraine-tension

25. Rosen, "China Is Buying Russian Energy with Its Own Currency."

26. Tsvetana Paraskova, "Belarus Threatens to Cut Off Transit Gas Flows to Europe," *OilPrice.com*, November 11, 2021, https://oilprice.com/Latest-Energy -News/World-News/Belarus-Threatens-To-Cut-Off-Transit-Gas-Flows-To -Europe.html

27. "South China Sea: Reserves and Resources," EIA, October 15, 2019, https:// www.eia.gov/international/analysis/regions-of-interest/South_China_Sea

28. China Power Team, "How Is China's Energy Footprint Changing?"

29. Country Profile: China, "Electricity Generation by Source, People's Republic of China 1990–2020," IEA, accessed February 5, 2022, https://www.iea.org /countries/china

30. China Power Team, "How Is China's Energy Footprint Changing?"

31. "China's Coal Imports from US in September Jump," International Centre for Sustainable Carbon, November 5, 2021, https://www.sustainable-carbon.org /chinas-coal-imports-from-us-in-september-jump

32. China Power Team, "How Is China's Energy Footprint Changing?"

33. Seulkee Jang, "Amid Coal Shortages, Chinese Traders on the Hunt for More North Korean Coal," *Daily NK*, October 7, 2021, https://www.dailynk .com/english/amid-coal-shortages-chinese-traders-hunt-more-north-korean -coal

34. Anna Henderson, Stephen Dziedzic, James Oaten, and Som Patidar, "China's *The Global Times* Appears to Confirm a Ban on Australian Coal Imports amid Perilous Trade Tensions," *ABC News*, December 14, 2020, https://www.abc.net .au/news/2020-12-14/global-times-reports-australian-coal-exports-blocked -by-china/12983336

35. Ken Silverstein, "China Snubs Australian Coal, Giving U.S. Coal Producers Breathing Space," *Forbes*, September 12, 2021, https://www.forbes.com/sites /kensilverstein/2021/09/12/china-snubs-australian-coal-giving-us-coal-producers -breathing-space/?sh=4f6aaa1815b0; Robert Vergara, "China's US Coal Imports Jump 748% in Q4'20 amid Australian Trade Dispute," *S&P Global*, February 23, 2021, https://www.spglobal.com/marketintelligence/en/news-insights/latest-news -headlines/china-s-us-coal-imports-jump-748-in-q4-20-amid-australian-trade -dispute-62766412

36. Jane Nakano, *The Changing Geopolitics of Nuclear Energy: A Look at the United States, Russia, and China*, Center for Strategic and International Studies, March 2020, p. 8, https://www.csis.org/analysis/changing-geopolitics-nuclear-energy-look -united-states-russia-and-china

37. Nakano, *The Changing Geopolitics of Nuclear Energy*, p. 15.

38. Ministry of Finance of the Russian Federation, *Annual Report on Execution of the Federal Budget*, updated October 28, 2021, https://minfin.gov.ru/en/statistics /fedbud/?id_65=119255-annual_report_on_execution_of_the_federal_budget _starting_from_january_1_2006#

39. "Country Analysis Brief: Russia," EIA, December 13, 2021, p. 1, https://www.eia.gov/international/analysis/country/RUS; Cory Welt and Rebecca M. Nelson, *Russia: Domestic Politics and Economy* (Washington, DC: Congressional Research Service, September 9, 2020), p. 29.

40. European Commission, "Trade Policy: United States," *Opening Foreign Markets* section, September 6, 2021, https://ec.europa.eu/trade/policy/countries-and-regions/countries/united-states

41. Andrew S. Bowen and Cory Welt, *Russia: Foreign Policy and U.S. Relations* (Washington, DC: Congressional Research Service, April 15, 2021), 40–42.

42. Gazprom, "Foreign Projects," accessed December 18, 2021, https://www.gazprom.com/projects

43. Rochelle Toplensky, "Western Oil Companies Leave Russia to Their State-Run Rivals," *The Wall Street Journal*, March 2, 2022, https://www.wsj.com/articles/western-oil-companies-leave-russia-to-their-state-run-rivals-11646229422

44. European Commission, "Diversification of Gas Supply Sources and Routes," *Energy Security*, accessed May 31, 2022, https://energy.ec.europa.eu/topics/energy-security/diversification-gas-supply-sources-and-routes_en

45. Sarah E. Garding, Michael Ratner, Cory Welt, and Jim Zanotti, *TurkStream: Russia's Southern Pipeline to Europe* (Washington, DC: Congressional Research Service, May 6, 2021); BP, "Southern Gas Corridor—Project of the Century," *News and Insights*, January 29, 2021, https://www.bp.com/en/global/corporate/news-and-insights/reimagining-energy/southern-gas-corridor-special-feature.html

46. Dave Keating, "Trump Imposes Sanctions to Stop Nord Stream 2—But It's Too Late," *Forbes*, December 21, 2019, https://www.forbes.com/sites/davekeating/2019/12/21/trump-imposes-sanctions-to-stop-nord-stream-2—but-its-too-late/?sh=6df922a35df1; News, "Ukraine Crisis: Germany Halts Nord Stream 2 Approval," *Deutsche Welle*, February 22, 2022, https://www.dw.com/en/ukraine-crisis-germany-halts-nord-stream-2-approval/a-60867443; Paul Belkin, Michael Ratner, and Cory Welt, *Russia's Nord Stream 2 Natural Gas Pipeline to Germany Halted* (Washington, DC: Congressional Research Service, March 10, 2022).

47. Organization of the Petroleum Exporting Countries, "Declaration of Cooperation," accessed December 18, 2021, https://www.opec.org/opec_web/en/publications/4580.htm

48. Alex Ward, "The Saudi Arabia–Russia Oil War, Explained," *Vox*, March 9, 2020, https://www.vox.com/2020/3/9/21171406/coronavirus-saudi-arabia-russia-oil-war-explained

49. Gas Exporting Countries Forum, "GECF History," accessed December 18, 2021, https://www.gecf.org/about/history.aspx; Bowen and Welt, *Russia: Foreign Policy and U.S. Relations*, p. 42. Despite its global status as a gas producing nation, the United States has not been invited to join this multilateral organization.

50. Country Analysis: Russia, "Figure 7: Russia's Total Primary Energy Consumption, 2020," EIA, December 13, 2021, https://www.eia.gov/international/analysis/country/RUS

51. Hilary Hooper, Justine Barden, and Tejasvi Raghuveer, "Europe Is a Key Destination for Russia's Energy Exports," *Today in Energy*, EIA, March 14, 2022, https://www.eia.gov/todayinenergy/detail.php?id=51618

52. Country Analysis: Russia, "Figure 7: Russia's Total Primary Energy Consumption, 2020," EIA, December 13, 2021, https://www.eia.gov/international/analysis/country/RUS

53. "Country Profile: Russia," International Hydropower Association, accessed June 1, 2022, https://www.hydropower.org/country-profiles/russia; Anastasia Lyrchikova, "Russia Doubles Electricity Exports to China to Help Ease Power Crunch," *Nasdaq*, October 1, 2021, https://www.nasdaq.com/articles/russia-doubles-electricity-exports-to-china-to-help-ease-power-crunch-2021-10-01

54. Nakano, *The Changing Geopolitics of Nuclear Energy*, p. 6.

55. Nakano, *The Changing Geopolitics of Nuclear Energy*, pp. 6–7; Rosatom, "Projects," accessed December 18, 2021, https://rosatom.ru/en/investors/projects

56. Adam Kredo, "Exposed: The Russian Companies That Will Get Billions from New Iran Nuclear Deal," *Washington Free Beacon*, April 8, 2022, https://freebeacon.com/national-security/exposed-the-russian-companies-that-will-get-billions-from-new-iran-nuclear-deal; Richard Nephew, "The Wisdom of Nuclear Carve-Outs from the Russian Sanctions Regime," *War on the Rocks*, March 17, 2022, https://warontherocks.com/2022/03/the-wisdom-of-nuclear-carve-outs-from-the-russian-sanctions-regime

57. Rosatom, "Nuclear Icebreaker Fleet," accessed December 18, 2021, https://rosatom.ru/en/rosatom-group/the-nuclear-icebreaker-fleet/index.php?sphrase_id=2543019

58. Oil and Petroleum Products Explained, "Oil Imports and Exports," EIA, April 13, 2021, https://www.eia.gov/energyexplained/oil-and-petroleum-products/imports-and-exports.php; Natural Gas Explained, "Natural Gas Imports and Exports," EIA, July 12, 2021, https://www.eia.gov/energyexplained/natural-gas/imports-and-exports.php

59. Coal Explained, "Coal Imports and Exports," EIA, May 27, 2021, https://www.eia.gov/energyexplained/coal/imports-and-exports.php

60. Augustine Kwon, "Today in Energy: California Imports the Most Electricity from Other States; Pennsylvania Exports the Most," EIA, April 4, 2019, https://www.eia.gov/todayinenergy/detail.php?id=38912; "Natural Gas Explained: Natural Gas Imports and Exports," EIA, July 12, 2021, https://www.eia.gov/energyexplained/natural-gas/imports-and-exports.php

61. Country Profile: United States, "Electricity Generation by Source, United States 1990–2020," IEA, accessed February 5, 2022, https://www.iea.org/countries/united-states

62. Hydropower Explained, "Where Hydropower Is Generated," EIA, April 8, 2021, https://www.eia.gov/energyexplained/hydropower/where-hydropower-is-generated.php

63. Westinghouse Electric, "New Plants and Operating Plants," https://www.westinghousenuclear.com; Bechtel, "Nuclear Power," https://www.bechtel.com/services/energy/nuclear, both accessed February 5, 2022

64. Fluor, "Nuclear & Civil," https://www.fluor.com/client-markets/mission/nuclear-civil, accessed February 5, 2022; NuScale Power, "U.S. Nuclear Regulatory Commission Issues Standard Design Approval for NuScale's SMR Design," September 14, 2020, https://newsroom.nuscalepower.com/press-releases/news-details/2020/U.S.-Nuclear-Regulatory-Commission-issues-Standard-Design-Approval-for-NuScales-SMR-design/default.aspx

65. American Petroleum Institute, "Oil & Natural Gas Contribution to U.S. Economy Fact Sheet," accessed January 22, 2022, https://www.api.org/news-policy-and-issues/taxes/oil-and-natural-gas-contribution-to-us-economy-fact-sheet; Energy Futures Initiative, "Wages, Benefits, and Change: A Supplemental

Report to the Annual U.S. Energy and Employment Report," April 6, 2021, https://www.usenergyjobs.org/wages

66. Bureau for Development, Democracy, and Innovation, "Energy Programs and Initiatives," U.S. Agency for International Development, accessed January 22, 2022, https://www.usaid.gov/energy/programs; Bureau of Energy Resources, "Energy Resource Governance Initiative," U.S. Department of State, accessed January 22, 2022, https://www.state.gov/key-topics-bureau-of-energy-resources/#ERGI

67. *Foreign Relations of the United States*, "1977–1980, Volume XVIII, Middle East Region; Arabian Peninsula, 45. Editorial Note," Office of the Historian, U.S. Department of State, Document 45, https://history.state.gov/historicaldocuments/frus1977-80v18/d45

68. Antonia Juhasz, "Why the War in Iraq Was Fought for Big Oil," *CNN*, April 15, 2013, https://www.cnn.com/2013/03/19/opinion/iraq-war-oil-juhasz/index.html. However, scholar Emily Meierding argues that oil wars are rare occurrences: "Countries may occasionally decide that it is worth initiating an oil spat to obtain desired resources, especially when targeted territories are contested and other issues are at stake. However, fighting major conflicts for oil does not pay." Emily Meierding, "The Exaggerated Threat of Oil Wars," *Lawfare*, August 2, 2020, https://www.lawfareblog.com/exaggerated-threat-oil-wars

69. George W. Bush, *National Security Strategy* (Washington, DC: The White House, March 2006), https://georgewbush-whitehouse.archives.gov/nsc/nss/2006/nss2006.pdf; Bureau of European and Eurasian Affairs, *Protocol of Intent Among the U.S. Agency for International Development and the Russian Energy Agency on Cooperation in Promoting Clean Energy in the Russia Far East*, U.S. Department of State, April 19, 2012, https://2009-2017.state.gov/p/eur/ci/rs/usrussiabilat/192412.htm

70. Nikolas K. Gvosdev, Jessica D. Blankshain, and David A. Cooper, *Decision-Making in American Foreign Policy: Translating Theory into Practice* (New York: Cambridge University Press, 2019), p. 294.

71. National Nuclear Security Administration, *123 Agreements for Peaceful Cooperation*, U.S. Department of Energy, January 10, 2022, https://www.energy.gov/nnsa/123-agreements-peaceful-cooperation

72. Bureau of European and Eurasian Affairs, *U.S.–Russia Energy and Energy Efficiency Cooperation*, U.S. Department of State, June 18, 2012, https://2009-2017.state.gov/p/eur/rls/fs/193091.htm

73. Bureau of European and Eurasian Affairs, *U.S.–Russia Bilateral Presidential Commission*, U.S. Department of State, accessed June 1, 2022, https://2009-2017.state.gov/p/eur/ci/rs/usrussiabilat/index.htm

74. Thomas O. Falk, "How Realistic Is Vladimir Putin's Nuclear Threat?" *Al-Jazeera*, March 3, 2022, https://www.aljazeera.com/news/2022/3/3/how-realistic-is-vladimir-putins-nuclear-threat; Max Hastings, "With Nuclear Threat, Putin Makes the Unthinkable a Possibility," *Bloomberg*, March 27, 2022, https://www.bloomberg.com/opinion/articles/2022-03-27/max-hastings-putin-s-nuclear-threat-against-ukraine-is-serious

75. Biden, *Interim National Security Strategic Guidance*, p. 21. President Biden's 2022 *National Security Strategy* includes similar language.

76. U.S. Department of State, "U.S.–China Joint Glasgow Declaration on Enhancing Climate Action in the 2020s," press release, November 10, 2021, https://www.state.gov/u-s-china-joint-glasgow-declaration-on-enhancing-climate-action-in

-the-2020s; Cecelia Smith-Schoenwalder, "Biden Rebukes Russia, China Leaders after Apologizing for American Inaction," *U.S. News & World Report*, November 2, 2021, https://www.usnews.com/news/national-news/articles/2021-11-02/biden-rebukes-russia-china-leaders-for-skipping-cop26-climate-summit-theyve-walked-away

77. International Renewable Energy Agency (IRENA), *Collaborative Framework on Green Hydrogen*, accessed January 29, 2022, https://www.irena.org/collaborativeframeworks/Green-Hydrogen

78. U.S. Geological Survey, *Critical Mineral Commodities in Renewable Energy*, U.S. Department of the Interior, June 4, 2019, https://www.usgs.gov/media/images/critical-mineral-commodities-renewable-energy

79. U.S. Geological Survey, *Mineral Commodity Summaries: Silicon, Copper, Rare Earths*, U.S. Department of the Interior, January 2021, https://pubs.usgs.gov/periodicals/mcs2021/mcs2021-silicon.pdf, https://pubs.usgs.gov/periodicals/mcs2021/mcs2021-copper.pdf, https://pubs.usgs.gov/periodicals/mcs2021/mcs2021-rare-earths.pdf

80. World Bank and Solargis, *Global Solar and Wind Atlases: China, Russia, and the United States*, accessed January 29, 2022, https://globalsolaratlas.info/map, https://globalwindatlas.info

81. David Feldman, Kevin Wu, and Robert Margolis, *H1 2021 Solar Industry Update*, U.S. National Renewable Energy Laboratory, June 22, 2021, 40, https://www.nrel.gov/docs/fy21osti/80427.pdf

82. Feng Zhao, "GWEC Releases Global Wind Turbine Supplier Ranking for 2020," Global Wind Energy Council, March 23, 2021, https://gwec.net/gwec-releases-global-wind-turbine-supplier-ranking-for-2020

83. Office of the U.S. Trade Representative, "WTO Panel Rejects China's Solar Safeguard Challenge," Press release, September 2, 2021, https://ustr.gov/about-us/policy-offices/press-office/press-releases/2021/september/wto-panel-rejects-chinas-solar-safeguard-challenge; Thomas Catenacci, "Biden Unveils Plan to Counter China Despite Paving Way for More Chinese Solar Imports," *Fox Business News*, June 27, 2022, https://www.foxbusiness.com/politics/biden-plan-counter-china-solar-imports

84. International Atomic Energy Agency (IAEA), *The Potential Role of Nuclear Energy in National Climate Change Mitigation Strategies* (Vienna: United Nations, November 2021), 8–9, https://www-pub.iaea.org/MTCD/Publications/PDF/TE-1984web.pdf

85. James Conca, "Why Was Nuclear Side-Lined at COP26?" *Energypost eu*, November 16, 2021, https://energypost.eu/why-was-nuclear-side-lined-at-cop26; Jeffrey Donovan, "Countries Detail Nuclear Power Climate Change Plans in COP26 Event with IAEA Director General," *IAEA News*, November 4, 2021, https://www.iaea.org/newscenter/news/countries-detail-nuclear-power-climate-change-plans-in-cop26-event-with-iaea-director-general

86. Marina Strauss, "European Commission Declares Nuclear and Gas to Be Green," *Deutsche Welle*, February 2, 2022, https://www.dw.com/en/european-commission-declares-nuclear-and-gas-to-be-green/a-60614990

87. IAEA, *International Project on Innovative Nuclear Reactors and Fuel Cycles (INPRO)*, accessed January 29, 2022, https://www.iaea.org/services/key-programmes/international-project-on-innovative-nuclear-reactors-and-fuel-cycles-inpro

88. Office of Nuclear Energy, *Bilateral Cooperation*, U.S. Department of Energy, accessed January 22, 2022, https://www.energy.gov/ne/nuclear-reactor-technologies/international-nuclear-energy-policy-and-cooperation/bilateral

89. Yuriy Humber and Jacob Adelman, "Russia Offers Fukushima Cleanup Help as Tepco Reaches Out," *Bloomberg*, August 26, 2013, https://www.bloomberg.com/news/articles/2013-08-25/russia-offers-to-help-clean-up-fukushima-as-tepco-calls-for-help; U.S. Department of Energy, "A Statement from U.S. Secretary of Energy Regarding Fukushima," press release, November 1, 2013, https://www.energy.gov/articles/statement-us-secretary-energy-ernest-moniz-regarding-fukushima

90. IRENA, *Regional Action Plan: Accelerating Renewable Energy Deployment in Latin America*, January 10, 2019, 7, https://www.irena.org/-/media/Files/IRENA/Agency/Regional-Group/Latin-America-and-the-Caribbean/IRENA_LatAm_action_plan_2019_EN.PDF?la=en&hash=12D7D12BF816911D9ED12AFEA0F34E73258B18F2;

91. IRENA, *Clean Energy Corridors*, accessed January 29, 2022, https://www.irena.org/cleanenergycorridors; IRENA, "IRENA Members Back New Blueprint for Hydropower Advancement," June 7, 2022, https://www.irena.org/newsroom/articles/2022/Jun/IRENA-Members-Back-New-Blueprint-for-Hydropower-Advancement

92. USAID, *Latin America and the Caribbean Energy Programs & Initiatives*, accessed January 29, 2022, https://www.usaid.gov/energy/super/cybersecurity-latin-america

93. Evan Ellis, "Russia's Latest Return to Latin America," *Global Americans*, January 19, 2022, https://theglobalamericans.org/2022/01/russia-return-latin-america

94. Energy Capital & Power, *US.–African Energy Forum 2021*, Houston, TX, December 9–10, 2021, https://energycapitalpower.com/event/us-africa-energy-forum

95. USAID, *Memorandum of Understanding between IRENA and USAID for Power Africa*, February 12, 2021, https://www.usaid.gov/sites/default/files/documents/IRENA-USAID-MOU-Jan-2021.pdf

96. Sebastian Ibold, "BRI Projects," Belt and Road Initiative, accessed January 29, 2022, https://www.beltroad-initiative.com/projects; African Energy Chamber, "Russia–Africa Energy Committee to Drive Investment and Deal-Making in the Energy Sector," *CNBC Africa*, March 24, 2021, https://www.cnbcafrica.com/2021/russia-africa-energy-committee-to-drive-investment-and-deal-making-in-the-energy-sector; USAID, *Power Africa*, October 1, 2021, https://www.usaid.gov/powerafrica

97. Stephen M. Walt, "China Wants a 'Rules-Based International Order,' Too," *Foreign Policy*, March 31, 2021, https://foreignpolicy.com/2021/03/31/china-wants-a-rules-based-international-order-too

98. "Nationally Determined Contributions," *Paris Agreement*, United Nations Climate Change Conference (COP21), December 12, 2015, article 4, paragraph 2, https://unfccc.int/files/meetings/paris_nov_2015/application/pdf/paris_agreement_english_.pdf

99. Cory Welt, *Russia's Invasion of Ukraine: Overview of U.S. Sanctions and Other Responses* (Washington, DC: Congressional Research Service, April 22, 2022); European Council, *EU Restrictive Measures against Russia over Ukraine (since 2014)*, June 7, 2022, https://www.consilium.europa.eu/en/policies/sanctions/restrictive-measures-against-russia-over-ukraine

PART II

Competing and Winning with Allies and Partners

CHAPTER 7

Fighting with Allies: A World War II Case Study

Kevin J. Weddle and Joel R. Hillison

INTRODUCTION

Winston Churchill once said that "There is only one thing worse than fighting with allies, and that is fighting without them."[1] With his deep knowledge of history, Churchill knew that fighting a war within military coalitions and alliances is not a guarantee of success. Indeed, there are many pitfalls and disadvantages of coalition warfare. Likewise, many benefits come with fighting with allies. Senior political and military leaders must be clear-eyed when entering into a coalition, understanding the potential difficulties that will inevitably surface along with the many benefits.

Operation OVERLORD marked the beginning of the end of Nazi Germany. On June 6, 1944, the Allies launched the long-awaited offensive to liberate occupied Europe. It was the largest and most complex military operation in World War II and arguably in all of military history. Following two months of hard fighting, terrible casualties, and operational failures and disputes, which threatened on several occasions to weaken the bonds in the Allied coalition, British and American armies regained the initiative. They drove Axis forces back to the Siegfried Line by the end of the summer. Allied decision-makers and planners drew upon the hard-won lessons learned from previous operations in North Africa, Sicily, and Italy.

The Normandy invasion and the subsequent campaign in France through the liberation of Paris represent an unmatched feat of arms and an essential study for practitioners of the art of warfare. But the cross-channel invasion had its origins well before the invasion of North Africa, and the strategic decisions that led to OVERLORD were not simple, nor were they easy. The decision to conduct the cross-channel invasion to open up a new theater of operations when they did was arguably the key military strategic decision of the war. The massive planning and deception, the vast logistical considerations, the alliance structure, and coalition warfare are but a few of the challenges that confronted Allied strategists, and Operation OVERLORD represented the culmination of their efforts from 1940 through the spring of 1944. Senior strategic leaders from the United Kingdom and the United States came at the strategic challenge of defeating the Axis powers and liberating the enslaved nations of Europe and Asia from very different perspectives. British and American military and civilian leaders were often at odds when it came to determining a viable military strategy for defeating Germany and liberating Europe.[2] These perspectives were products of history, culture, geography, politics, prejudices, professional judgment, and personalities. Their debates over the fundamental questions of coalition strategy formulation provide valuable lessons to leaders today and help the United States and its allies to maintain their strategic advantages.

We like to think that the strategic environment that we lived through over the past two decades and face today is much more complex and challenging than our predecessors could ever imagine. This, I think, is an incorrect reading of history. If there ever was a volatile, uncertain, complex, and ambiguous (VUCA) environment, World War II was it. Franklin D. Roosevelt, Winston Churchill, General George C. Marshall, and General Alan Brooke had to figure it out and make the best decisions they possibly could under the most trying of circumstances.

THE NATURE OF COALITIONS: ADVANTAGES AND DISADVANTAGES

Coalitions are temporary partnerships created to address a specific, immediate security threat. The operative word is "temporary." An alliance is typically a long-term relationship among nations that help enhance collective security. An example of an alliance is the North Atlantic Treaty Organization (NATO). The relationship between the United States and Great Britain (and later the USSR)—often referred to as the Grand Alliance—was a wartime coalition created gradually to address first the German Nazi threat and later Italy and Japan and their allies.

Coalitions offer participants many benefits often not available when unilateral military action is taken. Political legitimacy often attracts nations to coalitions, especially if one or more coalition partners is more respected and admired on the world stage. Even if one coalition member has sufficient resources to achieve its political objectives, recruiting other members with the same goals will enhance the validity of their actions. Perhaps the most apparent advantage of a coalition is shared sacrifice and, related to that, increased availability of resources and combat power. More resources will generally lead to more varied strategic options. Weaker nations are often drawn to coalitions for a big payoff for a small investment and to wield outsized influence on the world stage.

While the benefits of coalition members are many, there are significant and numerous challenges. The coalition partners might very well have different strategic cultures—the propensity of a nation to make certain, consistent kinds of strategic choices—that will complicate their ability to decide on strategic courses of action. Ensuring unity of command or at least unity of effort can be difficult if nations demand that their military assets be under national control. Member nations, especially weaker partners, must often subordinate their national interests and objectives to the coalition's goals, which might differ in many respects. To gain buy-in and keep the coalition together, the members might be driven to conduct suboptimal operational and strategic courses of action. Finally, while coalitions are often formed out of necessity due to the rise of a significant threat, they often become extremely fragile. They will collapse with the diminishment or elimination of the threat, negatively affecting successful conflict termination.[3]

LAYING THE STRATEGIC GROUNDWORK: THE ATLANTIC CHARTER AND GERMANY FIRST

The first step in developing any military strategy is determining the desired end state for which the conflict will be fought. British Prime Minister Winston Churchill and U.S. President Franklin Roosevelt knew this instinctively. In their first World War II meeting, the August 14, 1941, Atlantic Conference (code-named RIVIERA), they established what they wanted the world to look like after the allies won the war. Meeting on warships off Newfoundland, they developed and agreed to the so-called Atlantic Charter. The Charter put forth eight points, among which were "no territorial gains," "(universal) freedom from want and fear" (echoing FDR's famous January 6, 1941 "Four Freedoms" speech), "self-determination," and the "lowering of trade barriers."[4] The Atlantic Charter essentially described Churchill's and Roosevelt's shared vision for the post–World War II world. A cynic might point out that many of their points were never achieved during or after World War II, but this was an aspirational document almost unprecedented in history. It demonstrated that the two allies (this was almost five months before the United States entered the war) agreed on the most fundamental interests of the coalition. It set the stage for the more challenging strategic negotiations to come. It was a remarkable point of agreement that doesn't get enough attention.

The second point of fundamental strategic agreement came early in the coalition's life. This was the "Germany First" decision. The origin of arguably one of the most significant strategic decisions of the war came from the U.S. Chief of Naval Operations Admiral Harold R. Stark. On November 12, 1940, he wrote a memorandum, later called the "Plan Dog Memo," that was circulated at the highest political and military levels in the U.S. government. Stark, with no guidance, provided his assessment of the worldwide strategic environment in the wake of the German invasions of the West, including the capture of Poland, Norway, the Netherlands, Denmark, and France, as well as increasingly aggressive Japanese actions in China and the Pacific. In November 1940, Great Britain essentially stood alone against the Germans in Europe. Stark offered a full and bleak assessment of the challenges facing the United States if it entered the war, which he assumed would eventually happen. He also offered several options, but perhaps the most important passage in his memo stated that "if Britain wins decisively against Germany we could win everywhere; but if she loses the problem confronting us would be very great; and, while we might not lose everywhere, we might, possibly, not win anywhere."[5] The logical extension of Stark's statement was that the United States and Great Britain had to focus and prioritize their efforts and resources on Germany, their most dangerous adversary, no matter what other strategic threats existed elsewhere. This came to be known as Germany First. This bedrock foundational decision for Allied strategy was confirmed at the ABC-1 Conference of senior American, British, and Canadian staff officers in early 1941 and at the First Washington Conference, code-named ARCADIA, in late 1941 through early 1942.[6]

FDR, Churchill, and their staffs instinctively did exactly what strategic leaders should do with the Germany First doctrine and the Atlantic Charter. They established the coalition's first principles.

"GETTING TO KNOW YOU, GETTING TO KNOW ALL ABOUT YOU": DEVELOPING RELATIONSHIPS

At the strategic level, developing good personal and professional levels between and amongst civilian and military leaders is not only important, it is also essential. This is especially important in developing a successful coalition because the leaders do not have the ready-built advantage of common national identity. This is a time-consuming process, and leaders must devote the resources—mainly time—to this effort. The key members of the U.S.–Great Britain World War II coalition did just that. Their hard work at relationship building paid huge dividends and was a significant factor in the coalition's victory in Europe and the Pacific.

It must be remembered that the "special relationship" that the United States and Great Britain enjoy today did not exist in 1940. An Anglophobe in the ranks of the U.S. officer corps today would be almost impossible to find, but that was not the case at the beginning of World War II. Despite working together in World War I, the upper ranks of both the United States' and United Kingdom's senior leadership were filled with key personnel who were suspicious at best and contemptuous at worst of each other. These feelings could have easily derailed the coalition before it even got on its feet, but the key leaders of both countries made sure that did not happen.

The most important step they took to fight the mutual distrust that could fatally impede the coalition's ability to craft and execute a viable military strategy was to open communications channels, and indeed, they forced each other to talk and work out their issues through a variety of means. FDR had already made it quite clear that, despite American neutrality in the prewar period and immediately after the beginning of the war, he favored the Western European democracies over the German aggressors. He did this through his outstanding use of the "bully pulpit" through his Fireside Chats and things like the "Arsenal of Democracy" speech of December 1940. The Lend-Lease Act of 1941 and the decision to use U.S. naval warships to escort transatlantic convoys to Great Britain were additional steps toward U.S. direct intervention in the war. All of these concrete steps were important, but for the coalition's cohesion, the almost constant communication between FDR and Churchill ensured its ultimate success.

Before the ARCADIA Conference, Churchill and FDR had already communicated privately over 200 times using cables, letters, and transatlantic telephone calls.[7] Their respective staffs were doing the same thing. But they also realized that no other method of communication could replace face-to-face meetings, and these two leaders along with their senior military officials met eleven times during the war.[8] This was an extraordinary investment in time and resources for these senior leaders to meet so often, but the result was a war-winning military strategy.

The first major meeting took place in Washington, DC between January 29 and March 27, 1941, consisting of senior military staff from the United States, Britain, and Canada. As already discussed, the main decision that came out of the ABC-1 Conference was agreement by all parties to Admiral Stark's "Germany First" concept. The Atlantic Conference in August 1941, also mentioned previously, and the first face-to-face meeting between Churchill and FDR, took place off Newfoundland and resulted in the Atlantic Charter. Finally, the First Washington Conference, the second meeting between Churchill and FDR and the first after Pearl Harbor and the U.S. entry into the war, took place from December 11, 1941 to January 14, 1942. At this meeting the two groups of decision-makers agreed to Germany First, established the Combined Chiefs of Staff, agreed to the concept of unity of command for major theaters, and declared none of the United Nations would negotiate independently with the enemy.

While the first three major conferences resulted in important decisions and products, perhaps the most important result was creating a coalition team, building relationships, and developing trust. These first three wartime conferences were also, for the most part at least, remarkably free from disagreement and acrimony. Both sides worked hard to get along, and these early meetings were critically important for the long-term health of the coalition. The special relationship the two nations enjoy today was born, nurtured, and grew during the perilous early days of World War II, and the primary architects of that relationship were Winston Churchill, FDR, and their subordinates.[9]

ASSESSMENT AND REASSESSMENT

Strategy is very simple: the calculated application of ways and means to achieve a military or political objective. But as the great Prussian theorist of war and strategy, Carl von Clausewitz warns us, "Everything is very simple in war, but the simplest thing is difficult."[10] If military strategy was easy, we would be a lot better at it, as the recent debacle in Afghanistan reminds us. In fact, the United States and its coalition partners have not developed or executed military strategy in an effective manner since World War II. Throughout Clausewitz's *On War*, he also advises strategists that because war is never predictable and because conditions are constantly changing, senior leaders must continually adjust their strategies. If not, they will waste resources on what may very well be operations that do not contribute to the end for which the war is being fought.[11] The World War II coalition of the United States and Great Britain was successful at developing and executing military strategy in the most challenging strategic environment in history for two reasons: The senior leaders were able to develop effective personal and professional relationships, and they conducted periodic and extensive assessment and reassessments of their strategy. Those two points might well be the most significant lessons that strategists today can learn from the World War II Grand Alliance.

The senior political and military members of the coalition began to address specific military strategic options not long after the ARCADIA Conference. While the addition of American military power was a huge relief to Great Britain and Churchill, its potential had not yet been reached. American industrial production of aircraft, warships, weapons, ammunition, and military-related output was still two years from its peak, which meant that strategic options for the coalition in 1942 were limited in size and scope.

The American and British strategic leaders viewed the war and how to fight it in very different ways. As the eminent historian Maurice Matloff observed, these very different approaches were "because of their varying traditions, interests, policies, geography, and resources, the three [includes the Soviet Union] partners looked at the European war through different spectacles."[12] As an island nation with limited resources that relied on its significant sea power for survival, Britain was concerned about maintaining the empire and its overseas possessions and thus preferred Mediterranean operations. The United States, the world's leading industrial power, had abundant internal resources, was protected by two oceans, was suspicious of Britain's imperial goals and had not suffered grievously during World War I as their partners did. Indeed, it is impossible to overemphasize the impact that World War I had on the strategic approaches preferred by the respective coalition partners. For Britain and the Commonwealth, World War I was a devastating and almost catastrophic event that resulted in the deaths of one million young men—almost an entire generation—in the trenches of France, Belgium, Turkey, and other theaters. Virtually all the senior British military and civilian leaders had experienced the horrors of World War I. These combat deaths and the fact that Britain had already suffered significant land defeats in France and North Africa reinforced their desire to avoid direct confrontations with German land power. They preferred a peripheral Mediterranean-focused strategy combined with gradually weakening Germany with sea and air power.

The Americans on the other hand, had a very different experience in World War I. In the opinion of many American leaders, World War I had been a short, sharp fight won by the injection of thousands of U.S. troops that saved the day as the other Western European powers—France and Britain, in particular—began to falter in the face of relentless German attacks. It was the direct application of American combat power against the German army that won the War to End All Wars. The American experience in World War I reinforced their historic strategic preferences that emphasized a direct approach with overwhelming combat power backed by U.S. industrial might. These two very different strategic approaches meant that disagreement and conflicts over military strategy were almost inevitable, despite the efforts on both sides to build trust within the coalition.[13]

The first major disagreement between the members of the coalition took place in late June 1942 at the Second Washington Conference, code-named ARGONAUT. Weeks before this conferences, U.S. Army Chief of Staff and FDR's primary military advisor, General George C. Marshall, had visited the United Kingdom to propose several options developed by his new chief planner, Brigadier General Dwight D. Eisenhower. Marshall proposed the buildup of U.S. forces in Great Britain, code-named BOLERO, an early cross-channel invasion of occupied France, code-named ROUNDUP, and if the USSR faced imminent collapse, an emergency invasion of France, called SLEDGEHAMMER. Marshall left believing that he had the support of Churchill and the chief of the Imperial General Staff, General Alan Brooke. He did not. In reality, the British believed that Marshall was wildly optimistic about what was militarily possible at this point in the war. They knew that the American buildup in the United Kingdom was just getting started and it would be many months, even years before enough combat power was marshalled for an invasion. They also knew that the sea lines of communication between the United States and the United Kingdom had to be secure, and thus the Battle of Atlantic won before an invasion could take place. Finally, air supremacy over the channel was critical for an invasion's success. It was unclear when any, not to mention all, of these prerequisites for an invasion would be achieved.[14]

When the British delegation arrived in Washington in June, it was clear that Churchill instead advocated an allied invasion of North Africa, code-named TORCH. Churchill's significant persuasive skills convinced FDR that an early cross-channel invasion was folly at this point in the war and that the quickest way to get Americans into the fight against Germany was through North Africa. Marshall was appalled and voiced his strong disagreement to FDR and Churchill. FDR was firm and in no uncertain terms let Marshall know that that while he appreciated the chief's advice, as the commander in chief it was his decision, and the decision was for TORCH. After the successful Anglo–U.S. TORCH landings in North Africa in November 1942, the first real combined operation of the war, the leaders agreed to meet again at Casablanca in January 1943, code-named SYMBOL, to review the overall strategic situation and to consider their next major moves. A lot had happened in 1942, much of it bad, but by the second half of the year, the Allies began to push back with U.S. victories at the Coral Sea and Midway, and the invasion of Guadalcanal in the Pacific; the British victory at the Battle of El Alamein; and the Soviet isolation of Hitler's Sixth Army at Stalingrad. It was time to look ahead. While they made a number of key decisions such as increased resources allocated to the Battle of the Atlantic and the Combined Bomber Offensive, the most important choice was where to strike next after the North African operations wrapped up. They reviewed several options, including a cross-channel invasion, but also other locations in the Mediterranean including Sardinia, the Balkans, Sicily, and Greece. Once again, the Americans were outmaneuvered by their British counterparts, and all attempts to push an early cross-channel invasion were rebuffed. Churchill and FDR settled on Sicily as the next major operation. Once again, Marshall was disappointed as he felt that it was a diversion from what should be the main object, the shortest route to the heart of Germany. As a good strategist, he asked a very important question: "Is an operation against Sicily merely a means toward an end, or is it an end in itself? Is Sicily part of an integrated plan to win the war, or is it simply taking advantage of an opportunity?"[15] Marshall worried that the Mediterranean operations were a waste of combat power and only delayed a cross-channel invasion, and instead served to support Britain's imperial ambitions more than the rapid defeat of Germany. Churchill, on the other hand, believed that an invasion of Sicily would force Italy out of the Axis coalition and could act as a springboard for future operations. As it turned out, both were partly right, but in hindsight it is clear that the ultimate success of Operation OVERLORD in June–August 1944 was due in no small part to the lessons learned from the invasion of Sicily. Once again, the Allies had argued over the way ahead. The process was not neat and orderly, nor was it without bitter arguments, but while the Americans felt outmaneuvered once more, the coalition only became stronger and the military strategy sounder.[16]

The next meeting took place in Washington, DC in May 1943, code-named TRIDENT. At TRIDENT, compromise was once again the order of the day: The Americans finally got a date for a cross-channel invasion, code-named OVERLORD, and the British got the Americans to agree to an invasion of Italy after the conclusion of the Sicily campaign. Until TRIDENT, the British were clearly the senior members of the coalition, particularly in experience and staff preparation and competence. Going forward, however, the Americans with their clear predominance in resources and almost total control of the Pacific theater of operations had elbowed the British aside as the dominant coalition partner.

Subsequent conferences in 1943, the First Quebec Conference in August 1943, code-named QUADRANT, and the Teheran Conference in November and December 1943, code-named EUREKA, confirmed OVERLORD and authorized initial planning for the operation. At these conferences Churchill continued to argue for increased resources for the Italian campaign, and he was partially successful, but by mid-1943, the Americans, now wielding the most influence in strategy-making, ensured that OVERLORD would remain the priority for the coalition. The Tehran Conference was the first attended by Soviet Premier Joseph Stalin. Churchill and FDR had invited Stalin to join them at the Casablanca Conference, but he had refused, citing the Stalingrad crisis. At Tehran, it was clear that Great Britain and the United States had very different visions for the postwar period. For months Stalin had pushed the Allies for an early cross-channel invasion, so by the end of 1943, he was impatient. When assured that OVERLORD would take place the coming spring, he demanded to know the name of the commander. Stalin's inquiry forced FDR and Churchill to confront this key decision. They had both promised their respective senior military advisors—Marshall and Brooke—the job, but realized that they were too important in their current positions for their counties and for the good of the coalition. In the end, the obvious choice was Eisenhower, who had already commanded TORCH, HUSKY, and AVALANCHE. With the appointment of Eisenhower as the Supreme Allied Commander, the last major piece was in place for the planning, coordination, and execution of the greatest military operation in history and the most important decision made by the coalition.[17]

The preceding are just a few of the key meetings and decisions between the key military and civilian leaders from Great Britain and the United

States. Neither of the coalition members got everything it wanted. Almost all of the decisions on major military operations and the resources to execute them involved compromise with the attendant successes and failures. However, the coalition was able to achieve ultimate success due to the strategic and personal foundations laid early in the process and the continued efforts on both sides to maintain and improve relationships of trust and respect. They made mistakes along the way, and many of their decisions can be critiqued with 80-year-old hindsight, but their insistence on constant assessment and reassessment of their military strategy ensured Allied success. We can learn a lot from these leaders who developed and executed a war-winning strategy against the most dangerous regimes in human history.

It is not enough for military leaders alone to review and evaluate military strategy to determine if the objectives (ends) are still valid, and that the ways (strategic options) selected, developed, and executed are making progress toward the ends. Indeed, that path often leads to confirmation bias, choice-supportive bias, and the sunk cost fallacy as we experienced in Vietnam and Afghanistan.[18] Any comprehensive assessment and reassessment of a military strategy must include both military and civilian senior leaders in the process. The same energy expended in the formulation, development, and execution of a military strategy, must be spent on the evaluation of that strategy. If there is one lesson that the most senior Allied leaders in World War II can teach us and that we must take to heart, it is this one.

FRACTURES IN THE COALITION

While there were bitter disagreements over military strategy, especially over the timing and scope of a cross-channel invasion, the U.S.–British coalition was never seriously threatened from within in the first years of the war. However, once it became obvious that the Allies would defeat Germany and Japan, the coalition faced highly contentious post-war issues, many of which were never solved or only partially solved, and the coalition soon disintegrated over these disputes, with the United States and Britain and their close allies breaking into one camp and the Soviets and their satellites into the other. Two years after Victory in Europe (VE) and Victory in Japan (VJ) days, the Cold War began.

IMPLICATIONS FOR TODAY

While the allies certainly faced a complex and challenging threat during World War II, the United States today faces a multidimensional threat from great powers, such as China and Russia, and lesser, nuclear capable or aspiring countries such as Iran and North Korea and multiple nefarious nonstate actors. To face the many challenges in the contemporary security environment, the United States will have to rely upon its allies and partners in both permanent alliances and temporary coalitions. Contemporary U.S. allies and partners are much more numerous than they were during World War II, thus increasing the diversity of strategic perspectives and the complexity for developing and maintaining a united front in the face of those challenges. Yet this diversity is also a strategic advantage for the United States.

LAYING THE STRATEGIC GROUNDWORK: OLD FRIENDS AND NEW PARTNERS

In some ways, European allies are much easier to work with due to the institutional, political, and military arrangements within NATO and the over seventy years of working with each other for the original twelve members. It is no coincidence that NATO's first Supreme Allied Commander Europe (SACEUR) was General Dwight D. Eisenhower, who institutionalized many of the lessons he learned during World War II. Yet as the alliance has grown, so too have the problems of collective action and gaining consensus among more and more members. Four new members were added during the Cold War: Greece and Turkey in 1952, Germany (our former adversary in World War II) was added in 1955, and Spain was added in 1982. Fourteen new NATO members have been added since the fall of the Soviet Union, with many more hoping to gain admittance in the future. As of this writing, both Sweden and Finland have applied for NATO membership. Others, such as Ukraine, can only hope for membership in the future. Many of the NATO members are also members of the European Union, therefore opening greater room for cooperation with the EU, which is essential for the economic and diplomatic resources they can wield. As more allies were added, compromise and patience became more important to the ultimate survival and success of the alliance. Once all thirty members have consented to a decision, NATO can bring significant political legitimacy and military capabilities to bear.

In the rest of the world, the United States lacks a NATO-like alliance structure, making multilateral coalitions more difficult to create and sustain. Yet, these predominantly bilateral alliances and partnerships can in theory make decisions more quickly than NATO's cumbersome process. Though many of the U.S. alliance arraignments outside of NATO are also longstanding, there are fewer multilateral arrangements and even fewer integrated military structures than there are in NATO. The United States and the Republic of Korea (ROK) have had a strong alliance since 1950 and have a combined military structure in the ROK/U.S. Combined Forces Command. Similarly, the United States has strong military and political relationship with Japan but does not have a combined military command structure. The Australia, United Kingdom, and United States (AUKUS) countries have also established an alliance to confront the threats posed by China.[19] The United States also has longstanding mutual defense treaties with the Philippines and Thailand, which both have critical geostrategic locations. Finally, the United States has a longstanding though ambiguous commitment to the security of Taiwan. This arrangement, more than any other, is a friction point between the United States and China and could lead to a disastrous military conflict.[20] The U.S. security relationships with Taiwan and these other states anchor the U.S. position in the Indo-Pacific region.

The United States has also established a system of looser coalitions and relationships that can be a competitive advantage, for example, the Quadrilateral Security Dialogue (QUAD) between the United States, India, Japan, and Australia.[21] While United States has had longstanding political, economic, and military alliances with Japan and Australia, creating a more formalized partnership with India is an important development. Due to its long history of nonalignment and suspicions of U.S. foreign policy (e.g., past support of Pakistan), India is unlikely to align itself as closely with the United States as other allies. However, due to its size (soon to be the world's most populous nation), its democratic system of government, and its growing economic and military heft, India is an important partner for the United States, especially in the Indo-Pacific region.[22] Where the United States and India can act together, they will be able to do so with greater political legitimacy and resources. The United States will have to be more careful and willing to compromise where India is concerned, but the lessons from World War II will certainly help guide that developing cooperation.

There are significant disadvantages to the hub-and-spoke nature of U.S. alliances in the Indo-Pacific. Many of these allies and partners lack the interoperability and shared values found in NATO. In addition, there are often historic tensions that remain among many U.S. allies (such as between Korea and Japan). This makes collective action more difficult when domestic politics and historic injuries overshadow shared security concerns. Just as the series of allied conferences in World War II described previously helped to overcome mistrust, competing interests in the nature of the postwar structure, and differing strategic cultures, so too will the United States have to work to establish a shared vision with its contemporary allies and partner to confront the common challenges facing them.

CREATING A SHARED VISION FOR COOPERATION

Just as the Atlantic Charter established the framework for future cooperation between the United States and Great Britain during the second world war, the United States will have to develop similar frameworks to guide its relations with modern allies and partners. The Washington Treaty, which established the NATO alliance, provides an example of one such charter where alliance members agree to "share risks, responsibilities and benefits of collective defense."[23] The concept of collective defense was explicitly nested under Article 51 of the UN Charter. The NATO treaty goes on to describe how NATO members will improve relations with each other "by strengthening their free institutions, by bringing about a better understanding of the principles upon which these institutions are founded, and by promoting conditions of stability and well-being."[24]

While it is not realistic to expect a similar charter among other U.S. partners and allies, that does not mean that the United States cannot set the groundwork for developing a shared vision and approach for the threats facing these partners. In addition to the arrangements mentioned previously, the United States has a wide range of bilateral military alliances and partnerships as well as economic approaches such as the Indo-Pacific Economic Framework announce in 2022.[25] Just as the U.S. relationship with Britain and the Soviet Union developed over the course of the war through numerous conferences and agreements, so too will the United States have to continually assess and develop its vast network of partners and allies. Developing these relationships will rely heavily on extensive diplomatic efforts and patience in dealing with multiple strategic cultures with varying perceptions of threats and opportunities. These relationships will also continue to develop through economic interchange of trade and investment and military outreach such as exercises and other efforts to build partner capacity.

THE MORE THINGS CHANGE: DEMOCRACIES VERSUS AUTOCRACIES

In many ways, the United States faces a similar array of forces in its competition with Russia and China as the allies did in facing Nazi Germany, Fascist Italy, and Imperial Japan. As Michael Schuman noted in *The Atlantic* in 2022, "improving ties between China and Russia have security experts worried that the U.S. will need to contend with an unholy alliance of the world's two most powerful authoritarian states determined to reshape the global order in their favor."[26] The same could have been said of Nazi Germany and its Third Reich and Imperial Japan and their Greater East Asian Co-Prosperity Sphere.[27]

China is a powerful economic and military country. It has few meaningful allies, outside of North Korea, Cambodia, and Burma. Russia, though still militarily powerful, especially due to its nuclear weapons, boasts a weak set of supposed allies, such as Belarus, Kazakhstan, Cuba, Syria, Eritrea, and North Korea. China's "no limits" relationship with Russia is strong but based on a respect for each country going their own way. As

Chinese Foreign Minister Wang Yi said in a news conference, "The China-Russia relationship is valued for its independence."[28] Therefore, their relationship is not as intertwined as the United States' relations with its allies.

China and Russia bristle at the unique position of the United States in the world and its willingness to interfere in their areas of influence. China and Russia also have compatible economic relations with Russia supplying raw materials, especially energy, to China and China providing needed "investment and high-tech products."[29]

Yet there are some inherent tensions in that relationship as there was in the early Cold War relations between the USSR and China. In those early days, China was the supplicant and grew resentful of the unequal relationship. Now that the roles are somewhat reversed, it's easy to see Russia being uncomfortable with a junior partner role with China. Therefore, the United States and its allies might have to prioritize which of these two to focus on: the most immediate and dangerous threat (Russia), or the most likely threat in the future (China).

A STARK CONTRAST WITH WORLD WAR II: WHICH THREAT TO FACE FIRST

While the allies in World War II faced significant challenges, one of their earliest decisions—Germany first—helped focus efforts as the war progressed and the coalition developed. With the myriad of challenges facing the United States and its allies, it will be a much more difficult task gaining consensus on which threats to prioritize. U.S. partners and allies differ in how to approach such threats as climate change, terrorism, and nuclear proliferation. Perhaps more importantly, how does the United States prioritize its competition with China and its competition with Russia? Due to their significant conventional and nuclear military power, their vast energy resources, and their aggressive behavior (as shown by the devastation in the invasion into Ukraine in 2022), Russia is certainly an imminent danger to U.S. and NATO interests in the short term. Yet based on their economic growth, their dominance in certain sectors such as 5G, solar technology, and rare earth metals, China represents a longer term and perhaps more complex challenge to U.S. interests. Getting agreement on a Germany First–type agreement would therefore be extremely difficult, even among our closest allies.

For our NATO allies, especially those in Eastern Europe and Scandinavia, Russia looms as the most dangerous threat. Within the European Union, there is little consensus on whether China represents a threat, a competitor, or merely a partner for expanding trade and investment. While our allies in Asia are certainly leery of a growing and increasingly aggressive China, they also face nearer threats that consume much of their attention. India remains focused on the threat posed by a historically hostile

Pakistan, with whom it still has unresolved border claims. The Republic of Korea, and to a lesser extent Japan, focus much of their attention on North Korea in addition to China. Similarly, longtime allies such as the Philippines have also courted better relations with China despite disputes over territory, such as the Spratly Islands. Taiwan is certainly focused on the threat to its existence posed by China but is also economically interdependent with the mainland. Even the Five Eyes partners, Australia and New Zealand, have differing views on how to approach China, with Australia much more willing to balance against them, while New Zealand is content to hedge against China.

MAINTAINING RELATIONSHIPS

Unlike U.S. relations with Britain during World War II, which was heavily focused on military actions (at least early on), relations with the current network of allies and partners will depend more upon diplomatic, informational, and economic interactions. Relations with organizations such as the Association of Southeast Asian States (ASEAN) can provide needed political support for U.S. relations in the region and contribute to the narrative of the important role of the United States in achieving a shared vision for peace and prosperity in the region. Building closer economic relationships with key allies and partners will also be essential for the United States just as it was for rebuilding Europe in the aftermath of World War II. The Build Back Better World and Indo-Pacific Economic Framework are good starting points for this effort.

CONCLUSION

The lessons learned from the World War II alliance between the United States and Great Britain provide a roadmap for U.S. strategy in the current era. While the United States has a significant advantage in the number and strength of its alliances and partnerships, this alone will not guarantee a strategic advantage. The disadvantages and frustrations that come with dealing with allies have to be mitigated to reap the significant benefits of working with allies and partners, especially when they share your interests and values. Understanding the trade-offs that come with fighting and competing with allies will enable current military and political leaders to make the most of this unprecedented network.

NOTES

1. Churchill as quoted by in Field Marshal Lord Alanbrooke, *War Diaries, 1939–1945*, Eds. Alex Danchev and Daniel Todman (University of California Press, 2001), 680.

2. Note that this case study focuses on the British and American coalition. Of course, the Allied coalition in World War II also included many other counties, including the other senior partner, the Soviet Union. However, as Maurice Matloff has observed, "the Western powers fought their war and the Soviets theirs." Maurice Matloff, "Allied Strategy in Europe, 1939–1945," in Peter Paret, ed., *Makers of Modern Strategy from Machiavelli to the Nuclear Age* (Princeton University Press, 1986), 700.

3. I took this from notes that I've had for years but the origins for which are lost in the sands of time. Yet, the advantages and disadvantages are self-obvious.

4. Andrew Roberts, *Masters and Commanders: How Four Titans Won the War in the West, 1941–1945* (Harper Collins, 2009), 54, 125, 313; Matloff, 683–684; and Harry R. Yarger, *Strategic Theory for the 21st Century: The Little Book of Big Strategy* (U.S. Army War College Strategic Studies Institute, February 2006), 6–9.

5. Harold R. Stark, "Memorandum for the Secretary" (FDR Library, Marist College, November 12, 1940), 1, http://docs.fdrlibrary.marist.edu/psf/box4/a48b01.html.

6. Star, 1; Roberts, 45 and 50; Matloff, 683; and William T. Johnsen, *The Origins of the Grand Alliance: Anglo-American Military Collaboration from the Panay Incident to Pearl Harbor* (University Press of Kentucky, 2016), 136–137 and 251–252.

7. Robert Schmuhl, "Mr. Churchill in the White House," accessed May 2, 2022, https://www.whitehousehistory.org/mr-churchill-in-the-white-house-1

8. Four Freedoms Park Conservancy, "The Special Relationship: Franklin D. Roosevelt and Winston Churchill," 1–2, 10. https://fdr4freedoms.org/wp-content/themes/fdf4fdr/DownloadablePDFs/IV_StatesmanandCommanderinChief/08_TheSpecialRelationship.pdf (accessed May 2, 2022)

9. Roberts, 45, 50, 52–54, 66–67, 70–76, 79–84, 86–89, and 187–218; Matloff, 683–687; and Johnsen, 193–206 and 249.

10. Carl von Clausewitz, ed. and trans. by Michael Howard and Peter Paret, *On War* (Princeton University Press, 1976), 119.

11. Clausewitz, 92–93. See also Alan Beyerchen, "Clausewitz, Nonlinearity, and the Unpredictability of War," *International Security* 17, 3 (1992), 85–90.

12. Matloff, 679.

13. Roberts, 139; Matloff, 684–685.

14. Roberts, 140–166; Matloff, 685–686.

15. Albert N. Garland and Howard M. Smith, "United States Army in World War II: The Mediterranean Theater of Operations: Sicily and the Surrender of Italy," in *World War II: Sicily and the Surrender of Italy* (Center of Military History, 1993), 10.

16. Roberts, 313–345; Matloff, 687–688.

17. Roberts, 391–398, 401–408, 443–454; Matloff, 688–692.

18. All three of these biases and fallacies are closely related. Confirmation bias occurs when people ignore information that does not confirm their own thoughts. Choice-supportive bias is the tendency to apply success to a choice in the face of negative evidence, which can block consideration of other options. The suck cost fallacy is the tendency to continue something after committing significant resources to it, even in the face of evidence that it is not working. These definitions are from Paul Perry, "24 Forms of Bias: How to Identify and Avoid Them in Your Organization," April 8, 2020, https://blog.submittable.com/bias (accessed May 3, 2022) and Jamie Ducharme, "The Sunk Cost Fallacy Is Ruining Your

Decisions. Here's How," *Time Magazine*, July 26, 2018, https://time.com/5347133/sunk-cost-fallacy-decisions/.

19. Daniel Stewart, "Australia, United Kingdom and United States celebrate first anniversary of AUKUS," *360 News*, September 23, 2022. https://www.msn.com/en-us/news/world/australia-united-kingdom -and-united-states-celebrate -first-anniversary-of-aukus/ar-AA12bewc

20. Brendan Rittenhouse Green and Caitlin Talmadge, "The Consequences of Conquest: Why Indo-Pacific Power Hinges on Taiwan," *Foreign Affairs*, July/August 2022. https://www.foreignaffairs.com/articles/china/2022-06-16/consequences -conquest-taiwan-indo-pacific

21. Sumitha Narayanan Kutty and Rajesh Basrur, "The Quad: What It Is—And What It Is Not," The Diplomat, March 24, 2021. https://thediplomat.com/2021/03 /the-quad-what-it-is-and-what-it-is-not

22. "India Will Be Most Consequential Partner for U.S. in Indo-Pacific This Century: Esper," Hindu, October 21, 2020, https://www.thehindu.com/news /national/india-will-be-most-consequential-partner-for-us-in-indo-pacific-this -century-esper/article32905679.ece.

23. The North Atlantic Treaty Organization, Founding Treaty, Updated September 2, 2022. https://www.nato.int/cps/en/natolive/topics_67656.htm

24. The North Atlantic Treaty, Article 2, Washington, D.C., April 4, 1949. https:// www.nato.int/cps/en/natohq/official_texts_17120.htm

25. The White House, "Statement on Indo-Pacific Economic Framework for Prosperity," May 23, 2022. https://www.cnbc.com/2022/05/26/ipef-what-is -the-indo-pacific-framework-whos-in-it-why-it-matters.html

26. Michael Schuman, "China's Russia Risk," *The Atlantic*, March 9, 2022, https:// www.theatlantic.com/international/archive/2022/03/xi-putin-friendship-russia -ukraine/626973

27. Williamson Murry and Allan R. Millett, *A War to Be Won: Fighting the Second World War* (Cambridge: Harvard University Press, 2000), 162.

28. Evelyn Cheng, "China Upholds Its Relationship with Russia, Says Negotia-tions Needed to Solve Ukraine Conflict," CNBC, March 7, 2022, https://www.cnbc .com/2022/03/07/china-upholds-its-relationship-with-russia-says-negotiations -needed-to-solve-ukraine-conflict.html

29. Michael Schuman, "China's Russia Risk," *The Atlantic*, March 9, 2022.

CHAPTER 8

Competing with Allies: Benefits and Burdens of NATO and European Union Relations

Joel R. Hillison and Maryann F. Foster

Since the mid-twentieth century, the United States has enjoyed a multinational political and economic alliance with members of the North Atlantic Treaty Organization (NATO) and useful political and economic ties with the European Union (EU). U.S. competitors China and Russia lack the breadth and density of U.S. allies and partners. This gives America a comparative advantage in dealing with these and other competitors. America's oldest and most powerful foreign alliance is in Europe. Forged by common history, values, and interests, U.S. relations with NATO—and with its partner organization the European Union (EU)—have provided the United States an advantage in the battle of ideas, in terms of material capability, and in competition too, ranging from security and economic interests to more peripheral interests such as human rights and the promotion of democratic ideals.

Yet European security is in flux, especially considering the 2022 war in Ukraine. European states within NATO and the EU vary in their approaches toward Russia. This fact could erode their future solidarity if Eurasian security deteriorates further. Eastern European countries and their Scandinavian colleagues are clearly focused on the threat they perceive from Russia; thus, they expect the EU and NATO to continue their support to partners such as Ukraine to deter further Russian aggression. At the same time, older, larger NATO allies fear escalating any conflict with Russia and are eager to restore a more cooperative political and economic relationship with it, if some settlement can be found. Still other countries, such as Hungary, have close ties with Russia and can frustrate Alliance calls for movements against Russian aggression. Finally, Russian aggression's variable human and economic costs to other states influences how actors view its behaviors. As the United States seeks to maintain its global strategic advantage, it must strengthen its network of allies and partners in Europe and beyond to encourage or maintain solidarity among them—especially against current and future aggressors.

For more than seventy years, the United States has generally understood "competition with" its European allies and partners to mean working shoulder-to-shoulder with them, cooperating according to an indivisible transatlantic security bond—one largely of shared liberal democratic values and interests. It has also viewed the European Project as an effort in tandem with NATO's security goals, to promote greater prosperity and economic interdependence, within both Europe and the transatlantic community. Today, however, the phrase may be taking on another meaning, one that connotes competition *against* one another, especially as the EU aspires to wield incrementally greater supranational or federalizing powers rather than operating merely as an intergovernmental organization as NATO does.

In 2016, with its membership including 22 NATO allies, the EU published its first global strategy, in which it identified an ambition of "strategic autonomy."[1] In the years since, EU policies and interests have increasingly diverged from NATO and the United States'—NATO's de facto leader—whether regarding a nuclear agreement with Iran, transatlantic trade protocols, data protection, corporate taxation, climate change mitigation, or even relations with U.S. great power rivals such as Russia and China. Such differences raise questions about the nature of U.S. "competition with" European states and can strain transatlantic relations, suggesting a reassessment of the cost–benefit ratio of U.S. foreign relations with its transatlantic cohort.

Regardless of NATO's early solidarity upon Russia's 2022 invasion of Ukraine, the United States may still find itself vacillating between whether it and many of its allies in Europe share largely similar security interests or whether they are now in competition against one another. While the United States and its allies and partners share some similar security interests, they also find themselves competing in other areas. For example, the United States precluded a French submarine sale to Australia in 2021 by offering the latter U.S. nuclear-powered vessels instead. America vehemently objected to its ally Turkey's purchase of Russian S-400 missile defense systems and likewise to Hungary's close relations with Russian President Vladimir Putin, despite Russian aggression against NATO partners Ukraine and Georgia. In light of such diversions, this chapter reassesses the U.S. relationship with NATO and the EU. Discussions will consider the benefits and burdens of U.S. involvement with each organization, followed by a consideration of their futures and how the United States might proceed in its relations with each organization, given the changing security environment.

NATO

Since 1949, NATO has been the United States' preferred regional organization through which it pursues its national interests in Europe. With thirty allies as of June 2022, and more than forty global partners, NATO offers its members well-established intergovernmental political and military fora. Its integrated command structure facilitates joint, combined military action and enhances interoperability to safeguard allies and partners' collective and cooperative security, respectively. NATO's deterrent power hinges on Article 5 of its treaty, which states that an attack against one ally will be considered an attack against all. For more than seven decades, the U.S. military has contributed the largest and most expensive share of NATO's military power. With this burden has come the benefit of significant U.S. political influence in Europe, America's most important trade region. The following section discusses some of the benefits and burdens of America's NATO involvement in Europe, categorized by diplomatic, informational, military, and economic sources of power.

NATO's Benefits and Burdens

Diplomatically, NATO furthers enduring U.S. foreign policy interests, such as sustaining global alliances and partnerships and promoting liberal institutions, values, and norms. Belonging to a security community of likeminded states encourages mutual trust and cooperation on myriad issues. Having NATO members' solidarity behind certain U.S. foreign policy objectives can be a powerful source of legitimacy within the international community.

NATO cooperates with various organizations such as the United Nations, the European Union, the Organization for Security and Cooperation in Europe, and the African Union.[2] The largest partnership group that NATO hosts is the Euro-Atlantic Partnership Council (EAPC), comprising thirty allied and twenty nonallied European states, whose representatives dialogue regularly at NATO headquarters regarding security issues. EAPC partners include Armenia, Austria, Azerbaijan, Belarus, Bosnia and Herzegovina, Finland, Georgia, Ireland, Kazakhstan, the Kyrgyz Republic, Malta, the Republic of Moldova, Russia, Serbia, Sweden, Switzerland, Tajikistan, Turkmenistan, Ukraine, and Uzbekistan.[3]

From 1997 until 2021, the Russian Federation even participated within NATO's framework as a special partner under the Founding Act on Mutual Relations, Cooperation and Security between NATO and the

Russian Federation (or the "NATO-Russia Founding Act").[4] Georgia[5] and Ukraine remain special partners.[6]

Farther afield, NATO connects the United States to various NATO partner groupings such as the Mediterranean Dialogue ("Med-D"), comprising Algeria, Egypt, Israel, Jordan, Mauritania, Morocco, and Tunisia; the Istanbul Cooperation Initiative (ICI) of Bahrain, Qatar, Kuwait, and the United Arab Emirates; and NATO's Partners Across the Globe (or "Global Partners"), Afghanistan, Australia, Colombia, Iraq, Japan, the Republic of Korea, Mongolia, New Zealand, and Pakistan.[7] NATO partnerships' global reach enhances U.S. access.

As beneficial as NATO's size and reach may be to its legitimacy, having many allies can slow decision-making, which operates according to a tradition of consensus.[8] Any ally's veto can thwart a proposed action at any committee level.[9] Given diverging threat perceptions and policy priorities among so many allies,[10] getting consensus within NATO's supreme decision-making body, the North Atlantic Council (NAC), can take days, weeks, months, or even years, depending on an issue's urgency.

Crisis deliberation can signal tenuous solidarity, which damages a security organization's credibility. Yet the United States must endure the burden of slowed action, at times, in exchange for the operational legitimacy NATO consensus confers. Not consulting allies can harm others' trust in the United States and may incline some allies toward greater independence from U.S. security policies. Ignoring allied concerns is especially unwise at a time when U.S. great power rivals seek to sever transatlantic security bonds. Given past or present economic ties between key NATO allies such as Germany[11] and France[12] with U.S. peer competitors such as China and Russia,[13] the United States must be mindful of not distancing itself economically or diplomatically from its most important allies and partners.

Informationally, NATO evokes a positive narrative. Founded according to Article 51 of the United Nations Charter, which permits national and collective self-defense,[14] and backed by U.S. military power, NATO has supported democracy, human rights, individual liberty, and the rule of law for more than seventy years. As the Alliance has grown from twelve to thirty members, its appeal and expansion have spread these values throughout Europe.

At times, allies' behavior can burden NATO with charges of hypocrisy—especially if it appears to tolerate their illiberal domestic practices.[15] The risk of perceived NATO tolerance is high; for, were an ally were to flout NATO values, there is no mechanism in the Washington Treaty to censure or expel an offender.

As NATO's largest contributor, the United States must concern itself with its allies and partners' actual and perceived political behaviors. Among them are moral hazards they may exploit. So-called "free-riding"[16] poses not only a possible military and economic burden for the United States and its taxpayers, but it also risks a perception of U.S. weakness if America cannot persuade its allies to share collective security burdens.

Militarily, U.S. forces project power from their continuing presence in Europe. NATO channels further connect nearly eighty nations and thereby extend U.S. global reach. Through NATO, the United States can provide security assistance such as defense sector reform, training, materiel, or other resources.[17] NATO agency can make U.S. military aid more palatable for some countries' domestic politics than such direct aid may be.

America enjoys a standing leadership billet atop NATO's integrated command structure, in the person of the Supreme Allied Commander Europe (SACEUR). This leader is dual hatted as the commander of U.S. European Command (EUCOM). SACEUR coordinates with the Alliance's U.S.-based, but always European, Allied Command Transformation (ACT) commander to shape NATO forces' training, doctrine, strength, interoperability, and future capabilities.[18]

In developing national forces for NATO's potential benefit, Europeans spent $314 billion on defense in 2020.[19] This exceeded China's ($193 billion) and Russia's ($60.6 billion) spending combined.[20] Allies and partners' investments, combined with NATO's quest for members and partners' interoperability, can benefit the United States when allies and partners buy U.S. materiel.[21]

Much of NATO's current membership once lay east of the Iron Curtain (Albania, Bulgaria, Croatia, Czech Republic, the former East Germany, Estonia, Hungary, Latvia, Lithuania, Montenegro, North Macedonia, Poland, Romania, Slovenia, and Slovakia). These and others have benefited from NATO efforts to reform their militaries' organization, equipment, or doctrine, to put their forces under civilian control, and to make them defenders of the liberal rights and values that NATO promulgates.

Beyond enlarging the potential defense market, NATO's eastward creep has extended the United States' existing power projection options. Vast swaths of air, land, sea, and space belonging to U.S. allies and partners are now areas from which U.S. forces could possibly operate.[22] Such access is critical to U.S. global operations.[23] It provides a huge operational advantage, including deterrence potential.[24] This is vital, given great powers' Eurasian ambitions and China's interest in Halford Mackinder's "World-Island" concept, for example.[25]

A further benefit of U.S. membership in NATO is its access to both allies' and partners' expertise, partly resident in nearly thirty NATO-certified "Centers of Excellence" which explore best practices for a wide variety of security tasks, from cyber operations and hybrid warfare to military policing and more.[26]

Despite the benefits NATO membership accrues to the United States, it also entangles its security with that of many allies. Any could find itself in a security crisis requiring U.S. military assistance. Unfortunately, some security challenges arise from within the Alliance's own membership, which could also frustrate an enduring U.S. interest in a Europe which is "whole . . . free . . . and at peace."[27] For example, allies Greece and Turkey, and partners Armenia and Azerbaijan, have clashed in interstate disputes for decades. Russia poses risks to NATO candidates Finland and Sweden, and to aspirants and other partners such as Georgia, , Ukraine, Belarus, Moldova, and others. NATO has already responded to crises in the Balkans, where tensions nevertheless linger.

Alongside such perils, U.S. membership in the largely European alliance fuels Russia's security dilemma and its attempts at preserving a sphere of influence. For example, Russian leaders have long claimed NATO enlargements threaten their state.[28] On such claims, Russia staked its 2014 and 2022 Ukraine invasions.[29] Russian forces also occupy portions of NATO partners Georgia and Moldova's territories.[30] Other Russian provocations in Europe include air and sea incursions of allies' national boundaries;[31] violations of longstanding arms control treaties;[32] attacks on allies' cyberspace;[33] and threats to allies and partners' energy supplies.[34] To face myriad security threats, NATO members have long been urged by the United States to adopt a U.S.-led standard against which allies and partners should ideally develop and operate their militaries.[35] Beyond technical or logistical challenges inherent to harmonizing U.S.–NATO capabilities or operations, allies must contend with sensitive political–military constraints, such as 'caveats,' which certain allies' governments impose on their forces' use.[36]

No ally is obligated to provide forces for NAC-agreed operations,[37] thereby forcing a NATO commander's having to "take it or leave it," as far as volunteer force contributions go, caveats and all. Such restrictions can render a force contribution almost useless in combat. Tokenism raises the specter of allies "free-riding" on U.S. military might.[38] Though often tolerated, this further burdens U.S. soldiers with tasks the allies could ably perform—if only their nations would allow it.

Economically, NATO's expansion benefits international trade. Commerce follows security, and NATO enlargement extends the transatlantic market, including for defense goods.[39] Further, the cost of peace is incalculable. Were Europe to be plunged back into widespread war, the human and material costs would be staggering, and therefore worth the current security investments to avoid them. The U.S. financial burden for NATO's commonly funded expenses (though only 0.3% of total alliance spending) is higher than that of all other allies, except Germany. Both allies plan to fund around 16% of such outlays through 2024.[40] Under NATO's "costs fall where they lie" principle for supplying national forces for NATO operations, the United States (overwhelmingly) absorbs the highest cost share, given its huge defense budget.[41] Its financial contribution during twenty years of NATO-led operations in Afghanistan alone reached a staggering $2.3 trillion.[42]

Upon signing the Washington Treaty, America understood its ten European allies depended on its might for their security. America accepted that burden as a strategic necessity after the Second World War, when Europe lay in ruins. U.S. interests hinged on rebuilding Western Europe into a capitalist, transatlantic market with which America could trade. That plan's success yielded a collective European market (if the United Kingdom is included) which today nearly rivals U.S. wealth in gross domestic product (GDP) terms.

However, for all the European allies' wealth, as a bloc, they remain de facto members of a U.S. security "colony" of sorts—one still dependent on American might for its safety against great power adversaries. The latent military power which Europeans' wealth affords them has not yet been translated into any effective, independent, regional defense enterprise. This leaves the United States in its current position of shoring up Europe's defense, along with all the benefits and burdens this vital U.S. interest entails. The alternative, a strategically autonomous Europe, has long been a European ambition.

EU

Postwar European efforts at collective defense predate NATO. The Treaty of Brussels was signed in 1948 between France, the United Kingdom, and the Benelux countries (Belgium, Luxembourg, and the Netherlands) to promote mutual defense.[43] This treaty later became the basis for the Western European Union. Many of these countries, France in particular, saw that the region needed economic cooperation to rebuild after World War II's devastation. They also seized opportunities to cooperate with former rivals, especially Germany, to rebuild their economies; to prevent the kind of interwar social unrest that had enabled the rise of fascism; and to promote economic interdependence that might improve relations, increase the benefits of cooperation, and avoid the costs of future conflict.

In 1952, these same countries (less the United Kingdom), plus former belligerents (West Germany and Italy) established the European Coal and Steel Community (ECSC). Just five years on, these countries created the European Economic Community (EEC) under the Treaty of Rome. In 1973, the United Kingdom, Denmark, and Ireland also joined the EEC. In 1992, the EU emerged from it, formalized by the Maastricht Treaty. By 2007, the EU enshrined its own common defense guarantee, codified as Article 42.7 of its Lisbon Treaty.[44] Over time, the EU evolved into a supranational organization—one with ambivalent U.S. support. Today, the EU has twenty-seven members (the United Kingdom left the Union in 2020) and had a GDP of about $17.1 trillion in 2021 (compared to $22.9 trillion for the United States).[45] The Eurozone, a nineteen-member subset of the EU, shares a common currency and is even more closely integrated, given its common monetary policy and banking union. (By a special agreement with the EU, the Vatican also uses the Euro but is not an EU member.)

EU forefathers, such as France's René Pleven, envisioned the eventual creation of a European army. In 1950, he proposed today's European Defense Community (EDC), with strong U.S. support, at the time, for both the EDC and for greater European integration.[46] However, a 1954 "no" vote in the French National Assembly crushed Pleven's dream. It took almost fifty years for renewed movement on autonomous European defense.[47]

Since the end of the Cold War and the ensuing Balkans conflicts, the EU has increasingly become a security provider in Europe and its periphery. Embarrassed by the dismal failures of regional militaries to deter or stop the Balkan conflicts, some European states pushed for greater military capability, to include a collective European capacity to act independently of the United States. This ambition revived with the 1998 St. Malo agreement between Britain and France. Most recently, the EU proposed achieving strategic autonomy in its 2016 Global Strategy. Though this phrase means different things to different EU members, as well as to their external partners, it has called into question the nature of U.S.–EU relations and has spurred their reassessment. The benefits and burdens of U.S. relations with the EU are discussed in the following across the diplomatic, informational, military, and economic instruments of U.S. power they entail.

The United States has at times encouraged and then discouraged European forays into the autonomous security realm. Former U.S. Secretary of State Madeleine Albright expressed the U.S. perspective on European security efforts in terms of three "Ds," which the United States insisted upon: no *duplication* of existing capabilities, no *discrimination* as to non-EU countries' participation in defense capability development (read, no excluding the United States), and no *delinking* of North American and European security regimes.[48]

Diplomatically, when U.S. and EU interests align, the EU can provide another cooperative partner for international action on preventing threats to human rights, safeguarding the rule of law, upholding international law, promoting free and fair trade, and addressing climate change. For example, France's seat on the UN Security Council, and the sum of EU member states' voting shares in the International Monetary Fund (IMF) and the World Bank, all add to U.S. diplomatic power. The EU, as a supranational actor, also has an extensive diplomatic service, in addition to those of its individual member states. All these can help further U.S. diplomatic efforts, especially with countries where the United States does not have direct or formal diplomatic representation. The EU–U.S. relationship, when augmented by its allies in the United Kingdom and Canada, comprises the basis of an oft-called-for "league of democracies" to promote and protect global democratic values and human rights. EU states also retain certain legacy relationships from their colonial pasts, which can further U.S. interests. For example, longstanding ties in Francophone African and other countries can complement U.S. diplomatic efforts in those regions. The EU's Common Foreign and Security Policy (CFSP) is also highly popular with citizens of EU member states, with an almost 70% approval rating. It too can augment U.S. diplomatic efforts in times of need.[49]

As might be expected, EU foreign policy at times diverges from that of the United States. Such divergence exists even in important areas such as the EU's approach to China, in its energy relations with Russia, and regarding the Middle East and other issues. For example, the EU–China Comprehensive Agreement on Investment (CAI), signed at the end of the Trump administration, came as a shock to the United States.[50] While China's apparent insouciance over Russia's war on Ukraine has since put the future of the CAI at risk, this was a dramatic divergence from ongoing U.S. trade wars with China.[51] The United States further opposes some European countries' connections to the People's Republic of China's Belt and Road Initiative (BRI),[52] which endure despite the EU's having established its own Global Gateway fund to compete with the BRI.[53]

EU members, especially Germany, had already differed with the United States with regard to the harshness of sanctions on Russia after its 2014 invasion of Ukraine.[54] Perhaps the greatest divergence has been seen in the Middle East. For example, the EU remained in the Joint Comprehensive Plan of Action (JPCOA), aimed at reining in Iran's nuclear ambitions, even after the United States left the agreement. The EU then explored ways to get around U.S. sanctions on companies trading with Iran.

EU relations with Turkey, a NATO member since 1952, have also worked against U.S. interests. For example, the United States supported Turkey's entrance into the EU when many of its EU partners did not. In addition, EU policy is often at odds with U.S. policy in areas such as data privacy, climate change, capital punishment, and the jurisdictional reach of the International Criminal Court.[55]

The EU also has a collective action problem, which often results in longer decision times and less forceful positions, especially in matters of economics and national security. Even within the EU, member states retain a great deal of autonomy and therefore pursue their own interests, which can at times be at odds with those of the United States. For example, Germany pursued the Nordstream 2 energy pipeline project with Russia, which the United States opposed. The project nevertheless advanced until halted in response to Russia's 2022 war on Ukraine.[56]

Informationally, the U.S.–EU partnership offers the United States a significant advantage in its narrative competition with both Russia and China. Such competition depends upon a country's reputation—built on strength, reliability, and resolve.[57] The EU augments U.S. efforts at protecting against adversaries' information warfare and cyberattacks. U.S. and EU cyber collaboration helps them detect and address false narratives in a timely manner.

As a de facto partnership of democracies, the United States and the EU can boast a seven-decade-long relationship with significant, combined diplomatic, military, and economic strength. The EU has proven itself a reliable U.S. partner in rebuilding Europe after World War II, in coordinating recovery efforts after the 2008 global financial crisis, and in imposing sanctions on Russia in the aftermath of its 2022 war on Ukraine. Working together, the EU and the United States can better promote their shared values of democracy, individual freedom, human rights, and the rule of law. Such mutually reinforcing narratives from the world's most economically and politically advantaged actors lends credence to democracy as a superior model of governance and the preferred alternative to autocratic values and the power-based policies espoused by states such as China or Russia. While gaining consensus for the EU and United States to act in concert takes time and effort, these partners' agreement on certain issues amplifies and strengthens their combined message to the world.

Despite this, the partners' messages are sometimes directed at each other. For example, EU digital privacy regulations and concerns over U.S. dominance in the cyber economy have often led the United States to view the EU as obstructing the free flow of Internet data and of vilifying U.S. tech giants such as Google and Facebook to protect its own markets. Such disagreement impinges on a transatlantic solidarity narrative. The EU publicly criticizes the United States for permitting its states to exercise capital punishment.[58] It also condemned the U.S. withdrawal from the Paris Climate Accords.[59]

Militarily, a stronger EU defense force can benefit the United States. Increased EU military spending strengthens the "European pillar" within NATO, comprising those allies who belong to both organizations. As stated in the (de facto U.S.-led) NATO 2021 Brussels Summit Communiqué, "NATO recognizes the importance of a stronger and more capable European defence."[60] Additionally, EU involvement in regional crises allows the United States and NATO to focus on more conventional threats such as Russia (and China) in Europe and China in the Pacific. This was true in 1955, when the allies encouraged then West Germany to rearm and join NATO, to allow the United States to focus on the Korean War and on competition elsewhere with the USSR. It was also true in the Balkans, where the EU overtook NATO missions so the United States could focus its efforts on Iraq and Afghanistan. While such EU contributions were small, they did augment regional stability, in line with U.S. and EU interests.

Today, the EU is working to facilitate NATO member forces' movements throughout EU territory in the event of an emergency. It also developed its own battle groups. Should they ever become viable, EU Battle Groups could relieve some U.S. defense burdens.[61] So too a planned EU rapid reaction force could be useful, so long as Albright's "3 Ds" are considered.[62] Finally, the EU's version of coordinated defense planning could theoretically economize bloc-wide resources by standardizing more weapons systems across the EU, versus having up to twenty-seven individual national systems to fulfill certain capability requirements between the EU and NATO's missions.

Concerns over EU strategic autonomy appear overblown. The EU recognizes that, for the time being, a U.S.-underwritten NATO remains "the only viable framework to ensure the territorial defence of Europe."[63] A thus far elusive but necessary harmonization of EU acquisition processes perpetuates European reliance on key enablers which the United States largely supplies, such as intelligence, surveillance, and reconnaissance assets or precision-guided munitions.

Launched in 2017, the European Defense Fund (EDF) was designed to improve the EU's research and development capability. With some €7,953 billion in funding allocated from 2021 to 2027, the EDF was intended to jumpstart EU collective defense projects and to augment member state investment in future capabilities.[64] Nonmember states may also benefit from defense projects, under the EU's Permanent Structured Cooperation (PESCO) scheme, which theoretically permits U.S. participation.[65] PESCO's aim is to harmonize its twenty-five enrollees' defense efforts. Duplication and lack of standardization of current weapon systems among EU states now limits the capabilities each euro can buy. PESCO is perhaps the most promising and practical effort to address such inefficiencies.

One of the main U.S. objections to PESCO has been its exclusion of non-EU NATO countries from participating. However, the EU appears to have softened its stance on this issue. In May 2021, the EU Foreign Affairs Council approved U.S., Canadian, and Norwegian participation in the Netherlands-led Military Mobility Project. Such external partners' inclusion was a first for PESCO,[66] which may need all the help it can get, judging from current outcomes.

Despite PESCO's promise, fifteen of its initial projects were far behind their programmed milestones, according to a 2021 annual report.[67] By 2021, fully twenty-one of forty-six projects still lingered in a conceptual phase.

A downside to greater EU military capability is the finitude of resources. Given the large overlap between NATO and EU membership, most EU forces are dual-hatted against NATO requirements. This means that the same forces are assigned missions within both organizations. At times, this results in a competition for forces. In addition, the duplication of EU headquarters and staffs draws precious resources away from European states that are already lagging in fulfilling their commitments to NATO.

Depending on the EU's choices for developing and employing its own force structure, such ambitions could threaten EU–NATO interoperability—especially with NATO's most capable military, the United States. Technology transfers and intelligence sharing restrictions could further strain U.S. and EU military relations and these entities' ability to work together. Perhaps the greatest concern with the EU is over the question of its strategic autonomy. The EU's 2016 global strategy outlined the EU's ambition to achieve strategic autonomy, described as the "capacity to act autonomously when and where necessary and with partners wherever possible."[68] Such autonomy could undermine NATO's role and reduce U.S. influence in regional security issues. If the EU gains true strategic autonomy, it could, in time, become a U.S. rival.

Economically, the EU is collectively the United States' largest trading and investment partner. Both sides' commitment to the rule of law and transparency, as well as to regulated and stable financial and banking institutions, reinforce this strong economic relationship. The EU largely promotes the same economic interests as the United States, such as free markets, lowered trade barriers, private property protection, and subsidized agriculture, airlines, and other key domestic industries.

Though the EU slightly trails the United States in terms of both GDP and GDP per capita, it wields great economic power. For example, the U.S. GDP in 2021 ($22.9 trillion) and the EU's ($17.1 trillion) together equal about 45% of global GDP.[69] The EU is also the United States' largest trading partner, with trade estimated at $1.1 trillion in 2019, and its largest foreign investor (about $2.0 trillion in foreign direct investment in 2019).[70] The United States invested about $2.4 trillion in the EU during the same period. The two economies remain profoundly interdependent. Collectively, EU and NATO countries occupy six of the seven G7 seats, giving the United States additional leverage in establishing economic norms. The EU and the United States also coordinate with Japan in recurring trilateral discussions aimed at promoting market-oriented policies, World Trade Organization reform, and restrictions on state-owned enterprises.[71] The EU and the United States recently formed a U.S.–EU Trade and Technology Council to coordinate economic approaches to trade and technology and to address challenges posed by "non-market economies" (e.g., China).[72]

The EU also helps promote democratic and liberal economic values within its area of influence. States striving to join the EU must live up to its robust body of laws (the *Acquis Communautaire*), which include many U.S. priorities, such as protections for the rule of law, human rights, and open markets.[73] The EU can thus have a salubrious effect on those states seeking membership. In addition, the EU is the world's largest investor, which helps promote economic development and stability, especially in surrounding countries.[74]

Concerns about the EU's economic relations with China, though valid, are exaggerated. Both the United States and the EU have significant trading relationships with China. According to the U.S. Trade Representative's office, China was the largest trading partner of the United States (in goods) at just under $560 billion in 2020.[75] According to the European Commission, China is the EU's second largest trading partner (well behind the United States) at around $660 billion annually.[76] While the United States began to further scrutinize trade and investment with China much earlier, European states are finally ratcheting up their screening of Chinese investments in Europe to protect key industries from intellectual property theft and to undo foreign influence.[77] The EU is trying to balance cooperation with China, in areas such as climate change, while working with the United States on most issues where there are shared concerns over Chinese behavior.[78] The EU characterized China as a systemic rival in March 2019, a difference from the U.S. portrayal of China as a strategic rival.[79] This more nuanced approach to China has yielded some results and does not necessarily detract from a stronger EU response on political and security issues, which often align with U.S. interests.

However, the EU is not only a U.S. economic partner but also an economic competitor. Both economies produce comparable products and services; this has often led to contentious trade disputes. For example, in 2021, the United States won a trade dispute with the EU over Airbus subsidies. The EU countersued the United States for unfair protection of Boeing.[80] While this dispute was resolved peacefully, it lasted seventeen years and contributed to tensions and accusations, especially from former President Trump, of unfair EU trade practices. The United States also imposed tariffs on EU steel and aluminum producers to protect U.S. industries. These measures remained in effect until 2021, when they were replaced with voluntary quotas.[81]

In the past, the United States has also bristled at EU restrictions on U.S. products, for example, those developed with genetically modified organisms (GMOs) or chlorinated chicken, which the EU bars from its market.[82] In turn, the United States imposed tariffs on specific EU goods (e.g., French wines), in retaliation for France's enacting a digital services tax, which is seen as targeting U.S. big tech companies.

Meanwhile, some EU nations have allowed Chinese investment in sensitive industries such as 5G technology and Russian investment into controversial energy projects (e.g., Nordstream 2). China also owns large stakes of strategic infrastructure in EU states, such as the Greek port of Piraeus,[83] which gives China leverage over EU member states and which could pressure them to act against U.S. interests. For example, in 2017, Greece blocked a U.S.-backed statement condemning China for human rights abuses.[84] The EU even attempted to set up a payment transfer system to bypass that of the Society for Worldwide Interbank Financial Telecommunications (SWIFT) after the United States pulled out of the Joint Comprehensive Plan of Action (JCPOA)—a plan designed to hobble Iran's potential for developing nuclear weapons. The United States then threatened to retaliate against states which traded with Iran—though the JCPOA, developed jointly with the EU, would have allowed this.[85]

FOUR POSSIBLE FUTURES

So, where does all this leave the future U.S. relationships with NATO and the EU? The following section examines four possible futures, based on different developments which could seriously affect these relationships. We begin with the most unlikely, but also the most dangerous, potential future and end with the most advantageous developments for both sides of the transatlantic relationship.

NATO and EU Collapse

Waning political will, especially in France, Germany, the United States, or the United Kingdom, to support the NATO and the EU in a crisis could result in this worst of all possible outcomes. Jean Claude Juncker, the former Prime Minister of Luxembourg and president of the EU, famously described how this might happen: "They know what they need to do; they just don't know how to get re-elected once they have done it."[86] Thus domestic political realities often get in the way of collective aspirations of NATO and the EU. In an extreme situation, this pandering to domestic audiences could lead to the demise of the EU and/or NATO. A collapse of the EU and NATO would look much like the interwar period between the first and second world wars. Due to the significant interdependence among EU and NATO nations, the demise of these organizations would result in significant ill-will and numerous collective action problems within Europe and between America and its former allies and partners. The most likely cause of both NATO and the EU's demise would be a breach of NATO's Article 5 commitments or the EU's common defense commitments under Article 42(7). NATO's overextension, beyond a collective political will to come to the defense of its alliance members, is already a concern, given thirty allied states. That problem could be exacerbated if the Alliance were to expand to thirty-two members or beyond.

The lack of an institutional framework would force American diplomats to negotiate new bilateral security, investment, and trade agreements with European states to continue existing information sharing, coordinated action, or compliance monitoring regimes with them.[87]

The basis of the liberal international order would be in jeopardy if these pillars of the current order disbanded, ceding a values victory to China and Russia, who could exploit any vacuum to promote their own versions of collective action.

European nations' security would suffer most from any loss in collective defense capability, but this would also weaken the U.S. military's position.

Even if it could retain most of its allies through bilateral security arrangements, interoperability problems would worsen. Without NATO's security guarantee, many European states would question a U.S. willingness to intervene against threats, especially from Russia. This could lead Europeans to bandwagon with Russia, a critical regional energy supplier, and to further weaken U.S. interests. The United States would also lose the influence it now enjoys as NATO's de facto leader, embodied in the Alliance's supreme military commander. Removing NATO's security guarantee might encourage further Russian aggression, which, if tolerated, would end U.S. credibility as a superpower and would considerably weaken trust in the United States from its remaining European (and global) partners. That said, NATO's dissolution would free the United States from entangling security arrangements that could draw it into diverse allies' wars. Decreased security burdens would allow America to focus its efforts on selected or key allies' security.

It is difficult to quantify the economic effect this scenario would impose on America. U.S. bilateral bargaining power against former EU states would grow, but so would U.S. companies' cost of doing business, with a return to multiple currencies or national legal regimes. Reimposing national border controls would add cost too, resulting in less competitive transatlantic supply chains. Again, China and Russia could exploit this environment to strengthen economic and security ties with disaffected European states.

EU Collapse with Stronger NATO

Should the EU lose its current stature, NATO would inherit additional security burdens, especially given current Chinese and Russian ambitions. A loss of confidence in the Euro and a collapse of the EU monetary union is one possible scenario under which the EU might collapse while NATO survives. Other economic fragmentation is possible along geographic lines. In 2010's financial crisis, there were talks of separate unions—a frugal north and a profligate south. Political divide is also possible, if western EU states lift their sanctions on Russia—in exchange for affordable energy supplies, for example—while eastern ones insist on their continuation to deter Russian aggression. Absent a viable EU, NATO's appeal for coordinating collective military and political action would grow.

Though a stronger NATO (without an EU) could cost America more, it may be worth it. Tensions between NATO and the EU would cease. If NATO attracted more members and reformed its consensus-based decision-making protocol to avoid paralysis, it could boost European solidarity and U.S. influence.

Anxiety over costly EU duplication of NATO military functions would end. Former EU resources could go to NATO operations to alleviate U.S.–European burden-sharing friction.

A strong NATO benefits U.S. economic interests. A more secure, consolidated defense regime to protect the transatlantic region is likely to boost trade and investment. A Europe dependent on U.S. military protection will also be more likely to support U.S. interests in other realms.

Stronger EU with NATO Collapse

In 2019, French President Macron described NATO as "brain dead,"[88] upon perceiving U.S. disengagement from it. In response, he proposed developing the EU's geostrategic power in earnest. The surest route to NATO's demise would be a U.S. withdrawal from the treaty. Without U.S. security guarantees and leadership, NATO would likely dissolve for lack of credibility. Any failure to uphold Article 5 could sink NATO overnight.

NATO's loss would throw America's assumed goodwill among more than seventy likeminded allies and partners into doubt. A relatively meager U.S. diplomatic corps would face the daunting task of establishing tailored, bilateral military and economic relations with each of them.

The U.S. voice on the world stage would quiet somewhat, not least because America's adversaries could invoke NATO's dissolution as proof of U.S. weakness. Existing transatlantic divergences concerning climate change, terrorism, Middle East peace, and even how U.S. competitors such as Russia and China are characterized, could widen. U.S. credibility to lead or act could suffer, inviting adversaries' security challenges worldwide.

America could find itself increasingly isolated. Closer European ties with both Russia and China may be expected in the absence of U.S. pressure against them. U.S. defense industries would lose business, as the EU prioritized European defense industry investments to achieve some supranational interoperability.

However, if a powerful, autonomous EU did not oppose U.S. interests, it could prove essential to them. If the EU—instead of NATO—maintained European security and defense, this would benefit the United States by reducing current U.S. defense costs. As long as the EU upheld U.S.-shared liberal values, transatlantic cooperation would likely persist to mitigate the challenges any loss of direct U.S. influence in Europe might pose.

Stronger NATO–EU Partnership

The title of the EU's uplifting Beethoven anthem, "Ode to Joy," best characterizes this scenario. It expresses the goodness a unified Europe offers. A strong NATO with close EU ties would represent a significant ideological and security victory for U.S.-European relations, so long as they respected the final two of Madeleine Albright's "three Ds" (no duplication, no decoupling, and no discrimination), regarding non-EU NATO states.[89] A stronger NATO and EU partnership could minimize duplication and boost institutional interdependence for all members' mutual benefit. Success at this prospect depends upon relations among NATO's main power centers: the United States, the United Kingdom (UK), France, and Germany. The U.S.–UK relationship remains crucial for NATO power; France and Germany comprise the EU's twin engines. Where these four states go, others follow. This quad's ability to collaborate on critical issues such as Russian aggression will determine how likely a scenario this could be.

A stronger NATO–EU bond would increase U.S. and European states' global diplomatic influence, especially in areas of mutual agreement. The EU would benefit from NATO's credible security commitments, and NATO would benefit from the EU's economic heft and its ties throughout Europe and Africa.

A stronger relationship could also improve European states' military capabilities, if cooperation increased defense solidarity and resilience against outside threats. This could strengthen NATO's European pillar, enabling local forces to address regional problems without recourse to U.S. or NATO support. This supposes such partnership would not sideline non-EU NATO members (e.g., the United States, Turkey, and the UK). However, if the EU were to establish a supranational force generation process, this would complicate military command and control and could result in a NATO–EU competition for the same forces in a crisis.

Harmonized military capabilities should free resources for other needs. Greater integration of transatlantic defense industries and reduced procurement barriers would benefit both U.S. and European arms industries.

A secure and prosperous NATO–EU league of (ostensible) democracies would exemplify the appeal of Western values and ideas, and the benefits of free commerce and the rule of law for all to see and emulate.

CONCLUSION

In this chapter, we examined the benefits and burdens of U.S. relations with NATO and the EU to assess whether such relations present America a net burden or a net benefit over the long term. While both NATO and EU members sometimes work at cross-purposes to the United States, the best of all possible worlds is one is which NATO and the EU both thrive and collaborate closely with one another.

As the United States competes with China's influence in the Indo-Pacific, it will need strong European institutions such as NATO and the EU, with shared liberal values, to maintain regional security in its most important foreign market. It must also continue investing in its transatlantic relationships and bearing the burdens of its NATO leadership to keep the Alliance strong. The EU's role in European security, and its U.S. relationship, will depend upon how the former pursues strategic autonomy and what shape its harmonized military developments take. Whatever happens, the only transatlantic "competition" should be together—U.S. and European allies and partners on the same team, competing against outside players—*not* against each other. The challenge is to identify where U.S.–European interests align and are best served by cooperation, and where they diverge to be better addressed in constructive competition.

NOTES

1. European External Action Service (EEAS), *Shared Vision, Common Action: A Stronger Europe* (Brussels: European Union, 2016), https://www.eeas.europa.eu/sites/default/files/eugs_review_web_0.pdf.

2. "Partners," NATO, March 27, 2020, https://www.nato.int/cps/en/natohq/51288.htm.

3. "Partners," n.p.

4. "Founding Act on Mutual Relations, Cooperation and Security between NATO and the Russian Federation," NATO, November 5, 2008, https://www.nato.int/cps/en/natohq/official_texts_25470.htm?selectedLocale=en.

5. "NATO–Georgia Commission," NATO, December 4, 2012, https://www.nato.int/cps/en/natohq/topics_52131.htm.

6. "NATO–Ukraine Commission," NATO, January 25, 2018, https://www.nato.int/cps/en/natohq/topics_50319.htm.

7. "Partners," n.p.

8. Leo G. Michel, "NATO Decision-Making: The 'Consensus Rule' Endures Despite Challenges," in *New Security Challenges: NATO's Post-Cold War Politics,* ed. Sebastian Mayer (London: Palgrave Macmillan, 2014), 107–123.

9. "Consensus Decision-Making at NATO," NATO, October 2, 2020, https://www.nato.int/cps/en/natolive/topics_49178.htm.

10. Lucie Béraud-Sudreau and Bastian Giegerich, "NATO Defence Spending and European Threat Perceptions," *Survival* 60, 4 (2018), 53–74, https://doi.org/10.1080/00396338.2018.1495429.

11. Alan Crawford, "China's Ties at Hamburg Port Spotlight Germany's Economic Tradeoffs," *Bloomberg,* October 26, 2021, https://www.bloomberg.com/news/newsletters/2021-10-26/supply-chain-latest-china-deepens-germany-ties-at-port-of-hamburg.

12. Jonathan Eyal, "France, Germany and the 'Russia Engagement' Game," *Royal United Services Institute,* June 29, 2021, https://rusi.org/explore-our-research/publications/commentary/france-germany-and-russia-engagement-game.

13. Greg Myre, "Biden's National Security Team Lists Leading Threats, with China at the Top," *National Public Radio,* April 13, 2021, https://www.npr.org/2021/04/13/986453250/bidens-national-security-team-lists-leading-threats-with-china-at-the-top.

14. "United Nations Charter," United Nations, June 26, 1945, https://www.un.org/en/about-us/un-charter/full-text#:~:text=Article%2051,maintain%20international%20peace%20and%20security.

15. Celeste A. Wallander, "NATO's Enemies Within: How Democratic Decline Could Destroy the Alliance," *Foreign Affairs* 97, 4 (July/August 2018), 70–81.

16. Scott N. Siegel, "Bearing Their Share of the Burden: Europe in Afghanistan," *European Security* 18, 4 (2009), 469, https://doi.org/10.1080/09662839.2010.498824.

17. "Cooperative Security as NATO's Core Task," NATO, February 10, 2022, https://www.nato.int/cps/en/natohq/topics_77718.htm?selectedLocale=en.

18. Darrell Driver, "SACEUR, CJCS, and U.S. Military Influence in Transatlantic Security Policy," *Journal of Transatlantic Studies* (2021), 1–22.

19. "Europe," *The Military Balance* 121, 1 (2021), 66.

20. "Comparative Defence Statistics," *The Military Balance* 121, 1 (2021), 23.

21. Jenny Yang, "Smart Defense, Group Procurement, and Interoperability: Examining the Current State of Affairs," NATO Association of Canada, February 13, 2015, https://natoassociation.ca/smart-defense-group-procurement -and-interoperability-examining-the-current-state-of-affairs.

22. David Ochmanek, "Restoring the Power Projection Capabilities of the U.S. Armed Forces, Testimony presented before the Committee on Armed Services on February 16, 2017," RAND, February 16, 2017, https://apps.dtic.mil/sti/pdfs /AD1027356.pdf.

23. Mohammed Hussein and Mohammed Haddad, "Infographic: U.S. Military Presence Around the World," *Aljazeera*, September 10, 2021, https://www.aljazeera .com/news/2021/9/10/infographic-us-military-presence-around-the-world -interactive.

24. Juri Luik and Henrik Praks, "Boosting the Deterrent Effect of Allied Enhanced Forward Presence," International Centre for Defence and Security, 2017, https://icds.ee/wp-content/uploads/2017/ICDS_Policy_Paper_Boosting_the _Deterrent_Effect_of_Allied_eFP.pdf.

25. Francis P. Sempa, "Look to Classical Geopolitics to Understand China's Challenge," *Real Clear Defense*, January 26, 2019, https://www.realcleardefense .com/articles/2021/05/22/look_to_classical_geopolitics_to_understand_chinas _challenge_778283.html.

26. "Centres of Excellence," NATO, November 3, 2020, https://www.nato.int /cps/en/natohq/topics_68372.htm.

27. Robert E. Hunter, "A 'Europe Whole and Free and at Peace,'" RAND Corporation, September 9, 2008, https://www.rand.org/blog/2008/09/a-europe -whole-and-free-and-at-peace.html.

28. Sibel Kavuncu, "U.S. Russia Relations in the Context of NATO Enlargement," *Management and Education* 6, 4 (2010), 76, https://scholar.google.com/scholar _url?url=http://www.conference-burgas.com/maevolumes/vol6_2010/book4 /b4_r10.pdf&hl=en&sa=T&oi=gsb-ggp&ct=res&cd=0&d=13586607790903250577 &ei=aLKPYYT3FYGUy9YPkdKn2AQ&scisig=AAGBfm3HnlzhDctNdA _eG7OuAjqJ-I5dxg.

29. Robert Person and Michael McFaul, "What Putin Fears Most," *Journal of Democracy*, February 22, 2022, https://www.journalofdemocracy.org/what-putin-fears -most/?mc_cid=7dfa40c339&mc_eid=f3cb2aa7a1.

30. Henrik Larsen, "Ukraine, Georgia and Moldova: Between Russia and the West," Center for Security Studies, November 2021, https://css.ethz.ch/content /dam/ethz/special-interest/gess/cis/center-for-securities-studies/pdfs /CSSAnalyse293-EN.pdf.

31. "NATO Jets Intercept Russian Warplanes during Unusual Level of Air Activity," NATO, March 30, 2021, https://www.nato.int/cps/en/natohq/news_182897.htm.

32. Shannon Bugos, "U.S. Completes INF Treaty Withdrawal," Arms Control Association, September 2019, https://www.armscontrol.org/act/2019-09/news /us-completes-inf-treaty-withdrawal.

33. Lucian Kim, "Russian Cyberattacks Present Serious Threat to U.S.," *National Public Radio*, July 9, 2021, https://www.npr.org/2021/07/09/1014512241 /russian-cyber-attacks-present-serious-threat-to-u-s.

34. Arnold C. Dupuy, "Energy Security Is Critical to NATO's Black Sea Future," *The Atlantic*, May 12, 2022, https://www.atlanticcouncil.org/blogs/turkeysource /energy-security-is-critical-to-natos-black-sea-future/

35. Anthony H. Cordesman, "Rethinking NATO's Force Transformation," NATO, November 4, 2008, https://www.nato.int/cps/en/natohq/opinions _21953.htm?selectedLocale=en.

36. Per Marius Frost-Nielsen, "Conditional Commitments: Why States Use Caveats to Reserve Their Efforts in Military Coalition Operations," *Contemporary Security Policy* 38, 3 (September 2, 2017), 371–97, https://doi.org/10.1080/13523260 .2017.1300364.

37. Anthony Forster and William Wallace, "What Is NATO For?" *Survival* 43, 4 (December 2001), 107–22, https://doi.org/10.1080/00396330112331343155.

38. Siegel, "Bearing," 269.

39. Hans Binnendijk and Magnus Nordenman, "NATO's Value to the United States: By the Numbers," *Atlantic Council*, April 19, 2018, https://www .atlanticcouncil.org/in-depth-research-reports/issue-brief/nato-s-value-to-the -united-states-by-the-numbers.

40. "Funding NATO," NATO, August 12, 2021, https://www.nato.int/cps/en /natohq/topics_67655.htm.

41. Nick Routley, "Politics: This Is How Much NATO Countries Spend on Defense," *Visual Capitalist*, September 23, 2021, https://www.visualcapitalist.com /this-is-how-much-nato-countries-spend-on-defense.

42. Deirdre Shesgreen, "'War Rarely Goes as Planned': New Report Tallies Trillions US Spent in Afghanistan, Iraq," September 1, 2021, *USA Today*, https://www .usatoday.com/story/news/politics/2021/09/01/how-much-did-war -afghanistan-cost-how-many-people-died/5669656001.

43. Roy H. Ginsberg, *Demystifying the European Union* (Lanham, MD: Rowman and Littlefield, 2010), 9.

44. Joel R. Hillison, *The Relevance of the European Union and the North Atlantic Treaty Organization for the United States in the 21st Century* (U.S. War College Press, 2018), 28–29.

45. World Economic Outlook Database, IMF, October 2021

46. Max Bergmann, James Lamond, and Siena Cicarelli, "The Case for EU Defense," Center for American Progress, June 1, 2021, https://americanprogress .org/article/case-eu-defense/

47. Roy H. Ginsberg, *Demystifying the European Union* (Lanham, MD: Rowman & Littlefield, 2010), 48–49.

48. Ronald Tiersky and Erik Jones, *Europe Today*, 5th ed. (Lanham, MD: Rowman & Littlefield, 2015), Introduction.

49. Antonio Karlovi, Dario Cepo, and Katja Biedenkopf, "Politicisation of European Foreign, Security, And Defense Cooperation: The Case of the EU's Russian Sanctions," *European Security* 30, 2 (September 2021), 345.

50. Natalie Liu, "EU-China Investment Deal Threatens US-Europe Relations," *Voice of America*, January 1, 2021, https://www.voanews.com/a/east-asia-pacific _voa-news-china_eu-china-investment-deal-threatens-us-europe-relations /6200211.html.

51. William Yuen Yee, "Is the EU-China Investment Agreement Dead?" *The Diplomat*, March 26, 2022, https://thediplomat.com/2022/03/is-the-eu-china -investment-agreement-dead/

52. Jennifer Hillman and Alex Tippett, "The Belt and Road Initiative: Forcing Europe to Reckon with China?" Council on Foreign Relations, April 27, 2021, https://www.cfr.org/blog/belt-and-road-initiative-forcing-europe-reckon-china.

53. Henry Ridgwell, "Can Europe Compete with China's Belt and Road Initiative?" *Voice of America*, December 2, 2021, https://www.voanews.com/a/can -europe-compete-with-china-s-belt-and-road-initiative-/6337145.html.

54. Oliver Bilger, "US_EU differences over Russia," *Gulf Times*, August 8, 2017. https://www.gulf-times.com/story/559524/US-EU-differences-over-Russia.

55. Jonathan Olsen and John McCormick, *The European Union: Politics and Policies*, 6th ed. (Boulder: Westview Press, 2017), 330.

56. Holly Ellyatt, "Germany Halts Approval of Gas Pipeline Nord Stream 2 after Russia's Actions," CNBC, February 22, 2022, https://www.cnbc.com/2022/02/22 /germany-halts-certification-of-nord-stream-2-amid-russia-ukraine-crisis.html.

57. James C. McConville, "The Army in Competition: Chief of Staff Paper #2," Department of the Army, March 1, 2021, page v.

58. William A. Schabas, "The ICJ Ruling against the United States: Is it Really about the Death Penalty?" *The Yale Journal of International Law* 27 (September 12, 2013).

59. Alistair Walsh, "World Reacts to US Withdrawal from Paris Agreement," *Deutsche Welle*, June 1, 2017.

60. "Brussels Summit Communiqué," NATO Press Release (2021)086, June 14, 2021.

61. "EU Battlegroups," European External Action Service, April 2013, www .consilium.europa.eu/esdp

62. "EU to Establish Rapid Reaction Force with Up to 5,000 Troops," *Reuters*, March 21, 2022, https://www.reuters.com/world/europe/germany-offers -provide-core-eu-quick-reaction-force-2025-2022-03-21/.

63. "Why European Strategic Autonomy Matters," European External Action Service, December 3, 2020. https://eeas.europa.eu/

64. European Defence Agency, "European Defense Fund," accessed November 7, 2021, https://eda.europa.eu/what-we-do/EU-defence-initiatives /european-defence-fund-(edf)

65. Joel R. Hillison, *The Relevance of the European Union and the North Atlantic Treaty Organization for the United States in the 21st Century* (U.S. War College Press, 2018), 28–29.

66. Kingdom of Netherlands, "US, Canada and Norway Invited to Join EU PESCO Project Military Mobility," May 5, 2021, https://www.permanentrepresentations.nl/

67. Jacopo Barigazzi, "EU Military Projects Face Delays, Leaked Documents Shows," *Politico*, July 12, 2021.

68. "Why European Strategic Autonomy Matters," European External Action Service, December 3, 2020, https://eeas.europa.eu/.

69. World Economic Outlook Database, IMF, October 2021.

70. U.S. Trade Representative, "European Union," accessed December 13, 2021, https://ustr.gov/

71. U.S. Trade Representative, "Joint Statement of the Trilateral Meeting of the Trade Ministers of Japan, the United States and the European Union," January 14, 2020, https://ustr.gov/

72. United States Trade Representative, "U.S.–E.U. Trade and Technology Council (TTC)," accessed December 13, 2021, https://ustr.gov/

73. Roy H. Ginsberg, *Demystifying the European Union*, 2nd ed. (Lanham, MD: Rowman & Littlefield, 2010), 13.

74. European Commission, "Investment," accessed June 6, 2022, https://policy.trade.ec.europa.eu/help-exporters-and-importers/accessing-markets/investment_en.

75. U.S. Trade Representative, "U.S. China Trade Facts," accessed December 13, 2021, https://ustr.gov/.

76. European Commission, "Trade, Policy, China," accessed December 13, 2021, https://ec.europa.eu/.

77. John R. Deni, "The United States and the Transatlantic Relationship," *Parameters* 50, 2 (Summer 2020), 21.

78. Volker Perthes, "Dimensions of Rivalry: China, the United States, and Europe," *China International Strategy Review*, February 2021, https://doi.org/10.1007/s42533-021-00065-z.

79. Bruno Maçães, "Surprise! The EU Knows How to Handle China," *Politico*, June 22, 2021.

80. Silvia Amaro and Leslie Josephs, "U.S. and EU Resolve 17-Year Boeing-Airbus Trade Dispute," *CNBC*, June 15, 2021, https://www.cnbc.com.

81. Ana Swanson and Katie Rogers, "U.S. Agrees to Roll Back European Steel and Aluminum Tariffs," *The New York Times*, October 30, 2021, https://www.nytimes.com.

82. Valentina Pop, "EU 'Will Not Compromise' on Food Safety in US Trade Pact," *euobserver*, February 12, 2014. https://euobserver.com/world/123091.

83. Momoko Kidera, "Sold to China': Greece's Piraeus Port Town Cools on Belt and Road," *Nikkei Asia*, December 10, 2021, https://asia.nikkei.com/.

84. Robin Emmott and Angeliki Koutantou, "Greece Blocks EU Statement on China Human Rights at U.N.," *Reuters*, June 18, 2017, https://www.reuters.com/.

85. Katie Lobosco, "Trump's Tariffs on European Wine Have American Businesses Begging for Relief," *CNN*, January 7, 2020. https://www.cnn.com/2020/01/07/politics/trump-tariff-french-wine/index.html.

86. Ronald Tiersky and Erik Jones, *Europe Today*, 5th ed. (Lanham, MD: Rowman & Littlefield, 2015), Introduction.

87. Robert O. Keohane, *After Hegemony: Cooperation and Discord in the World Political Economy* (Princeton, NJ: Princeton University Press, 1984).

88. "Emmanuel Macron Warns Europe: NATO Is Becoming Brain-Dead," *The Economist*, November 7, 2019.

89. Nick Ottens, "EU Defense Union Worries Americans, Social Democrats Rally the Troops," *Atlantic Sentinel*, February 15, 2018. https://atlanticsentinel.com/2018/02/eu-defense-union-worries-americans-social-democrats-rally-the-troops

CHAPTER 9

Expanding Partnerships and Alliances in the Indo-Pacific

Jerad I. Harper

As President, I have, therefore, made a deliberate and strategic decision—as a Pacific nation, the United States will play a larger and long-term role in shaping this region and its future, by upholding core principles and in close partnership with our allies and friends.

. . . Our enduring interests in the region demand our enduring presence in the region. The United States is a Pacific power, and we are here to stay.

—*U.S. President Barack Obama, Speech to Australian Parliament, November 17, 2011*[1]

The United States has long been an Indo-Pacific power. However, with the rise of the People's Republic of China (PRC), the United States now finds its security dominance increasingly challenged in this vital region. This challenge will become even more difficult as China moves along its planned path to develop a military capable of countering the United States in the region by 2027 and becoming a "world class" military by 2049.[2] This will give the PRC the military means commensurate with its existing status as one of the two leading global economic powers and expand its already extensive ability to influence events in the region. Maintaining the security advantage necessary to achieve American interests in the Indo-Pacific will require the United States to develop new partnerships, expand existing ones, and change its existing security architecture from the current "hub-and-spoke" system to one instead composed of a number of stronger and overlapping multilateral relationships.

A large network of partners and treaty allies has been a traditional strength of the United States in the Indo-Pacific, as in other regions around the world. However, in contrast to the dense, multilateral security architecture built around the North Atlantic Treaty Alliance (NATO) in Europe and the North Atlantic, the United States in the Indo-Pacific instead has a "hub-and-spoke" system with the United States at the center and relatively weak ties between its partners. To compete effectively with a confident and increasingly aggressive China, this must change. Although duplicating a single, all-encompassing security alliance like NATO in the Indo-Pacific is probably unrealistic, a multilayered system composed of overlapping multilateral relationships and alliances is possible and would produce significant security gains. Additionally, the U.S. needs to expand and improve upon its existing individual security partnerships and tie them closer to these multilateral structures to facilitate "coalitions of the willing" when and if necessary.

Expanding partnerships and improving its alliance structure is critical to the emerging U.S. strategic concept of "integrated deterrence." Rather than relying so heavily upon the U.S. role—as was common during the "unipolar moment" of American hyperpower in the 1990s and early 2000s—this refined concept of deterrence instead relies much more extensively upon a closely interwoven network of allies and partners in addition to what U.S. Defense Secretary Lloyd Austin said on July 13, 2021 was "the right mix of technology, operational concepts, and capabilities . . . so credible, and flexible, and formidable that it will give any adversary pause."[3]

In building and expanding these partnerships, America's narrative needs to move off of its perceived "anti-China" policy to one that is instead "pro-Asia." The entire region is well aware that there is an increasingly tense environment between the United States and the People's Republic of China, but a conflict between these two great powers—likely fought in their waters, airspace, and potentially on their land—would be extremely damaging to all. Building partnerships and pragmatically focusing on the strength and usefulness of the U.S. relationship is more likely to produce results than a singular focus on countering China.

This chapter assesses the origins of the existing security architecture in the Indo-Pacific, America's treaty alliances and partnerships, and provides recommendations for both expanding and strengthening multilateral relationships and expanding individual partnerships to better posture the United States to compete for influence with the People's Republic of China.

ORIGINS OF THE CURRENT HUB-AND-SPOKE SECURITY ARCHITECTURE AND SEATO'S ATTEMPT AT MULTILATERALISM

In contrast to the multilateral Cold War security architecture that it developed in Europe, the United States created a post–World War II security architecture in the Indo-Pacific emphasizing a series of bilateral relations. In postwar Western Europe, the threat of invasion by the Soviet Union and its Warsaw Pact alliance system was a significant concern requiring the creation of an integrated collective defense system with joint command and planning systems. In order to create this system, the U.S. surrendered a degree of its power to a multilateral alliance of democratic states with whom it closely identified—the North Atlantic Treaty Organization.[4] In contrast, postwar Asia lacked the severity of the threat in Europe, and its societies at the time seemed far different to that of America, so the United States instead developed a web of bilateral alliances and partnerships.[5]

A series of bilateral relationships offer their members greater freedom of action than they would have in a multilateral organization requiring consensus among multiple members. The trade-off is that bilateral relations lack the pure military advantage offered by more integrated multilateral structures such as NATO.[6] Additionally, such bilateral alliances also give the more powerful partner—in this case, the United States—a greater degree of control over its partners alongside greater freedom of action. Thus, alliances with the Republic of Korea, Republic of China, and Japan were created not only to defend against communism but also to prevent America from being pulled into a land war in Asia. The United States gained increased influence over the Republic of Korea and Taiwan, as well as over the development of largely pacifist postwar Japanese security institutions. At the time, American planners were far more concerned with the threat of a resurgent Japan than potential gains through their inclusion in multilateral Indo-Pacific security structures.[7]

Although the region did see the temporary existence of a more multilateral security coalition—the Southeast Asia Treaty Organization (SEATO) created in 1954 by the United States, Australia, New Zealand, Pakistan, the Philippines, Thailand, France, and the United Kingdom—this coalition was far less effective than NATO and ultimately fell apart. Despite SEATO's goal of collective defense, U.S. leaders such as John Foster Dulles envisioned it as a much looser security pact and never saw it as a parallel to NATO, despite the similarities of their titles.[8] SEATO had no integrated military structure, and its members were not constrained by the same collective defense provisions as NATO—they were only bound to consult with each other.[9]

SEATO did prove an important vehicle for regional dialogue and collective action. However, it suffered from the incompatible views of its members states, one of whom was never very involved in the region (Pakistan) and two colonial powers (the United Kingdom and France) who became increasingly less invested in the region. Ultimately, the organization fell apart in the aftermath of the failure to hold South Vietnam. SEATO failed its first test when European members opposed intervention in the Laotian crisis of 1961–1962, the same situation for which SEATO was created. After this first crisis of confidence, European members again opposed intervention in Vietnam, where Australia, Thailand, and New Zealand contributed significant forces while the Philippines sent only a small contingent.[10] The fall of South Vietnam in 1975, Australian/New Zealand political backlash from the war, and Thai/Philippine reconciliation with the PRC ultimately led to the dissolution of the organization in 1977.[11]

The SEATO experience, may, however, provide lessons for multilateral structures in a world of increased great power competition. Building upon them requires an assessment of the current U.S. network of partners and allies and analysis of how these can be expanded.

U.S. ALLIES, EMERGING MAJOR PARTNERS, AND OPPORTUNITIES FOR INCREASED MULTILATERAL RELATIONSHIPS

The United States has five treaty allies in the region. Three of these—Australia, Japan, and the Republic of Korea—are strong and vibrant alliances that hold opportunity for the United States both in terms of their usefulness in increasing security partnerships as well as their economic and diplomatic influence. These first three alliances can be thought of as "alliances to build on," because of their significant opportunity for multilateral frameworks with the United States and will be discussed in this next section. The other two—the Philippines and Thailand—are important for U.S. basing and access, but more challenged by these countries' internal politics and relations with the PRC. This second group can be thought of as "alliances to maintain" and will be discussed in the following section looking at other critical relationships.

Treaty Alliances to Build On

Australia

The U.S.–Australian relationship is a partnership between liberal democracies with a multitude of overlapping interests cemented in the blood of shared military commitments spanning more than a century. This multifaceted and extremely close alliance benefits both countries enormously. Australia's military and intelligence forces are highly integrated with those of America, regularly practicing combined operations with the United States in both peacetime and war, and the two nations work closely together in the international diplomatic arena. Australians have participated alongside Americans in virtually every major U.S. conflict since (and including) World War I. Meanwhile, Australia and its small English-speaking population of 25 million spread around the edges of an American-sized continent benefit from the protection of the U.S. security umbrella.

Australia is a member of two of the emerging multilateral relationships in the region—the Quadrilateral Security Dialogue ("the Quad") between the United States, Australia, Japan, and India; and the AUKUS nuclear submarine technology-sharing agreement between the United States, Australia, and United Kingdom. In January 2022, Australia and Japan signed a Reciprocal Access Agreement to facilitate joint military training in each other's territory, logistical integration to support military exercises, and continued dialogue.[12] This was the first Japanese defense pact with any nation other than the United States since World War II, reflecting the strength of the evolving trilateral relationship.[13]

Other than its status as a fully integrated U.S. security and intelligence partner, Australia provides a major opportunity as a force multiplier for U.S. efforts. Since the majority of their interests closely overlap with those of the United States, Australians are able to go places due to their lower profile and engage other countries successfully on behalf of these shared interests when American activity would draw a greater response from China. This is only the first of several countries offering the potential to assist the United States in advancing security integration and strengthening the liberal regional order.

Australia's decision to pursue the development of advanced nuclear attack submarines capable of projecting critically needed military capability into flashpoints such as the South China Sea and Straits of Taiwan shows how seriously it takes the threat of China. The fact that the United States and United Kingdom were willing to share such closely held technology with the Australians shows the trust that is placed in this strategic partner. Finally, Australia's unique position provides it access to the western Pacific on its east coast, to maritime Southeast Asia on its north coast, and the Indian Ocean on its west coast. This makes it an important basing location as well as a key ally to potentially flex military forces into the South China Sea or the Indian Ocean to threaten China's vital lines of supply running through the strategic chokepoint presented by the Strait of Malacca (see the discussion on Singapore later in the chapter).

Japan

The U.S.–Japan alliance is universally considered one of the key cornerstones of peace and security in the Indo-Pacific. Japan has been a vital base of operations for U.S. power projection in the region since the conclusion of World War II. It is an active player in development assistance and economic investment throughout the region, particularly in Southeast Asia. Japanese concern with the threat of a rapidly expanding and increasingly confrontational Chinese military in the East China Sea and in the vicinity of Taiwan in combination with the ballistic missile/weapons of mass destruction challenge from the Democratic People's Republic of Korea (North Korea) has led to an increased focus on security issues. Consequently, Japan is simultaneously pursuing increasing security coordination with the United States, bilateral and trilateral cooperation with Australia (and the United States), and participation in the Quadrilateral Security Dialogue with the United States, Australia, and India. Alongside these efforts, Japanese Self Defense Forces are in a period of transition and redefinition toward becoming a military more capable of playing a larger role in world affairs and of defending Japan's interests.[14]

While Japan has made significant strides in improving its security capabilities, it continues to face significant hurdles. Domestic public opinion and continued constitutional hurdles limit its ability to project military power. In addition, joint integration within the services of the Japanese Self Defense Forces remains a challenge.[15] Japan's harsh early twentieth-century occupation of Korea and the brutal legacy of World War II provide significant challenges to its relationship with the Republic of Korea. Despite the fact that both countries are liberal democracies with close trade relationships, and the fact that each have integrated defense plans with the tens of thousands of U.S. military forces permanently stationed on their territory, strong nationalism constrains the political leaders of both countries from rapprochement, and close security cooperation has long remained out of reach.[16]

Japan is increasingly becoming a much more active player in the region. Its trilateral security relationship with the United States and Australia offers the opportunity for further development. Moreover, Japan has significantly evolved its stance on Taiwan—in July 2021, Japanese Deputy Prime Minister Taro stated that Japan would come to the aid of Taiwan if it were invaded by China.[17] This likely reflects Tokyo's view that any conflict in and around Taiwan would impact Japan's nearby southern islands as well as a more strategic concern with being outflanked by the PRC from the south.

Finally, Japan maintains extensive economic and development relationships in Southeast Asia and is beginning to enter the realm of security cooperation. The Japanese-led Asian Development Bank, established in 1966, committed some $22.8 billion to the Pacific region in 2021 and predates the Chinese-led Asian Infrastructure Investment Bank by almost fifty years.[18] Japanese direct infrastructure investment in Southeast Asia in 2019 exceeded that of the PRC by over $100 billion ($367 billion for Japan compared to $208 billion for the PRC). Almost two-thirds ($208 billion) of this went to Vietnam, but the rest was spread across maritime and mainland states.[19] Additionally, Japan has been a security assistance provider to the Philippine Coast Guard, providing significant numbers of critically needed small craft and surveillance radars along with training and exercises to enable their use.[20] Together, Japan's security advances and economic development assistance are becoming major advantages to leverage in expanding U.S. security relationships in the Indo-Pacific.

The Republic of Korea

As with Australia, the U.S.ROK alliance is built on shared sacrifice. U.S.-led intervention saved the ROK in the Korean War, and Americans have stood firm alongside their South Korean allies against the threat from the North for more than a half-century since. Bolstered by this protective shield, the ROK has transformed itself in the ensuring decades into a booming economic giant that today is a technology leader and liberal democracy much like its Japanese neighbors. Due to the continuing threat from the North, U.S. and ROK military forces on the peninsula today have a tightly integrated defense and intelligence relationship involving a wartime combined command relationship, shared planning staffs, and frequent large-scale combined exercises.[21]

Although the U.S.–ROK alliance is robust, it is singularly focused on the DPRK, rather than as a deterrent against the PRC. The PRC is an increasing challenge in the Indo-Pacific. But the long-running aggressive posturing and saber rattling of the Democratic People's Republic of Korea's (DPRK or North Korea) Kim family dynasty with their large conventional military forces and nuclear arsenal is an ever-present security threat cementing the alliance of their southern neighbor, the Republic of Korea (South Korea), with the United States. Thus, the U.S.–ROK relationship provides South Korea with a security guarantee and the United States with excellent basing and access to the mainland of North Asia. Nevertheless, this access appears to have limits, and South Korea's leaders have learned the tremendous economic cost of incurring China's wrath.

The 2016–2017 deployment of the U.S. Terminal High Altitude Area Defense (THAAD) ballistic missile defense capability to the ROK as a counter to the DPRK nuclear threat was met with immense political and economic coercion by the PRC against the ROK.[22] Although these systems were primarily designed to defend against the DPRK, China was extremely concerned that these systems' powerful X-band radar would allow the United States to see deep into the PRC. Such intelligence would threaten Chinese nuclear deterrence against the United States and otherwise constrain Chinese power in the region.[23] Although eventually lifted, the PRC's rapid imposition of stringent trade regulations and the restriction of Chinese tourists cost the ROK over $7.5 billion in economic losses—3% of South Korea's $219 billion trade with the PRC.[24] While these American air defense systems subsequently remained deployed in the ROK, this incident showed the vulnerability of the ROK to Chinese economic coercion. This will likely remain a factor in any potential attempts to enlist the ROK in security coalitions or other direct attempts that could be construed as attempts to counter China.

However, simply because direct opposition to the PRC is challenging doesn't mean that the ROK doesn't exert positive regional influence in its own way. One of the positive results of the 2017 THAAD crisis with the PRC was increased ROK efforts to diversify its market access, leading the ROK to increase political relations with the Association of Southeast Asian Nations (ASEAN) and India (see more on both of these later in the chapter). The ROK is an extensive economic and development player in Southeast Asia, with Vietnam (46.8% in 2019) and Singapore (31.7% in 2019) by far the largest recipients of ROK direct investment. The ROK also contributes about 25% of its international development assistance to Southeast Asian countries. This gives South Korea extensive influence in this subregion, particularly in two of Southeast Asia's most dynamic economies: Singapore and Vietnam.[25] Along with Japan, the ROK's extensive investments in the region are an important counter to Chinese influence.

Building Multilateral Structures around Existing U.S. Alliances

America's alliances with Australia, Japan, and the ROK offer significant opportunity for building multilateral structures. The rejuvenation of the Quadrilateral Security Dialogue ("the Quad") between the United States, India, Australia, and India in 2019 is the best place to begin this discussion. But before examining the Quad and potential "Quad-Plus" arrangements, it's first necessary to briefly examine India and the renewed interest of European countries such as the United Kingdom and France in the region.

India

The massive growth of China's power and influence in South Asia has prompted India to turn away from decades of nonalignment toward greater outreach to the United States and its allies. Disputed territory along India's northern border with Tibet has long been a source of friction between the PRC and India, resulting in a short war where Indian forces were quickly defeated by China in 1962. This boundary has seen increasingly tense standoffs from 2014 onward, including a 2020 clash resulting in the loss of lives on both sides.[26] India also faces an increasingly strong relationship between China and India's arch-rival, nuclear-armed Pakistan, which might eventually put a Chinese naval base in Gwadar, Pakistan on the Indian Ocean. Alongside of this direct confrontation on India's border with the PRC is increasing Chinese influence in states such as Nepal and Sri Lanka that were once firmly inside India's South Asian sphere of influence. The increasing power imbalance with an expanding China to the north has clearly changed India's strategic calculus away from its traditional path of nonalignment.[27]

To balance China, India has pursued an increasingly close relationship with the United States and its allies. India is a major customer for U.S. weapons. The annual Malabar naval exercises that it began with the

United States in 2014 have now expanded to become trilateral (with Japan) and then quadrilateral engagements (detailed further subsequently).[28] Perhaps most importantly, India is a key part of a renewed Quad along with Australia, Japan, and the United States.

Despite this, there are significant limits to the U.S.–Indian relationship. India still relies heavily on the Russian Federation for weapons supplies— a source of significant consternation for Washington. Leaders in New Delhi see no challenge in maintaining the long-running and beneficial relationship with Russia that it has maintained since early in the Cold War in parallel with its increasing partnership with the United States and its allies as India seeks to balance the PRC. But India's refusal to join Western-led sanctions against Russia during the 2022 Ukraine war and its decision to purchase highly discounted Russian oil (until then, India was not a major consumer of Russian oil) helped keep the Russian economy afloat, running counter to a major U.S. foreign policy objective.[29]

These decisions will likely serve as a warning of Indian coalition reliability for leaders in the other Quad countries in the future. The key point is that nonalignment remains an important part of India's strategic culture. India is clearly committed to strengthening its ties with the United States and other Quad members to balance China. But further strengthening deterrence will require greater moves to increase interoperability. Interoperability in the twenty-first century requires a high degree of command and control network integration, which in turn requires a great deal of trust between partners and allies. It remains to be seen if India is willing to make the kind of commitments necessary for its partners to extend that level of trust.

Finally, India's tense relationship with Pakistan is another challenge. Any gains in the U.S.–Indian relationship steadily push Pakistan closer to China. Other Quad members should question whether an alliance or deepened security cooperation with India would potentially risk the United States and/or its allies and other partners becoming embroiled in a future Indian–Pakistani conflict, particularly considering the possibility that such a conflict could go nuclear.

European Powers with Indo-Pacific Interests

India is not the only new player on the strategic scene in the Indo-Pacific. After decades where their attention was turned away from the region and where China was seen largely as an economic partner and less as a threat, the former colonial powers of Europe—in particular the United Kingdom and France and to a lesser degree Germany and the Netherlands—are once again taking a more active role in the region.

A post-Brexit United Kingdom is looking for a role on the world stage to maintain its status as an important foreign policy actor and to justify the United Kingdom's relevance in its special relationship with the United States. It demonstrated this in 2021 with British commitment to the AUKUS agreement to develop Australian nuclear submarine capabilities in partnership with the United States.[30] This was soon followed by the extended inaugural deployment of the United Kingdom's new Queen Elizabeth carrier strike group (with a U.S. Marine F-35 squadron on board) to the region and multiple combined exercises with the United States, Japan, Singapore, and other partners. Similarly, France retains possessions and 1.5 million citizens in the South Pacific, which it secures with a permanent naval, ground, and air presence as well as deployments to the region such as sending the Charles de Gaulle carrier strike group into the Pacific in 2019 and Indian Ocean in 2021.[31] Finally, in 2021 Germany sent a naval expedition to the South China Sea, and the Netherlands sent a destroyer to the region along with the British carrier task force.[32]

These are positive developments, and such activities should be encouraged by American leaders. All of this increased activity sends a message to China that the states of the liberal world order are concerned with the perception of increasingly aggressive Chinese military activity. However, the experience of SEATO shows the potential danger of bringing out-of-area players without significant interests in the region into important multilateral security structures. Any increased activity is positive for signaling, but today probably only the United Kingdom and France have enough significant interests in the region that offer opportunities for deeper multilateral partnerships.

The Quad

The Quadrilateral Dialogue first emerged as the brainchild of Japanese Prime Minister Shinzo Abe for a network of partnered states across Asia. Leaders from Japan, Australia, India and the United States came together for the first Quadrilateral Security Dialogue on the sidelines of the ASEAN Regional Forum in May 2007. Initial discussions focused on areas of common interest such as disaster relief, but soon expanded into the Malabar 07-02 exercise with the four countries' navies joined by Singapore in multicarrier sea control exercises in the Bay of Bengal. The Quad immediately faced heavy opposition in the form of PRC protests, and with the election of new Japanese and Australian prime ministers, it fell apart at the end of the year. At the time the four countries lacked a consensus on the primary threats to the region or means of addressing them.[33]

More recently, however, after a major growth in Chinese power, the leaders of the four nations resumed the meetings of what has sometimes become known as Quad 2.0 in November 2017 at another ASEAN summit in Manila. This new grouping—which has again been vigorously protested by China as a containment attempt—has been much more cohesive. The restored Quad has also benefited from a growing convergence of interests and an increasing concern with China shared by its members.[34]

The Quad is not a military alliance. It is still largely exactly what its name says—a largely political dialogue mechanism between countries with shared interests. Nevertheless, it has been a vehicle for significantly increased multilateral security cooperation. Such regularized cooperation, particularly when this involves increasingly complex military exercises, serves multiple purposes:

1. signaling increased commitment between participants, and beyond this on a military level;
2. building increasing trust between regular participants over time; and
3. regularizing interaction leads to increased interoperability—a critical requirement for effectiveness in the event of future combined operations, particularly high-end combat operations.

The Quad's Malabar naval exercises bring major surface units and air components together in rotating locations that have included the Bay of Bengal, Guam, and off the coast of Japan. What began as an annual bilateral exercise between the United States and India in 2006 was briefly expanded to a quadrilateral exercise (plus Singapore) in 2007. It began a more sustained expansion with the inclusion of Japan in 2014 and Australia in 2020. Such complex multilateral exercises are critical to improving interoperability.[35] The group's potential to balance China appears to be welcomed by many Southeast Asian elites.[36]

The Quad is certainly not an Asian NATO—and the previous discussion about India and Japan shows the challenges to ever becoming one. However, that hasn't prevented the Chinese government from labeling it as such. In October 2020, PRC Foreign Minister Li called the Quad an "Indo-Pacific NATO" stemming from an "old-fashioned Cold War mentality."[37] From 2020 onward, Chinese leadership statements and other strategic communications demonstrated that the PRC is clearly concerned that the expanded relationship between four of the Indo-Pacific's most powerful democracies could become the foundation for a future regional or potentially global anti-China coalition.[38]

Although the Quad's present contribution is principally as a signal of aligned interests, it has the potential to be evolved into something stronger. The original "1.0" version of the Quad formed as a shared commitment to humanitarian aid and disaster relief. Future Quad efforts in these perennial requirements in the Indo-Pacific is a good place to start. Efforts to coordinate critically needed infrastructure and economic development activities in Southeast Asia and around the region is another way to increase trust while signaling shared commitment without running into major divergence between the four partners. Alongside these easier efforts, a steady increase in the quantity, size, frequency, and complexity of quadrilateral military exercises, both with the four Quad nations and with other regional partners, is a must for developing the Quad into a viable deterrent. Given India's caution with formal military commitments, this expansion of security cooperation will likely need to proceed slowly to increase trust while building increased interoperability.

"Quad Plus" Dialogue and Security Cooperation Mechanisms

As Quad 2.0 has evolved into a slowly strengthening partnership, the concept of expanding into "Quad-Plus" coalitions has come into increasing use, paralleling Shinzo Abe's original intent for a larger network of nations to balance a rising China.[39] Leveraging the Quad relationship as a building block to tie together like-minded countries with shared interests provides a valuable opportunity to achieve the goal of integrated deterrence in the region.

Some argue for one single Quad Plus framework.[40] However, while that might be important for the sort of robust framework that could eventually be expanded into a potential NATO-like alliance structure, in the interim a series of Quad Plus coalitions may be more easily achievable. While less robust, these more limited partnerships may be more acceptable to countries that are willing to explore some level of relationship but not willing to incur the wrath of China that a more tightly integrated partnership with the United States would bring.

The United States has already begun such Quad Plus dialogues aimed at one group of countries. In March 2020, U.S. Secretary of State Pompeo began a series of weekly conversations with his foreign secretary counterparts from the Quad nations plus the ROK, New Zealand, and Vietnam. Beginning with attempts to deal with the COVID-19 pandemic, the talks expanded to trade facilitation, technology transfer, and migration.[41] This follows the point raised previously—humanitarian assistance and disaster relief are universally shared interests and easily lead to the discussion of mutually advantageous economic topics.

The 2020 Quad Plus talks provided the opportunity to tie in three critical partners in the region. As discussed previously, the ROK is a treaty ally which has nevertheless been reluctant to develop a trilateral U.S.–Japan–ROK partnership due to the historic and contemporary challenges between it and Japan. But bringing the two into a larger grouping on topics of shared interests allows these barriers to be overcome. Continuing such engagement offers the potential to slowly improve trust over time and help to build a more permanent bridge between these two critical (and neighboring) allies. Similarly, New Zealand is an important partner in the expanding competition in Oceania while Vietnam is increasingly becoming a critical partnership in the South China Sea region. Both of these partners will be discussed subsequently, but embedding them in regularly occurring dialogue mechanisms provides the opportunity to further coordinate their actions.

Some might question whether expanding the dialogue away from issues directly related to China waters down the deterrent value of the organization as a signal demonstrating united resolve to counter increasing Chinese assertiveness in the Indo-Pacific. But this misses the point. Linkages are crucial in building international partnerships. Regularized interaction in one foreign policy area has long been recognized to build trust useful for advancing partnerships in other areas.

While South Korea and others may share many similar interests with the Quad countries for issues such as disaster relief or economic development, they may be averse to signing up for an organization with a clear anti-China narrative—certainly in the short term.[42] Others, such as Vietnam, may be crucial partners in the event of a conflict but require an extended build-up of the relationship because of the immense potential repercussions of incurring conflict (whether kinetic or nonkinetic).

Issues that are "pro-Asia," that is, which clearly benefit Indo-Pacific countries, build important trust that can later be leveraged to other areas should Chinese assertiveness continue to increase and threaten those countries. A pro-Asia narrative could be very useful for bringing states such as Vietnam into a strengthened Quad Plus dialogue that could be leveraged in the future. And bringing in a key Southeast Asian nation such as Vietnam also threatens China's narrative that the Quad is a group of "extra-regional powers" seeking to contain China.[43]

But envisioning Quad Plus solutions should not be limited only to states in the Indo-Pacific region. Both the United Kingdom and France have strong interests in maintaining the liberal democratic order and have demonstrated a renewed commitment to the region. By joining in the AUKUS agreement, the United Kingdom demonstrated that it was willing to commit some of its most closely held secrets to help improve the capabilities of its Australian partners. This is not something done lightly and has been backed up by an uptick in extended deployments of British military force to the region. Similarly, France has been quite clear about its desire to preserve the current world order.[44]

The United Kingdom and France should also be brought into regular combined dialogue with the Quad countries as well as security cooperation efforts, particularly military exercises. They may not fit directly into the detailed dialogue that is possible to address regional internal matters with countries such as Vietnam and the Philippines. However, they could easily fit into a Quad Plus partnership focused on slightly different issues than the dialogue with South Korea, Vietnam, and New Zealand mentioned previously. A Quad Plus with the United Kingdom and France could focus more on hard-power coordination in military exercises and other areas of security cooperation as well as on soft-power issues such as the coordination of disaster relief.

Despite their promise, the Quad and Quad Plus solutions clearly have limits in terms of becoming a full military alliance, at least in the short to mid-term future. However, this doesn't mean that there aren't possibilities built around three quarters of the Quad.

Trilateral Defense Alliance—United States, Australia, Japan

One of America's major Indo-Pacific goals as it heads into the second quarter of the twenty-first century should be to expand its existing bilateral defense alliances with Japan and Australia into a trilateral alliance. Achieving a truly integrated deterrence to contain or balance an increasingly competent Chinese military requires a core alliance at its heart—not a series of bilateral ones. People's Liberation Army (PLA) leaders need to see that they face a determined and capable opposition in the event of conflict—modern high-end combat operations require a high level of integration and interoperability that isn't gained by simply exercising annually.

Achieving this alliance will provide two firm democratic anchors running from Northeast Asia through Southeast Asia. It would not only provide a capability for potential future combat operations but also help to further increase unity of effort across all the instruments of power (diplomatic, informational, military, and economic) across the entire region. Both of these countries are already strong actors in the region whose interests closely overlap with the United States. Both also exercise extensively with the United States, have increasingly begun to engage in trilateral exercises, and as mentioned previously, Japan and Australia have now begun to expand the systems to enable a closer bilateral security cooperation relationship of their own.

Achieving this will require Japan to continue its present trajectory slowly overcoming entrenched domestic constraints, including its Constitution. However, it's a virtual certainty that the continued expansion of PRC military power will continue, with more of the increasingly aggressive Chinese behavior in the East China Sea and Taiwan Straits that have already spurred the ongoing changes in Japanese strategic culture. Many of the political and economic measures that will help build the trust necessary for such an endeavor have already begun, particularly with the Quad's second incarnation. Expanding the size and frequency of trilateral exercises will lay the foundation for potential trilateral operational deployments in critical areas such as the South and East China Seas as well as in vicinity of Taiwan. While this will take time, it is potentially an achievable goal and one worth expending significant effort to achieve.

Trilateral Alliance Plus—United States, Australia, Japan, UK . . . France?

A trilateral U.S.–Australian–Japanese alliance could soon be tied in with other U.S. military allies, particularly the United Kingdom and later potentially France. As mentioned above, the United Kingdom has already committed itself to deepening its Indo-Pacific partnership with Australia and the United States through the AUKUS agreement. While the three countries have long fought together on battlefields largely outside the region (since World War II), this new commitment is clearly aimed at the Indo-Pacific. While AUKUS is largely a security assistance (technology development, training, and potentially weapons sales) agreement, the ties to exercises and combined operations is a natural expansion, particularly with the United Kingdom again clearly willing to commit major combat forces to the region. Adding Japan to the existing U.S.–UK–Australian partnership (absent the release of highly sensitive nuclear technology) should be a fairly easy lift and could potentially be expanded into a formal alliance structure. With the United Kingdom on board, gaining French participation is also a potentially achievable outcome and one worth pursuing due to their largely shared interests.

An expanded "trilateral plus" alliance might be seen as sufficiently capable of balancing an increasingly capable China and could help buoy more vulnerable U.S. allies, such as the Philippines, and other potential key partners, such as Vietnam. However, alongside efforts to develop a multilateral alliance network, the United States needs to continue to focus its efforts to develop a web of strengthened and expanded partnerships, whether those are tied into an existing multilateral alliance network (ideally) or not.

Another Trilateral Arrangement—United States, Japan, ROK

Despite the significant barriers to Japanese–South Korean security cooperation, the continuing threat of North Korean ballistic missiles (potentially tipped with nuclear weapons) is an enormous shared survival interest that holds the possibility to an improved relationship. On June 29, 2022, U.S. president Biden and the new leaders of Japan and the ROK—Fumio Kishida and Yoon Suk-yeol—met alongside a NATO summit in Europe and agreed to renew trilateral missile warning and tracking exercises and to explore the potential for further combined coordination efforts and exercises between the three countries.[45] Although only an early step, such structures and processes offer the potential to bridge the nationalistic barrier between Japan and South Korea—countries that otherwise share many interests.

The shared partnership with the United States offers the means to begin bridging this gap. Although any such security structures stemming from such cooperation would likely be limited to facing North Korea due to the challenges discussed previously, they provide the building block for further improved cooperation and potential security integration.

A WEB OF PARTNERS: ASEAN AND OTHER CRITICAL RELATIONSHIPS

While many larger powers in the Indo-Pacific may be more resistant to China, others are much more vulnerable to Chinese economic or military coercion and require a more nuanced approach in today's highly competitive world. This is particularly true in Southeast Asia, which is the most dynamic Indo-Pacific area of competition between China and U.S. allies and partners. Many of this crucial subregion's countries remain "at play" and are heavily involved in efforts to balance the United States and China. Although some of them—such as Thailand and the Philippines—are even treaty allies, the power imbalance with the PRC creates many more significant challenges that aren't faced by the larger and more powerful countries described above.

Southeast Asia is highly strategic because of the critical South China Sea shipping routes and the critical chokepoints through which transit almost two-thirds of global maritime trade—crucial lifeblood for the economies of the PRC, Japan, South Korea, and Taiwan and thus to the world at large.[46] In addition to Southeast Asia, the PRC has more recently begun to significantly increase its influence in the Southwest Pacific, and the United States' long-running partnership with New Zealand will be a critical factor in balancing Chinese efforts in this less critical but still important subregion.

Managing relations with these states will require pragmatic and concerted efforts by the United States in concert with the allies and partners discussed previously. Building a stronger web of partnerships will require careful management and attention to relationships with the Association of Southeast Asian Nations, several key individual states in Southeast Asia, and New Zealand in Oceania.

The Association of Southeast Asian Nations (ASEAN)

Although not an alliance, ASEAN is a critical player in Southeast Asian nations' relations with each other and their larger regional and international allies. The organization is composed of ten member states (Brunei, Cambodia, Indonesia, Laos, Malaysia, Myanmar, Philippines, Singapore, Thailand, and Vietnam) and bridges enormous disparities in government types and numerous regional disputes.[47] ASEAN's processes recognize and are designed to bridge these significant differences, making the organization a critical vehicle for dialogue. Its consensus-based decision-making process makes it is slow to act, but this also prevents hasty actions and makes it harder for outside players to dominate the organization, as they would have to secure the agreement of every player.

Great power influence over the organization's individual members exists on a spectrum. Although some ASEAN countries (such as Laos and Cambodia) are heavily influenced by China, others (such as Vietnam, Thailand and the Philippines) lean more toward the United States. However, for the majority, including most of the states mentioned previously, foreign policy consists of a careful balancing act between a strong and aggressive Chinese neighbor and a largely supportive but geographically distant and sometimes distracted United States.[48] *The impact of balancing these relationships will be explored further in the next chapter examining China's "Maritime Silk Road" and the struggle for influence.* But in terms of this chapter's focus on alliances and partnerships, ASEAN is an extremely important player as its unifying features provide a buffer for regional states against outside influence.

For the United States, engaging with either ASEAN as an organization or individually with its member-states is best done in concert with the players mentioned in the first section above—particularly with Australia, Japan, South Korea, and India. All of these states have strong involvement in Southeast Asia and together this provides a supportive buffer against an increasingly aggressive China. Engaging ASEAN will always be somewhat frustrating and may often seem like little is being achieved. Nevertheless, such engagement is important as it provides a clear signal of an alternative to an ever-present PRC. Many of the overall interests of most regional states—particularly for stability and economic growth—are largely parallel with those of the United States and its partners, offering important ground to build on. ASEAN's processes are deliberately slow, but sustaining its multilateral institutions provides not only a balance against Chinese influence across all of its members, but an inroad to key states in this strategic subregion.

Critical Individual Relationships

Treaty Alliances to Maintain—The Philippines and Thailand

Although they have many important individual differences, America's remaining two alliance partners in the Indo-Pacific—one in mainland (Thailand) and one in maritime (the Philippines) Southeast Asia—also share crucial similarities in terms of the state of these alliances. Most importantly, both are important strategic partners offering critically needed basing and access to the region. However, while the U.S. alliance with each state provides an important factor in the relationship, domestic political complications combined with their power imbalance and close proximity to the PRC creates a situation rendering both vulnerable to a geopolitical struggle for influence.

For both states, domestic politics has been an issue. Military intervention in Thailand's government has brought diplomatic pressure and legal restraints on U.S. security cooperation. Similarly, human rights abuses by the Philippines' Duterte government (2016–2022) brought U.S. diplomatic pressure for change. In both cases a Chinese government unconcerned by democratic backsliding was able to take advantage of the situation and gain political influence.[49] All of this occurred alongside the extensive increase of Chinese economic influence over both countries—giving the PRC far greater economic leverage than the United States.[50]

In America's favor, however, the long-running United States security cooperation relationships and frequent exercises provide important counterbalances. Throughout the region, the U.S. military is an important balancing factor against Chinese economic strength. This is particularly true with these two treaty allies, although in the case of the Philippines, the United States security alliance provides a much stronger benefit. America's close military and intelligence support has been a key factor in the Philippines long-running struggle with Islamist insurgents. Additionally, America's defensive alliance is a vitally important factor buoying the Philippines as it struggles with China to enforce its sovereignty over disputed territory in the South China Sea.

Moving forward, security cooperation will again be critical to retaining a close relationship with these two countries. Here again, the United States can combine its efforts with Japan and Australia, who also have close security cooperation relationships with the Philippines. America's goal in both of these relationships should generally be to maintain the status quo against strong Chinese pressure and make small gains in terms of basing and logistics access where possible. Access to the airbases and harbors of the Philippines would be critical in the event of a regional conflict and the access to similar facilities in Thailand would also be important. Their ability to serve as critical logistics nodes to support the forward deployed forces of the United States and other allies and partners would be crucial to success and thus these relationships deserve concerted and combined engagement efforts.

Vietnam

The United States relationship with Vietnam is one of the fastest growing and most critical Indo-Pacific relationships in recent years—certainly in terms of nonalliance members. As China has increased its efforts to expand its control over the South China Sea, this fiercely independent South China Sea claimant has made efforts to carefully push back against the PRC and to pragmatically reach out to other states to balance the PRC.

Vietnam is the last country to fight a war with the PRC (the month-long Sino-Vietnamese conflict in 1979) and harbors strong nationalistic anti-Chinese resentment. While the United States fought an extended conflict with Vietnam from the 1960s and 1970s during its consolidation into one single country, Vietnam's thousand-year struggle for independence from China has now taken on a greater priority for the country. If pursued carefully, this relationship provides a vital opportunity for the United States and its allies and partners to leverage this shared interest of balancing Chinese and develop a strong partnership.

Vietnam is a leader in efforts to counter China in ASEAN, and it has experienced increasing friction with China over disputed claims in the South China Sea—most recently over the resource-rich waters around Vanguard Bank.[51] However, despite its outreach to the United States in a carefully managed attempt to balance China, this does not mean a strong U.S. partnership is a certain outcome. Vietnam's ruling communist party—particularly its older members—still retains affinity with its Chinese counterpart. Vietnam continues its long-running history of weapons purchases from Russia, a hold-over from its strong previous relationship with the Soviet Union. Finally, the newly enhanced U.S.–Vietnam relationship is beginning from an entry level. In Vietnamese diplomatic parlance, the relationship with the United States is still only a "comprehensive partnership"—below that of Myanmar ("comprehensive cooperative partnership"), and Vietnam's sixteen "strategic partnerships" which include China, Russia, India, Japan, South Korea, and Australia. Despite this diplomatic fig-leaf, however, which has likely been followed to avoid the wrath of China, the U.S. relationship has become increasingly vital to Vietnam.[52]

Given Vietnam's close proximity to China and the complexities of negotiating a partnership with a communist former enemy, engaging with Vietnam will require careful, sustained, and pragmatic efforts. A critical factor in improving the relationship will be an integrated approach with U.S. allies in partners. As noted above, Japan and South Korean investment in Vietnam exceeds that of China and India and Australia also have strong partnerships that can be leveraged. Vietnam's agreement to explore potential Quad-Plus partnerships alongside South Korea and New Zealand have already been mentioned previous and demonstrate that Vietnam is receptive to at least limited multilateral engagement, with the potential for more to come.

Given its strategic importance, a subsequent chapter will provide a case study exploring the U.S.–Vietnam relationship in depth and assessing realistic opportunities—and limits—to expanding this crucial partnership.

Singapore

With its critically strategic location at the eastern entrance to the Strait of Malacca, the city-state of Singapore is a global transshipment hub whose anchorages, economic wealth, and stability are critical to both the United States and China. The Strait of Malacca is a critical chokepoint between the South China Sea and Indian Ocean relied on by the PRC, Taiwan, ROK, and Japan for 75–90% of their annual energy shipments.[53] Loss of access would be immensely damaging for any of these countries. Unless the PRC's Belt and Road Initiative is able to secure overland routes through central Asia, this is unlikely to change, and any BRI successes in building new land routes will likely only minimally impact this trade.

Singapore has a long-running security cooperation relationship with the United States but it has more recently chosen to balance this by affording equal access to the People's Liberation Army. The city-state is home to the U.S. logistical command unit coordinating warship deployments and logistics throughout the Western Pacific, is a base for U.S. littoral combat ships and naval surveillance aircraft, and Singapore's Changi Naval base (like Yokosuka, Japan) is one of the few bases capable of handling U.S. aircraft carriers.[54] Despite this, however, Singapore has carefully tried to maintain equal relations with the PRC, with which it has signed agreements for joint military training, visiting forces, and ministerial dialogue.[55]

Maintaining access to Singapore as a vital U.S. logistics node is highly important for the twenty-first century and will require careful management and effort. As a former British colony, Singapore has a long-running security cooperation with the United Kingdom. as well as with Australia. Additionally, its location next to the Malacca Straits also makes it vitally important to India. The United States should work closely with these countries to maintain a close relationship with Singapore. America's goal should be to maintain access to this logistics hub even in the event of a future conflict with China. This will require leveraging its partners to emphasize their many shared interests with Singapore.

New Zealand

In addition to the Southeast Asian relationships mentioned previous, the United States will need to expand its partnership with one final country to counter China's increasing influence in Oceania—the far-flung islands of the southwest and central Pacific spanning the lines of communication between Hawaii and Australia. Although small and geographically isolated, New Zealand brings a long history of working with the United States and Australia and will be an important force multiplier in countering Chinese influence in the region. Since the United States does not have major resources to expend in restoring its traditional influence in the region, it will need to leverage partners such as New Zealand to successfully balance China's increasing efforts to become the dominant influence in the subregion.[56]

Although New Zealand today maintains a strong partnership with the United States, this has not always been the case. In 1951, Australia and newly independent New Zealand signed the ANZUS treaty with the United States, formally replacing these countries' previous security relationship with the United Kingdom.[57] Australia remains an ANZUS member today. In 1987, however, New Zealand's antinuclear stance brought its government to ban nuclear weapons carrying vessels from its waters and the United States suspended its alliance commitments to New Zealand. Following New Zealand's military contributions to the wars in Iraq and Afghanistan alongside Australia, bilateral relations were resumed in 2010 and today the two countries maintain a strong security cooperation relationship.[58] Another sign of the strong trust that New Zealand has regained is its membership in the "Five Eyes" intelligence-sharing relationship with the United States, Australia, Canada, and United Kingdom.

While a close partner with the United States and Australia, New Zealand's small size renders it vulnerable to Chinese economic coercion. As with the ROK, New Zealand has been subject to both threats and the actual imposition of trade and tourism restrictions, making challenges to the PRC potentially problematic for this small nation.[59] In general it thus seeks inclusiveness and dialogue with China rather than confrontation due to concern that irritating China could threaten New Zealand's beneficial trade relationship with the PRC.[60]

New Zealand's most critical contribution as a U.S. partner comes from its unique access to (and location in) Oceania. Despite its small size, New Zealand maintains significant influence in the islands of the southwest Pacific and regularly conducts military exercises with these states. While this area is largely an "economy of force" effort for the United States, it is New Zealand's primary operating area. With the PRC's signing of a security and basing agreement with the Solomon Islands in March 2022 (allowing the basing of Chinese police, ground forces, and warships) alongside simultaneous efforts to secure an airbase in Kiribati, efforts to counter or contain increasing Chinese activity have now become critical.[61]

Additionally, New Zealand has already been part of efforts to establish a Quad Plus relationship along with South Korea and Vietnam, as mentioned previously. As with the ROK, the United States and its allies need to work to lessen New Zealand's economic dependence on China for key elements of its economy. Providing options for New Zealand to diversify its trade markets to Quad Plus markets will be important to building its resiliency.

New Zealand's long and close partnership with Australia will also be critical to leveraging its assistance. The two have long worked closely together and given their shared interests in Oceania, the United States should seek the combined assistance of both of these partners in a combined diplomatic and security cooperation effort to avoid ceding influence with these states to China. The nations of Oceania would likely be highly receptive to such increased outreach given their long history of interactions with the United States and its partners. The United States simply can't be everywhere, but by enlisting partners such as New Zealand to lead these efforts and then supporting them in combination with Australia, it can still achieve its interests with a much smaller expenditure of resources.

CONCLUSION

As the twenty-first century continues, the post–World War II hub-and-spoke system of alliance and partner relationships that the United States has long maintained in the Indo-Pacific will be insufficient to maintain U.S. interests in this vital region in competition against an increasingly powerful and confident People's Republic of China. Yet America's strong network of relationships is still a key competitive advantage over the PRC and a strength that can be further built upon. To sustain its competitive advantage and achieve integrated deterrence, the United States needs to expand and deepen these relationships into a many-layered web of multilateral alliances, multilateral relationships, and individual partnerships. As discussed previously, these different relationships will be mutually supporting and should thus be pursued simultaneously by the United States and key allies and partners.

The region's previous SEATO experience demonstrates the importance of building multilateral relationships and individual partnerships around shared interests—an important lesson to remember as these expanded relationships are pursued. The countries of the renewed Quad and several Quad Plus groupings should be key focus areas. Within these are key force multipliers—countries like Australia, Japan, the ROK, and India which the United States can lean on as partners to pursue their own coordinated efforts aimed at deepening this web of relationships supporting a liberal world order. In doing so, the United States should be pragmatic and seek to initiate these efforts from a pro-Asia narrative emphasizing shared gains from the partnership rather than a more anti-China narrative that might force states to choose sides before they are more thoroughly supported and integrated.

Alongside less stringent relationship building, the United States should seek to evolve its existing bilateral defense alliances to develop more capable and resilient multilateral alliances. Formalizing a truly integrated structure with Japan and Australia should be a major U.S. foreign policy goal, followed soon by the United Kingdom and potentially later by France and even South Korea. Such a core alliance at the heart of this web of relationships will be critical in providing stronger and more integrated deterrence against the PRC. In the event of conflict with China this sort of tighter, integrated, and closely networked core will be necessary to fight and win in high-end combat operations. Additionally, it provides the central framework upon which other partners can be tied to build coalitions of the willing depending on the nature of the particular future crisis situation.

America's strategic planners know that existing bilateral relationships will not be enough to fight and win against a great power adversary like the PRC. Hopefully the layered alliance and partner web described above will be enough to deter such a conflict. If not, however, it will be a critical component in assuring that a desired outcome is achieved.

NOTES

1. The White House, "Remarks by President Obama to the Australian Parliament," November 17, 2011, https://obamawhitehouse.archives.gov/the-press-office/2011/11/17/remarks-president-obama-australian-parliament.

2. U.S. Government, Office of the Secretary of Defense, "Developments Involving the People's Republic of China 2021: Annual Report to Congress," 36.

3. U.S. Department of Defense, "Secretary of Defense Austin Remarks at the Global Emerging Technology Summit of the National Security Commission on Artificial Intelligence (As Delivered)," July 13, 2021, Secretary of Defense Lloyd J. Austin III, https://www.defense.gov/News/Transcripts/Transcript/Article/2692943/secretary-of-defense-austin-remarks-at-the-global-emerging-technology-summit-of/.

4. Victor Cha, "Powerplay: Origins of the U.S. Alliance System in Asia," *International Security* 34, 3 (2009), 158–159; Christopher Hemmer and Peter Katzenstein, "Why Is There No NATO in Asia? Collective Identity, Regionalism and the Origins of Multilateralism," *International Organizations* 56, 3 (Summer 2002), 588.

5. Cha, 158–159.

6. Hemmer and Katzenstein, 580.

7. Cha, 158–159, 161.

8. Hemmer and Katzenstein, 579.

9. Aaron Bartnick, "Asia Whole and Free? Assessing the Viability and Practicality of a Pacific NATO," Harvard Kennedy School: Belfer Center for Science and International Affairs Paper, March 2020, 8–10, https://www.belfercenter.org/index.php/publication/asia-whole-and-free-assessing-viability-and-practicality-pacific-nato.

10. Leszek Buszynski, "SEATO: Why It Survived until 1977 and Why It Was Abolished," *Journal of Southeast Asian Studies* 12, 2 (September 1981), 287–288.

11. *Ibid.*, 290–291, 296.

12. Thomas Wilkins, "Another Piece in the Jigsaw: Australia and Japan Sign Long-Awaited Reciprocal Access Agreement," Australian Institute of International Affairs, January 20, 2022, https://www.internationalaffairs.org.au/australianoutlook/another-piece-jigsaw-australia-japan-sign-long-awaited-reciprocal-access-agreement/.

13. Haruka Nuga and Steve McMorran, "Australia, Japan Sign Defense Pact as China Concerns Loom," *The Diplomat*, January 7, 2022, https://thediplomat.com/2022/01/australia-japan-sign-defense-pact-as-china-concerns-loom/.

14. U.S. Department of State, "US Security Cooperation with Japan," January 20, 2021, https://www.state.gov/u-s-security-cooperation-with-japan/.

15. John Wright, "Solving Japan's Joint Operations Problem," *The Diplomat*, January 31, 2018, https://thediplomat.com/2018/01/solving-japans-joint-operations-problem/#:~:text=Solving%20Japan's%20Joint%20Operations%20Problem%20Japan%20needs%20its,Jan.%2011%2C%202018%2C%20at%20Kadena%20Air%20Base%2C%20Japan.

16. The National Bureau of Asian Research, "Interview with Jennifer Lind: The Next Steps for U.S.-ROK-Japan Trilateralism," September 4, 2020, https://www.nbr.org/publication/the-next-steps-for-u-s-rok-japan-trilateralism/.

17. Bertil Lintner, "How Far Would Japan Really Go to Defend Taiwan," *Asia Times*, July 19, 2021, https://asiatimes.com/2021/07/how-far-would-japan-really-go-to-defend-taiwan/.

18. Asian Development Bank, "Who We Are," accessed July 3, 2022, https://www.adb.org/who-we-are/about.

19. Alongside $208 billion in infrastructure investment in Japan, Japan also invested $74 billion in Indonesia, $43 billion in the Philippines, $19 billion to Singapore, and $15 billion in Thailand in 2019. All figures are in U.S. dollar; Andy Brown, "Japan Investing US$367 Billion in Southeast Asia Infrastructure," *International Construction*, June 27, 2019, https://www.international-construction.com/news/Japan-investing-US-367-billion-in-Southeast-Asia-infrastructure/1139032.article#:~:text=Japan's%20investment%20is%20heavily%20weighted%20in%20favour%20of,billion%20into%20Singapore%20and%20US%2415%20billion%20in%20Thailand.

20. Frances Mangosing, "Japan to Keep Aid Flowing to PH Coast Guard," *Inquirer.net*, December 7, 2021, https://globalnation.inquirer.net/200922/japan-to-keep-aid-flowing-to-ph-coast-guard.

21. Bryan Port, "Defense Readiness and the U.S.–ROK Alliance," Carnegie Endowment for International Peace, March 18, 2020, https://carnegieendowment.org/2020/03/18/defense-readiness-and-u.s.-rok-alliance-pub-81234.

22. Ethan Meick and Nargiza Salidjanova, "China's Response to U.S.-South Korean Missile Defense System Deployment and Its Implications," U.S. Economic and Security Review Commission, July 26, 2017, 8, https://www.uscc.gov/sites/default/files/Research/Report_China%27s%20Response%20to%20THAAD%20Deployment%20and%20its%20Implications.pdf.

23. Meick and Salidjanova, 5–8.

24. Victoria Kim, "When China and U.S. Spar, It's South Korea That Gets Punched," *Los Angeles Times*, November 20, 2020, ttps://www.latimes.com/world-nation/story/2020-11-19/south-korea-china-beijing-economy-thaad-missile-interceptor.

25. Oscar Petrewicz, "South Korea's Growing Economic Involvement in Southeast Asia," The Polish Institute of International Affairs, December 2, 2021, https://www.pism.pl/publications/South_Koreas_Growing_Economic_Involvement_in_Southeast_Asia#:~:text=To%20support%20its%20investment%20presence%20in%20the%20region%2C,the%2010%20largest%20recipients%20of%20South%20Korean%20ODA.

26. Tanvi Madan, "China Is Losing India: A Clash in the Himalayas Will Push New Delhi Toward Washington," *Foreign Affairs*, June 22, 2020.

27. Kanti Bajpai, "India's Emerging Grand Strategy after Galwan: Bridging the Power Gap," *Asia-Pacific Regional Security Assessment: Key Developments and Trends, 2021* (IISS: London, 2021), 63–64.

28. Krisn Kaushik, "Explained: The Malabar Exercise of Quad Nations, and Why It Matters to India," *The IndianExpress*, August 31, 2021, https://indianexpress .com/article/explained/malabar-exercise-of-quad-nations-why-it-matters-to -india-7472058/.

29. Jack Dutton, "U.S. Tells India There Will Be 'Consequences' for Dodging Russian Sanctions," *Newsweek*, April 1, 2022, https://www.newsweek.com/us -india-russia-sanctions-consequences-ukraine-invasion-1694076; Shruti Menon, "Ukraine Crisis: Why Is India Buying Russian Oil," *British Broadcasting Corporation*, June 10, 2022, https://www.bbc.com/news/world-asia-india-60783874.

30. Jennifer D. P. Moroney and Aland Tidwell, "Making AUKUS Work," *TheRANDBlog*, March 22, 2022, https://www.rand.org/blog/2022/03/making -aukus-work.html#:~:text=On%20September%2015%2C%202021%20President %20Biden%20announced%20the,deliver%20nuclear%20powered%20 submarines%20to%20Australia%20by%202039.

31. Abhijnan Rej, "French Joint Commander for Asia-Pacific Outlines Paris' Indo-Pacific Defense Plans," *The Diplomat*, April 13, 2021, https://thediplomat .com/2021/04/french-joint-commander-for-asia-pacific-outlines-paris-indo -pacific-defense-plans/#:~:text=As%20a%20nation%20of%20the%20Indo-Pacific %2C%20France%20operates,reinforced%20by%20specialized%20units%2C%20 coming%20from%20France%20mainland.

32. Richard Javad Haydarian, "Germany Wades Warily into South China Sea Fray," *Asia Times*, August 4, 2021, https://asiatimes.com/2021/08/germany -wades-warily-into-south-china-sea-fray/; "UK's First Carrier Continues Long Deployment in South China Sea," *The Maritime Executive*, October 12, 2021, https:// maritime-executive.com/article/uk-s-carrier-strike-group-exercises-in-south-china -sea#:~:text=The%20UK%20and%20Netherlands%20are%20among%20several%20 nations%2C,Sea%20after%20China%20enacted%20its%20new%20maritime%20law.

33. Patrick Gerard Buchan and Benjamin Rimland, "Defining the Diamond: The Past, Present, and Future of the Quadrilateral Security Dialogue," *Center for Strategic and International Studies*, March 2020, 2–3.

34. Buchan and Rimand, 4–5.

35. Krisn Kaushik, "Explained: The Malabar Exercise of Quad Nations, and Why It Matters to India," *The IndianExpress*, August 31, 2021, https:// indianexpress.com/article/explained/malabar-exercise-of-quad-nations-why-it -matters-to-india-7472058/.

36. Buchan and Rimland, 5.

37. Jagannath Panda, "Making 'Quad Plus' a Reality," *The Diplomat*, January 13, 2022. https://thediplomat.com/2022/01/making-quad-plus-a-reality/.

38. Kevin Rudd, "Why the Quad Alarms China," *Foreign Affairs*, August 6, 2021, https://www.foreignaffairs.com/articles/united-states/2021-08-06/why-quad -alarms-china.

39. Buchan and Rimland, 2.

40. Jagannath Panda, "U.S-China Competition and Washington's Case for 'Quad-Plus,'" *The National Interest*, September 28, 2020, https://nationalinteres t.org/feature/us-china-competition-and-washington's-case-'quad-plus'-169751.

41. Indrani Bagchi, "India, Quad-Plus Countries Discuss Covid-19 Battle, Economic Resurgence," *The Times of India,* March 20, 2020, https://timesofindia .indiatimes.com/india/india-quad-plus-countries-discuss-covid-19-battle -economic-resurgence/articleshow/74861792.cms

42. Derek Grossman, "Don't Get Too Excited, 'Quad Plus' Meetings Won't Cover China," *The Diplomat,* April 9, 2020, https://thediplomat.com/2020/04/dont -get-too-excited-quad-plus-meetings-wont-cover-china/#:~:text=Don't %20Get%20Too%20Excited%2C%20'Quad%20Plus'%20Meetings %20Won't,counter%20China's%20growing%20assertiveness%20in%20the %20Indo-Pacific.%20By.

43. Derek Grossman, "Don't Get Too Excited, 'Quad Plus' Meetings Won't Cover China."

44. Abhijnan Rej, "French Joint Commander for Asia-Pacific Outlines Paris' Indo-Pacific Defense Plans."

45. "Leaders of U.S, South Korea and Japan Agree Closer Cooperation over North Korean Threat," *Reuters,* June 29, 2022, https://www.reuters.com/world /asia-pacific/us-skorea-japan-cite-ongoing-concerns-over-nkorea-missile-tests -2022-06-29/; "North Korea Slams U.S.–South Korea–Japan Military Cooperation," July 7, 2022, https://www.politico.com/news/2022/07/03/north-korea-slams-us -japan-military-cooperation-00043872.

46. "How Much Trade Transits the South China Sea," *ChinaPower,* https:// chinapower.csis.org/much-trade-transits-south-china-sea/.

47. Ben Dolven, "The Association of Southeast Asian Nations (ASEAN)," Congressional Research Service, July 26, 2022, https://crsreports.congress.gov /product/pdf/IF/IF10348.

48. Viet Hoang, "The Code of Conduct for the South China Sea: A Long and Bumpy Road," *The Diplomat,* September 28, 2020, https://thediplomat.com /2020/09/the-code-of-conduct-for-the-south-china-sea-a-long-and-bumpy-road/.

49. Michael J. Green and Gregory Poling, "The U.S. Alliance with the Philippines," December 3, 2020, Center for Strategic and International Studies, https:// www.csis.org/analysis/us-alliance-philippines.

50. John Lee, "China's Economic Influence in Thailand: Perception or Reality," July 11, 2013, *Institute of Southeast Asian Studies,* 1–2, https://www.iseas.edu.sg /images/pdf/ISEAS_Perspective_2013_44.pdf; Anders S. Corr and Priscilla A. Tacujan, "Chinese Political and Economic Influence in the Philippines: Implications for Alliances and the South China Sea Dispute," *Journal of Political Risk* 1, 3 (July 3, 2013), https://www.jpolrisk.com/chinese-political-and-economic-influence-in -the-philippines-implications-for-alliances-and-the-south-china-sea-dispute/.

51. Lucio Blanco Pitlo III, "The Vanguard Bank Standoff: A New Tempest in the South China Sea and What It Represents," *China US Focus,* September 2, 2019, https://www.chinausfocus.com/society-culture/the-vanguard-bank-standoff-a -new-tempest-in-the-south-china-sea-and-what-it-represents.

52. Xuan Loc Doan, "U.S.–Vietnam Strategic Partners in all but name," *Asia Times,* April 10, 2019, https://asiatimes.com/2019/04/us-vietnam-strategic-partners-in -all-but-name/.

53. George Lauriat, "The Strait of Malacca and the Indo-Pacific Region: Between Regionalization and Maritime Trade," Italian Institute for International Political Studies, October 27, 2021, https://www.ispionline.it/en/pubblicazione

/strait-malacca-and-indo-pacific-region-between-regionalization-and-maritime
-trade-32052.

54. Ben Dolven and Emma Chanlett-Avery, "U.S.–Singapore Relations," Congressional Research Service, April 7, 2022, https://crsreports.congress.gov
/product/pdf/IF/IF10228.

55. William Choong, "Chinese-U.S. Split Is Forcing Singapore to Choose
Sides," *Foreign Policy*, July 14, 2021, https://foreignpolicy.com/2021/07/14
/singapore-china-us-southeast-asia-asean-geopolitics/.

56. Eddie Walsh, "Why Oceania Matters," *The Diplomat*, August 3, 2011, https://
thediplomat.com/2011/08/why-oceania-matters/.

57. Amy Catalinac, "Why New Zealand Took Itself Out of ANZUS: Observing 'Opposition for Autonomy' in Asymmetric Alliances," *Foreign Policy Analysis*
6, 4 (October 2010), 317, https://scholar.harvard.edu/files/amycatalinac/files
/catalinac_fpa.pdf.

58. Bruce Vaughn, "New Zealand: Background and Relations with the United
States," Congressional Research Service, May 12, 2021, 7, https://crsreports
.congress.gov/product/pdf/R/R44552.

59. Fergus Hanson, Emilia Currey, and Tracey Beattie, "The Chinese Communist Party's Coercive Diplomacy," Australian Strategic Policy Institute (August
2020), 31, 40–41, https://www.aspi.org.au/report/chinese-communist-partys
-coercive-diplomacy.

60. Derek Grossman, "Don't Get Too Excited, 'Quad Plus' Meetings Won't Cover
China."

61. Katie Lyons and Dorothy Wickham, "The Deal That Shocked the World:
Inside the China-Solomons Security Pact," *The Guardian*, April 20, 2022, https://
www.theguardian.com/world/2022/apr/20/the-deal-that-shocked-the-world
-inside-the-china-solomons-security-pact; Craig Singleton, "Beijing Eyes New
Military Bases across the Indo-Pacific," *Foreign Policy*, July 7, 2021, https://
foreignpolicy.com/2021/07/07/china-pla-military-bases-kiribati-uae-cambodia
-tanzania-djibouti-indo-pacific-ports-airfields/.

CHAPTER 10

Lessons from Our Adversaries and Partners on the Maritime Silk Road

Heather Levy

Competition with the People's Republic of China (PRC) in the Indo-Pacific has been a growing national security concern over the last several presidential administrations.[1] Much of this competition is along the Maritime Silk Road (MSR), a series of critical geographic locations that provide competitive advantages in diplomatic, military, economic, and informational competition to the countries that operate there. Physically, the Maritime Silk Road consists of multiple PRC infrastructure projects in different countries, deliberately placed along sea lines of communication and at critical maritime choke points. These projects are of strategic concern to the United States in its competition with the PRC.[2]

Despite these expressed concerns, there has been little published analysis regarding the extent to which Maritime Silk Road investments have contributed to the success of the PRC strategy in Southeast Asia or limited the U.S. freedom of action in the region. This chapter uses a deliberate analysis across the instruments of national power to assess the extent to which the MSR is facilitating Chinese access or limiting U.S. access to countries in the Association of Southeast Asian Nations (ASEAN), and in what cases Chinese and U.S. access is mutually exclusive. This evaluation of access drives strategic recommendations in support of the United States' Free and Open Indo Pacific Strategy and generates recommendations for continuing to sustain America's competitive advantage in the region.

The Maritime Silk Road is a subset of the PRC's Belt and Road Initiative (BRI) that focuses primarily on the sea lines of communication connecting

China with key economic corridors. In contrast with the overarching BRI, which includes significant overland components, the Maritime Silk Road is centered on expanding deep-water port access, energy and cyber connectivity close to port locations, and infrastructure (including road, rail, and airports) connecting maritime locations to land capabilities.

Varying analysts often approach the Maritime Silk Road (and the Belt and Road Initiative in general) from a strictly competitive perspective. Alarmists argue that the United States has already lost the strategic initiative to the PRC and has fallen behind in the race to compete for influence within Southeast Asia.[3] Other Western scholars of Chinese thought focus on how the PRC crafts geopolitical strategy through the integration of instruments of national power across domains.[4] Both these lenses of analysis, however important, ignore the most critical measures of MSR success in terms of access and influence: how the projects and the soft power they project are perceived by the host nation. The reaction of Indo-Pacific countries to Chinese interventions is the true measure of the Silk Road's success (or lack thereof).

The Maritime Silk Road is a key strategic line of communication for the People's Republic of China. If the overall Belt Road and Initiative is globally focused, the Maritime Silk Road is the more geographically constrained subset of BRI centering on Southeast Asian countries and access to sea lines of communication. The MSR is a means for improving the PRC's global access to and from critical resources and markets globally and in Southeast Asia, rather than an end in itself.[5] The five stated goals of the MSR are policy coordination, infrastructure connectivity, unimpeded trade, financial integration, and connecting people.[6] In practice, the PRC uses these objectives to gain access rights to energy resources (including hydro-electric and oil/natural gas), strategic minerals, markets for Chinese goods, and diplomatic influence in regional and global governing bodies. Many of these goals are centered around infrastructure projects, which are the focus of this chapter.[7] Increasing access to ASEAN states along these five lines of effort does not require limiting U.S. access.

By analyzing reactions to the MSR by the countries of Southeast Asia, and their effect on Chinese and U.S. access to a given country, this chapter identifies critical lessons learned that help define the complex scope of the geopolitical challenges in the region. Notably, the United States must enlist partners and allies to counter Maritime Silk Road investments while operating within domestic economic constraints. Equally important, it is the quality of work and interaction with participating countries that provides a strategic advantage, rather than the amount of money spent or the simple possibility of diplomatic leverage. The analysis discussed in this chapter suggests that neither the existence of MSR projects, nor the amount of money spent, equated to improved access for PRC in the hosting countries. In addition, they have not yet limited access for the United

States nor allies such as Japan to participate in either actual infrastructure construction or financial investments in the MSR participating nations. Rather, ASEAN nations have been content to participate in both Chinese and Western initiatives as a hedging strategy.[8] Western competition with Maritime Silk Road projects is largely self-limited by a lack of economic willingness to participate, and further hampered in Southeast Asia by domestic restrictions on lending that render some governments and economies ineligible for grants or loans. None of these aspects of this equilibrium, whether foreign or domestic, are likely to change in the future.

In a case highlighting the complexity of Silk Road project access, in January 2022, Cambodia banned future U.S. visits to Ream Naval Base after Washington levied targeted sanctions against senior military officials for corruption connected to Chinese companies. Both the United States and the PRC have paid for construction of various buildings on the base, but Ream officially belongs to Cambodia alone. Cambodia allowed new construction of additional Chinese-funded buildings on the base and deep-water dredging has begun in the area.[9] At the same time, however, Cambodia also cancelled annual military exercises with the PRC, citing a variety of concerns from COVID to flood reconstruction requirements to funding issues. Opposition leaders, however, point to a desire to appease the new U.S. administration.[10] Both Chinese and Cambodian leaders continue to deny that Ream will become the second overseas People's Liberation Army–Naval (PLA-N) base after Djibouti, and the first in the Indo-Pacific. But Western officials point to mounting evidence that the new construction will be for exclusive use of PLA-N forces and expect that the United States will never regain the access lost.[11]

This type of hedging between the interests of the two major powers is a technique seen throughout the nations of ASEAN.[12] The strategy of simultaneous positive and negative engagement with both major powers is exemplified by military strengthening without partner aligning and multilateral engagements without definitive mutual defense alliances.[13] Kuik and Rozman describe hedging as a spectrum of weaker states between "light hedgers," which want to maximize the benefit of alignment with the PRC while continuing to minimize subservience, and "heavy hedgers," which ally more strongly with the United States but continue to cultivate "balance-of-power" hedges across economic, diplomatic, and military lines of effort.[14] The countries of ASEAN are not a universal bloc and range from strong associations with the PRC to alliances with the United States. From an international relations perspective, they do not see either the United States or the PRC as a sufficiently immediate and hostile threat to require balancing or bandwagoning in terms of a formal mutual defense pact.[15] The United States must be prepared to operate within an environment of hedging, rather than expecting strict allegiance to one power across all domains, as most ASEAN countries do not see it as in their best interest to align solely with one of the major geopolitical powers.[16] The challenge facing the United States will be to build sufficient strength in the relatively weak ASEAN countries so if they do sense an immediate threat from the PRC they are sufficiently resilient to form a balancing coalition with the United States rather than feeling compelled to bandwagon with the PRC.

The United States and the PRC are competing for similar geographic access and influence in the Indo-Pacific. The U.S. National Defense Strategy has described the U.S. strategy as one of "integrated deterrence," which Secretary Austin has further refined as including multiple domains and instruments of national power integrated to deter the PRC from acts of violence or coercion.[17] The U.S. Indo-Pacific Command has identified milestones across the region that include both conventional deterrence and "gray zone" competition objectives as well, including economic and diplomatic goals.[18] Both the United States and China are currently working increase their own advantages in the region through competition while minimizing their adversary's influence. Analyzing the impacts of MSR in the Indo-Pacific region, focusing particularly on the countries in ASEAN, provides insights on some of the successes and shortcomings of a multi-faceted approach to access in the region. Through this analysis, the United States can identify where to focus critical deterrence efforts, learn from the PRC's experiences in economic outreach, and recognize both opportunities and threats in the region. Though the specific cultural and political factors are different, many of the basic frictions between influencing powers and the influenced are the same.

This chapter analyzes U.S. and PRC access to ASEAN countries from diplomatic, informational, military, and economic perspectives, and evaluates that access compared to the Maritime Silk Road investments and projects that the PRC has conducted in conjunction with those countries. The primary question is whether MSR investments increase access for the PRC, decrease access for the United States, prove neutral, or have some other effect on the participating countries. The intent is to use this analysis to guide the United States in developing a strategy for improving access to ASEAN countries to better compete with the PRC. Beginning with a brief discussion of the Maritime Silk Road and assessment of how the PRC has prioritized investments within ASEAN, the comparison of access between the two superpowers offers both guidance and gateposts on how to focus efforts to fill any gaps and maintain any strategic advantages.

MARITIME SILK ROAD PROJECTS IN ASEAN COUNTRIES

The initial cornerstone of the Maritime Silk Road is the concept of connecting regional nations to the PRC—economically, physically, culturally, and with informational ties. The PRC has stated that the MSR initiatives (and counterpart Silk Road Economic Belt and Digital Silk Road) include more than just economics and infrastructure, but also elements of soft power, diplomacy, and security support aspects as well.[19] The Maritime Silk Road includes projects ranging from basic extraction of metals and ores, to cell phone and cyber infrastructure, port construction, power generation capabilities, and development of transportation and other logistics functions.[20] Many of the Maritime Silk Road connected influence aspects, including diplomatic and cultural exchanges, are even more challenging to quantify but will be discussed briefly. Potentially excepting Ream naval base, the remainder of the MSR projects are either civilian or dual use in nature.[21]

Figure 10.1 shows the PRC's MSR-specific investments in ASEAN nations ranked from most to least, in billions of U.S. dollars. As seen in the figure, the PRC's three largest investments focus on strategically placed countries such as Indonesia, Malaysia, and Vietnam, with developing relationships, rather than with their closest partners (such as Cambodia or Myanmar).

Chinese investment in Indonesia's high-speed railway projects provides an illustrative example of these investments.[22] In 2015, when the PRC's favorable terms and promise of quick delivery won the contract bid, the project appeared likely to strengthen the relationship between the PRC and Indonesia. As of early 2022, construction had been delayed by multiple years, a Japanese competitor had been invited to support construction efforts, and the "influence" outcome was much less assured.[23] Indonesia itself was reluctant to engage in predatory lending schemes and continued negotiations until they were satisfied with the results of the terms of their agreement, a plan that was particularly effective due to Chinese and Japanese competition for the Jakarta–Bandung rail project.[24]

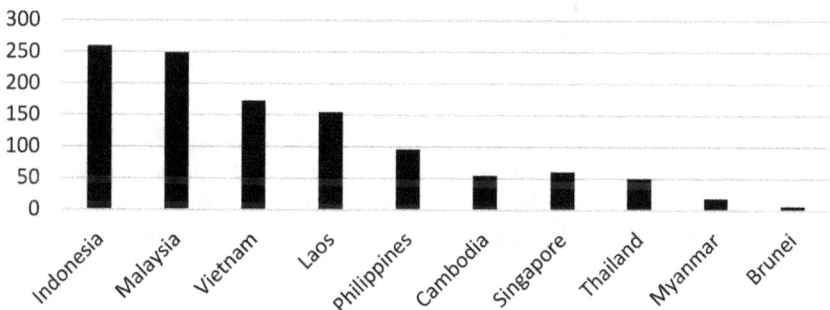

Figure 10.1   China's MSR Investments in ASEAN (U.S.$bn) as of 2019

Source: OECD, "The Belt and Road Initiative in the Global Trade, Investment and Finance Landscape," in *OECD Business and Finance Outlook 2018*, 61–102. Paris: OECD Publishing, 2018. Data compiled from larger statistical grouping.

Emerging Issues and Challenges

Other MSR projects have faced a series of disputes, both internationally and with hosting nations; these include accusations of "debt-trap diplomacy" and other unscrupulous banking practices,[25] issues with corruption, ignoring local and environmental concerns, disrupting national development, and construction cost and quality problems.[26] One of the most well-known cases centers around Chinese loans to develop Hambantota Port in Sri Lanka, which resulted in a Chinese company acquiring a majority stake in the Sri Lankan port. In concert with government officials in Colombo, China issued significant development loans that the United States and India were unwilling to grant, given the likelihood of return on investment.[27] This case generated Western fears of debt traps, though it actually appears to be more a case of China using economic engagement to its strategic advantage and then Sri Lanka defaulting rather than a deliberately developed plan.[28] Combined, all of these concerns contribute to lack of support for conducting business with the PRC and for the reduction of positive impact for projects that are conducted. Identifying the nature of these challenges in detail, however, is critical for understanding if the United States is going to effectively compete with Maritime Silk Road investment projects. As a result of this case and similar predatory lending practices, most host countries are now well-aware of the double-edged sword of Chinese infrastructure investments.[29]

The U.S. Secretary of State has denounced the PRC for predatory lending that "mires nations in debt and undercuts their sovereignty."[30] A 2021 Center for Global Development study reviewed hundreds of Chinese contracts and indicated that they included the following elements that were unusual for international lending contracts:[31]

- strict confidentiality clauses, even regarding the existence of the debt
- immediate repayment clauses
- prohibitions against loan restructuring with other creditors

In contrast, others have emphasized the benefits of the PRC's lending and suggested that concerns about harsh terms and a loss of sovereign freedoms are greatly exaggerated.[32] Though these practices could be perceived as levers for diplomatic or economic coercion, it is worth noting that the PRC has not overtly used them in this manner. It is not possible to determine whether the existence of these clauses changes the behavior of the participating countries, though the independent state behavior in ASEAN countries, and diversity among them, would indicate that it does not.

Despite the immense concern with potential debt traps, the reality appears to be more complex. Many of the MSR projects in dispute, such as Hambantota Port and the two Malay developments, were initiated by the host nations (rather than by the PRC), and money received by the host nations did not go back to the PRC for loan repayment but was used to pay down other sovereign debt.[33] As Malaysian Prime Minister Mahathir described when cancelling some of the corrupt and ill-considered projects, "this is our own people's stupidity. We cannot blame the Chinese for that."[34] In actuality, there is sufficient blame to be shared between the predatory lender and the greedy recipient of the loans in question. Accepting some of the international concern, the People's Republic of China has displayed some global efforts toward reducing impacts on countries with whom it contracts as well. In the past two years, the PRC has forgiven several multimillion-dollar loans (mostly with African countries) and four Chinese banks have restructured loans to defer interest payments as a part of the G20 Debt Service Suspension Initiative.[35]

There are two cautions for the United States when it comes to trumpeting the call of debt traps. First, the West is not without censure regarding its development assistance and terms for investment. ASEAN nations register complaints about the lack of investment from the United States, or about the rigid strictures that the United States places on countries using grants or other forms of foreign direct investment (FDI)—such as human rights guarantees or governmental transparency.[36] In addition, the constant warnings about Chinese debt traps ring as hypocrisy when the United States offers no viable alternative to poor nations with bad credit and worse records on treatment of their citizens.[37]

Complaints of corruption have also tarnished the impact and reputation of MSR projects in ASEAN. Public opinion survey respondents have argued for less corruption from Chinese companies, more transparency, and better opportunities for competition from local companies and local workers.[38] In some cases, this may be the fault of local business practices, where Chinese companies adapt to compete, even if this means using illegal or illiberal practices. In the course of this style of competition, especially where permitted by host governments, Chinese firms have on occasion violated even the host state's own national procedures, environmental restrictions or laws.[39] Given the reputation that Chinese companies have among many Southeast Asia nations, "crony corruption" might be a better strategic narrative for the United States to propagate as it places the blame on the Chinese businesses, not on the "victim" nation walking into the trap of interest rates and underwater loans.[40] In a 2021 global public opinion survey, the assessment of Chinese companies' performance and trustworthiness (35% trusted) remains well behind that of American companies (48% trusted) operating worldwide.[41] These indicators demonstrate that merely operating in a region is not sufficient to gain trust, a lesson that China is learning and that American companies can use to maintain a strategic advantage.[42]

Western policy analysts have also expressed alarm that the People's Republic of China's investment in development systems undercuts the technological development of hosting nations. Advocates of the PRC's practices make the inarguable point that there is little Western interest in competitive investing, and that the short-term losses by participating nations are worth the cost for the long-term benefits of the investments. Chinese investment in the Lao power grid illuminates this dissonance. In March 2021, a Chinese state-owned enterprise (SOE) signed a contract with Laos to develop a national power grid. Electricite du Laos Transmission Company Ltd. (EDLT), a joint venture between a Chinese company and the Lao state-owned company, will invest $2 billion in the Lao power grid and will manage large parts of it for a period of twenty-five years, after which the infrastructure will be ceded to the Lao government.[43] This move was necessary for Laos after drops in tourism and additional COVID-19 impacts on an economy that was already struggling and in debt to others, including the PRC. To Western observers, however, it demonstrates an expansion of Chinese control over regional energy resources that also grants diplomatic leverage over a sovereign nation.

Finally, MSR projects average a 20% cost overrun and often mandate the use of imported Chinese workers, materials, and equipment.[44] Many projects import Chinese workers and materials to reduce expenses, which impacts quality and imposes human costs as well, with imported workers often treated as victims of human trafficking.[45] The cost issues, though not unusual in the industry, contribute to an overall perception of Chinese companies and by extension the Chinese government.[46]

Overall, the most significant lessons from the MSR projects in ASEAN illustrate that both the United States and the PRC can learn from Chinese experiences. The PRC has primarily focused on the nations with the most strategic potential rather than their closest partners. This gives the United States and its allies, including Japan and South Korea, the opportunity to gain strategic advantage in the region through competition for investment opportunities. In addition, Chinese SOEs that are conducting business in hosting nations are seen as an extension of the Chinese state, and their performance can result in a backlash of unfavorable public opinion much as individual American misbehavior overseas can cause a similar cultural backlash.

The United States and potential partners must identify these lessons learned if they choose to increase competition with the PRC in infrastructure investment in Southeast Asia. In fact, the United States has committed many of these same errors internationally, as observed in the findings from the U.S. Agency for International Development's Office of the Inspector General (IG) in multiple semiannual reports to Congress between 2018 and 2021. These issues include $66 million in questioned costs that the IG identified in the second half of fiscal year 2021 regarding the $2 billion of foreign development funds audited. The nonquantified findings, like the PRC's MSR issues, include expecting local leaders to present information on behalf of the people and the nation, importing materials rather than using locally generated options, and balancing between local business practices and home-nation standards.[47] The United States should consider this carefully if it intends to emulate the PRC's Maritime Silk Road investment program.

MEASURES OF ACCESS

Maintaining access to the global commons is a national security objective in the U.S. National Security Strategy (2017) and Interim National Strategic Guidance (2021). As applied to national security, *access* is defined as the unhindered use of the global commons across land, water, airspace, and cyberspace.[48] The instruments of national power—diplomacy, information, economic, and military—are the means by which the various domains can be accessed to improve U.S. strategic advantages in the region. Evaluating the relative access of the United States and the People's Republic of China with respect to each of the ten ASEAN countries is the first step in gaining strategic advantages in the Indo-Pacific in accordance with the strategy of integrated deterrence as discussed in the National Defense Strategy of 2022.

Diplomatic Access

Literature describing diplomatic access has evolved in the past two decades from narrowly discussing formal alliances and appointed personnel to now including measures of public diplomacy and even evaluations of public opinion.[49] An overall measure of diplomatic power should include all three of these aspects of access—measures of standard diplomacy, public diplomacy and "new public diplomacy," or public opinion.[50] Moyer et al. (2019) discuss measures of standard diplomacy as meeting different conditions—including the level of diplomatic relations and agreements. From this perspective, the Philippines and Thailand's status as formal allies gives the United States stronger diplomatic access to those nations than to what the INDOPACOM Strategic Guidance (2019) describes as "key partners" such as Indonesia, Malaysia, Singapore, and Vietnam. Nations that are involved in more public diplomacy efforts (including the Philippines) grant more access from a diplomatic/political perspective than those that are not, such as Myanmar. And nations where the public, especially the elite, has a higher degree of trust of a country can aid in gaining access for those countries. Thus, both official public diplomacy and also public opinion efforts in concert with official diplomacy are important in gaining access to national channels.

Another indicator of potential diplomatic access is alignment with a superpower's vision, which predicts not only increased physical access (in terms of meetings or shared committees) but also whether countries participate in joint efforts based on shared vision, as described in detail by a 2018 RAND study. Alignment with shared vision describes the extent to which a given nation is committed to a leading nation's shared vision of the world, particularly if is culturally shared or a consensus-based model. It is important to note that, according to Lin's study at RAND, some countries may be similarly aligned with both the PRC and the United States as they do not see those visions as mutually exclusive.[51]

Figure 10.2 provides an analysis of two different measures of diplomatic access—comparative diplomatic ties with the United States and the PRC, as denoted in the bar graphs, and alignment or commitment to a shared vision with those two nations, as described by the graphed lines. In a detailed RAND study, Lin and others ranked diplomatic influence through a series of expert polls, surveys and analyses of published reports.[52] The x axis reflects these rankings on a scale from 1 to 5, with 5 indicating the highest possible shared ties or alignment, and 1 the lowest alignment or commitment and sharing of diplomatic ties with the United States or PRC.[53] The blue line demonstrates

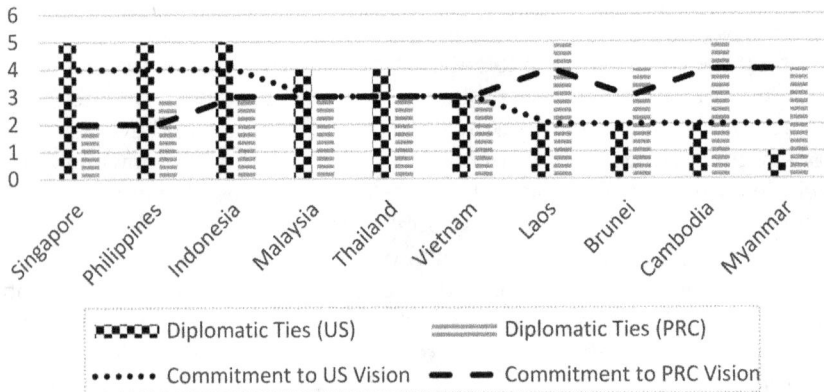

Figure 10.2 Comparative Measures of Diplomatic Access—Diplomatic Ties and Shared Commitment

Source: Data for first six countries from Bonny Lin, Michael S. Chase, Jonah Blank, Cortez A. Cooper III, Derek Grossman, Scott W. Harold, Jennifer D. P. Moroney, Lyle J. Morris, Logan Ma, Paul Orner, Alice Shih, and Soo Kim, *U.S. Versus Chinese Powers of Persuasion: Does the United States or China Have More Influence in the Indo-Pacific Region?* Santa Monica, CA: RAND Corporation, 2020. Extrapolated using same methodologies as used in Appendix B for remaining four countries.

alignment or commitment to overarching U.S. goals in the region, and the red line describes shared alignment with the PRC's vision and goals. As indicated by the figure, many respondents did not see the United States and the PRC as having mutually exclusive visions for the region.[54]

U.S. diplomatic access ranges from close ties (as in with allies such as the Philippines) with a high degree of cooperative efforts with the United States, including cooperative efforts, partnerships and agreements, and established diplomatic ties in the region to nations with limited ties such as Myanmar. Nations with more bilateral agreements with the PRC such as Cambodia, Myanmar, and Laos are more challenging for the United States to access from a political perspective.[55] David Shambaugh describes the alignment of ASEAN nations with the PRC ranging from closely aligned nations such as Cambodia to more independent nations like Indonesia. Though he admits that the orientation is not static, it is based on both historical, cultural, and more recent changes that have moved nations such as the Philippines, Thailand, and Malaysia closer to the PRC as they embraced more autocratic governments, and closer to the United States when those governments were rejected by the population in the case of the Philippines.[56]

In addition to standard measures of diplomacy, measures of public diplomacy or trust are also an important criterion of access-granting. In a public opinion survey across ASEAN, the United States was consistently rated as more likely to "do the right thing" in terms of upholding international law and good governance.[57] Figure 10.3 describes how survey

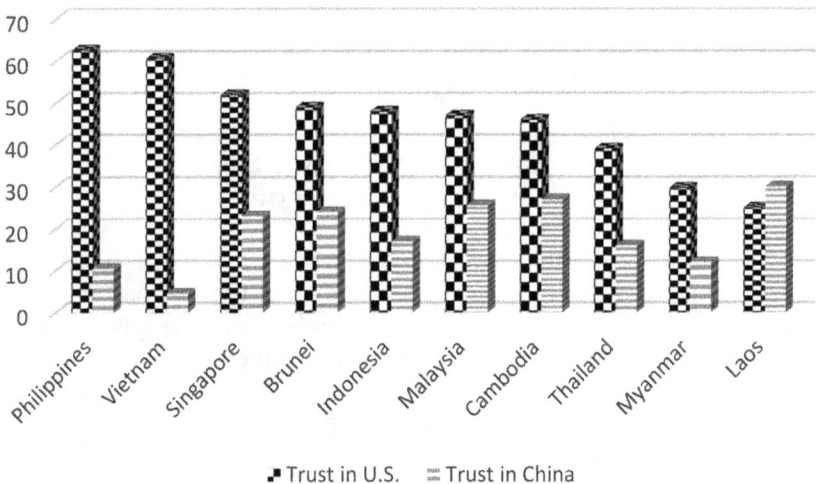

■ Trust in U.S.   ☰ Trust in China

Figure 10.3 Comparative Trust in the United States and China

Source: Iseas–Yusof Ishak Institute, "Survey of Southeast Asia Report 2021." Accessed January 31, 2022. https://www.iseas.edu.sg/articles-commentaries /state-of-southeast-asia-survey/the-state-of-southeast-asia-2021-survey-report/

respondents in each ASEAN country reported their level of trust that the United States or the PRC was a positive force in international relations and was likely to continue to act to uphold international law.[58] Since this data was collected after the 2020 election in the United States, but before significant policy updates from the Biden administration, events could have provided an optimistic poll bump rather than one achieved by a particular strategy.

Only Laos showed a higher degree of trust in the PRC than in the United States, demonstrating that even those countries where formal diplomatic access skews toward the PRC, the United States still retains influence on public opinion and access in the political realm. The potential advantage of public opinion toward the United States might be limited by some governments but remains a potential strategic advantage as the United States considers deterrence strategies in the region. The United States still maintains significant diplomatic access, both formal and informal, and shared vision with most of the ASEAN nations, including those where the PRC has made significant MSR investments. This demonstrates that, as of 2022 at least, the governments and people of ASEAN remained open to diplomatic overtures from the United States no matter the level of investment from the PRC in their infrastructure projects. This corresponds with what Kuik described as, at a minimum, light hedging, which includes seeking balance of political power by courting diplomatic efforts from both major powers in the region.

Informational Access

This chapter considers two metrics for access from an information perspective: one comprising citizen ability to freely receive and transmit information, and the other evaluating access to secure network infrastructure. For the first metric, nonprofit research organization Freedom House provides a detailed analysis of Internet freedom for citizens of different world countries. Though none of the ASEAN nations rank higher than "partly free," in terms of their ability to access open content without surveillance or censorship, a distinct ranking emerges that is useful from an access point of view.[59] For the second metric, this chapter evaluates governmental cyber policies that permit business access in competition with the PRC's restrictive monopoly on cyber infrastructure. From a cyber perspective, nations range from Singapore's extremely supportive cyber alliance with the United States to the restrictive PRC-aligned policies of Burma.[60]

In addition to Internet freedom from a content perspective, Shahbaz and many others from a national security perspective (including retired Lieutenant General H. R. McMaster) argue that providing informational infrastructure that is secure and free from governmental interference is an essential requirement of free democracies.[61] The access to secure cyber infrastructure is included as a key measure of access in the informational domain, given that there are no current options for many nations to the PRC-controlled Huawei 5G network. The United States and several allies do have the ability to incentivize cyber infrastructure competition.[62]

Based on Internet freedom and cyber access policies, information access is a challenge in Southeast Asia, partially due to the PRC's information and communications technology infrastructure monopoly in the region.[63] Within ASEAN, only Singapore, Thailand, and Vietnam use or are planning to use network infrastructure outside the Chinese-owned Huawei for 5G access. Figure 10.4 depicts the two key measures of information access: Internet freedom and cyber infrastructure.[64]

Freedom House's metric of Internet freedom (shown by the blue line on the graph) analyzes obstacles to access, limits on content, and violations of user rights in assessing a country's Internet freedom, which provides access in the information domain to the population and decision makers.[65] The *x*-axis scores reflect the Freedom on the Net index, as developed by Freedom House, which measures each country's level of Internet freedom based on a set of close to 100 methodology questions that cover an array of relevant issues to human rights online.[77] A lower number on the scale demonstrates limited access including to any government outside the ruling party, so it would apply relatively equally to the PRC and the United States, unless the specific message or info information is synchronized between autocracies.

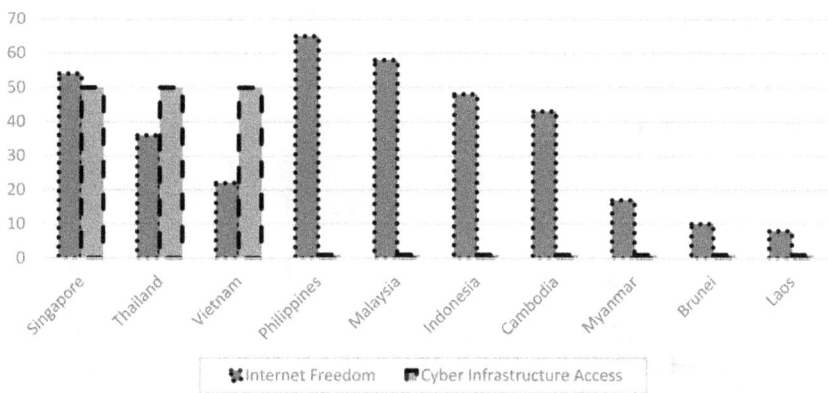

Figure 10.4   Access (Cyber Domain)

Sources: Internet freedom information from Shahbaz, Funk, Slipowitz, Vesteinsson, Baker, Grothe, Vepa, Weal eds. Freedom on the Net 2021, Freedom House, 2021, freedomonthenet.org. Cyber infrastructure information from Kentaro Iwamoto, "Huawei 5G Dominance Threatened in Southeast Asia," *Nikkei Asia*, July 20, 2020 and "Transmissions and Broadcasting," United States Agency for Global Media, March 23, 2022. https://www.usagm.gov/our-work/transmissions-and-broadcasting/

The second measure looks at cyber infrastructure. Specifically, those countries that are utilizing Huawei infrastructure will have limited access from a cyber perspective to U.S. networks due to U.S. sanctions, bureaucratic and policy limits on interoperability, and reluctance to work with the Chinese company.[66] ASEAN countries working with European providers Ericsson or Nokia, on the other hand, can still be used to provide cyber access to the United States as they develop the ability to access 4G or 5G networks.

As seen previously, the lack of Internet freedom limits both the PRC and the United States equally. It prevents people from gaining unfettered access to the Internet, whether from government controls, electricity or data issues, or other obstacles. The U.S. Agency for Global Media provides anticensorship technology and services to journalists and citizens across the world to ensure access to uncensored information, but ensuring access at the source provides a wider audience and wider access to information both from recipient and broadcast perspectives.[67]

As Figure 10.4 reveals, informational/cyber access varies widely among ASEAN countries. In this domain, future Chinese access could be enhanced by MSR projects as fiber and Internet connectivity infrastructure are included in economic corridor developments. The limits on domain access for the United States are largely self-imposed barriers based on U.S. suspicions of surveillance and the difficulty of inspecting cyber hardware for state-installed malware.[68] These domain access constraints are increased by the limited competition available in the area to Chinese-owned Huawei, which is also about 30% cheaper than European competitors.[69]

Military Access

The U.S. Department of Defense's *Joint Publication 3-0, Operations,* describes military access as being comprised of three critical components across multiple domains. These are freedom of navigation, land basing requirements (that are essential for force projection) and combined exercises with ASEAN partners (that ensure interoperability in case of large-scale conflict).[70] U.S.-led exercises in INDOPACOM range from small bilateral expert exchanges to large multilateral events such as Rim of the Pacific (RIMPAC), a major naval exercise combining the navies of multiple countries including the PRC on several occasions, or Southeast Asia Cooperation and Training (SEACAT), the multiday littoral-based counterterrorism event.

Military cooperation, especially in terms of combined exercises, has been a priority for the United States, sometimes to the chagrin of ASEAN nations who have stated their preference for economic engagement.[71] The U.S. focus can be seen in the relative number of U.S. cooperative exercises compared to PLA exercises. Just as the United States has limited appetite

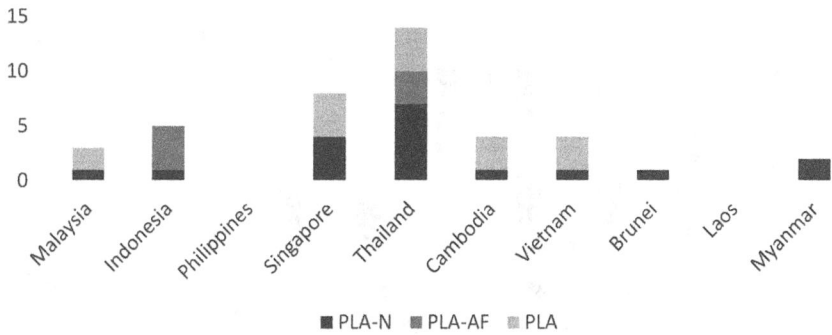

Figure 10.5 People's Liberation Army–ASEAN Countries Military Excercises (Total), 2002–2018

Source: Data derived from Kenneth Allen, Phillip C. Saunders, and John Chen, *Chinese Military Diplomacy, 2003–2016: Trends and Implications.* Washington, DC: National Defense University, 2017.

for overseas economic competition, so the PRC has been both late and limited in its military outreach both bilaterally and multilaterally as seen in Figure 10.5. Though RAND researchers expressed their concern that U.S. security cooperation is no replacement for Chinese economic cooperation, the converse is true as well—as Joseph Nye reminded the international relations community in 2011, economic and military resources and the power they produce are not interchangeable.[72] To close the gap in strategic advantage, the United States must expand economic policies, while maintaining the advantage it currently holds in military and security cooperation.

From a military access perspective, the PRC's slowly growing military cooperation goals have been supported by their diplomatic and economic strengths. Philippine President Duterte's turn toward the PRC for economic development was accompanied by cooperation in the security realm. In October 2016, Manila and Beijing signed economic deals worth $13.5 billion, and the two sides agreed to a memorandum for coast guard cooperation.[73] This would prove to be short-lived, as was most of the Filipino lean toward China. In late 2016, Duterte began threatening to revoke the U.S.–Philippines Visiting Force Agreement (VFA). In 2021, after recognizing that his deals with the PRC had not produced either economic success or a domestic legacy, and recognizing new opportunities with the Biden administration, Duterte stopped objections to the VFA.[74] Despite this five-year shift toward a more neutral stance, the Philippines has a tendency toward heavy hedging, maintaining long-held military and diplomatic arrangements with the United States even as it pursues more diversified economic hedging through preferential trade agreements with the PRC.[75]

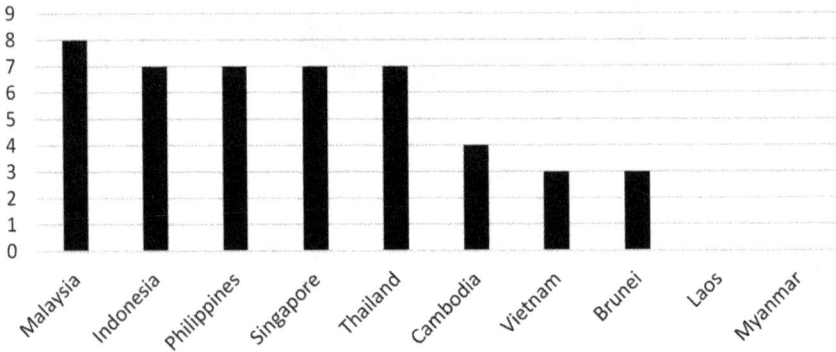

Figure 10.6   United States–ASEAN Military Excercises (Annual), 2002–2018

Source: Data from Satu P. Limaye, Alexander C. Feldman, and US-ASEAN Business Council, *ASEAN Matters for America/America Matters for ASEAN*. Washington, DC: East-West Center, 2020.

Compared to the PRC, the United States has significantly more military and security cooperation exercises, both bilateral and multilateral, and even including countries that are otherwise more closely aligned with the PRC. Note in Figure 10.5 and Figure 10.6 that the United States conducts approximately as many annual engagements as the PRC did with each country over a 16-year period. However, expanding military access with ASEAN may be challenging for the United States as even the nations most interested in balancing the rising hegemon want to remain unaligned from a military perspective.[76] The main challenges to expanding agreements and access with ASEAN partners are the reluctance of countries to anger China as the nearer superpower, the desire to remain publicly independent from allegiances, the United States' potential reputation for reluctance to engage militarily in the region, and Western limits on how much they want to reward such agreements.

Economic Access

Economic access is one of the PRC's highest priorities in Southeast Asia, and China's focus and regional ascendance is reflected in their balance of trade with ASEAN countries. Relative trade between ASEAN nations and the PRC as compared to the United States is depicted in Figures 10.7 and 10.8. The red bars indicate the percentage of a country's exports that are to the PRC, the blue bar indicates the percentage of exports that are sold to the United States, and the gray bar represent exports to the rest of the world. These percentages indicate how dependent a country's economy is on the buying power of the PRC compared to that of the United States, and how significantly the country would be impacted by sanctions

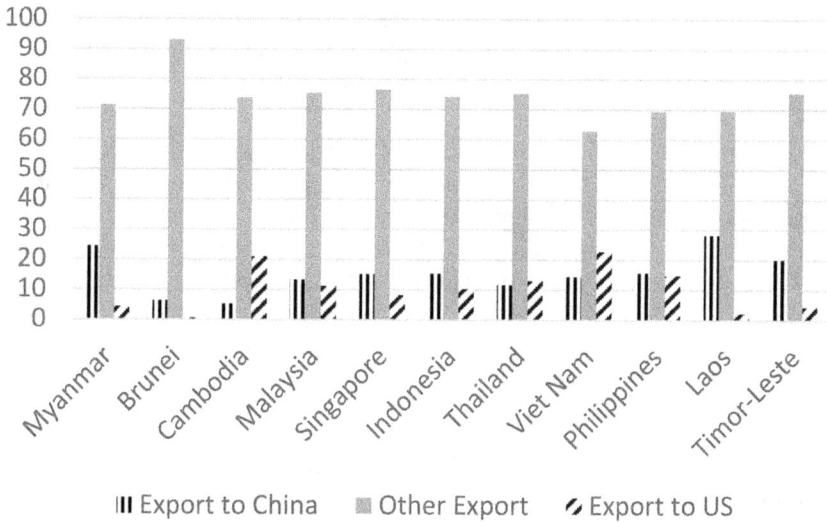

Figure 10.7   ASEAN Exports to the PRC, Other Nations, and the United States
Source: Data from https://oec.world/ as of December 19, 2021.

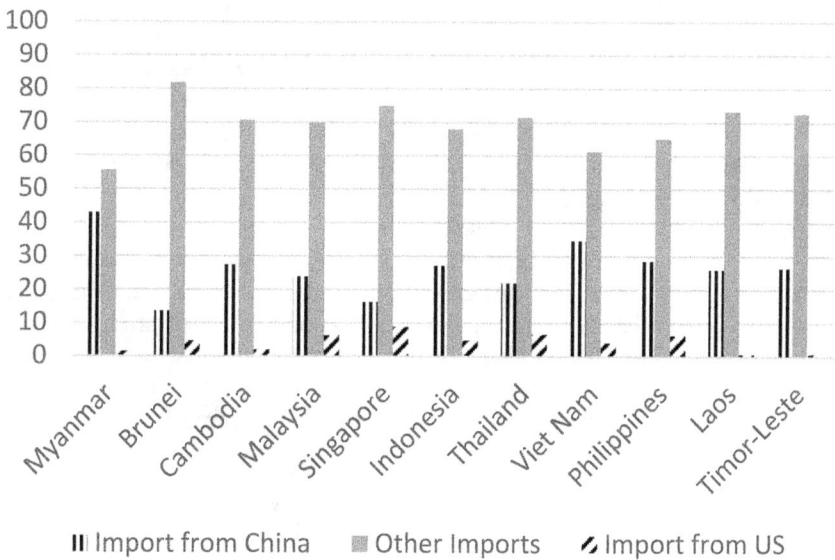

Figure 10.8   ASEAN Imports from the PRC, Other Nations, and the United States
Source: Data from https://oec.world/ as of December 19, 2021.

or informal threats to trade. Lessons from the first one hundred days of world sanctions on Ukraine demonstrate the importance of the weaponized dollar in a global economy. To nations that are similarly dependent on China, regional sanctions could be as dangerous as military instruments of power.

From an export perspective, as seen in Figure 10.7, both the PRC and the United States support ASEAN countries through purchases, and many of those nations are similarly dependent on the United States and the PRC as markets for their goods and products. But the PRC is the preferred market between the two—only Cambodia and Vietnam export more to the United States than they do to the PRC. Laos and Brunei only produce insignificant amounts of goods for export to the United States, though most of the other Southeast Asian nations are closer to parity in the export market.[77] This does demonstrate economic dependency outside MSR lending. The PRC is not afraid to use coercive economics for diplomatic effect. Given the regional trade dependency on the PRC, this is a broader concern than debt-trap diplomacy in particular.[78] The United States currently has sanctions against 26 countries.[79] In contrast, the PRC's nuanced and targeted use of economic leverage tends to be less transparent, shorter term, and more targeted at specific companies or individuals and has been increasing with the PRC's increasing foreign policy goals.[80] It is also important to note despite the PRC's willingness to use economic coercion, it recognizes that coercion could result in a negative counterreaction, especially in ASEAN.[81]

Imports to ASEAN countries are less balanced than exports, as seen in Figure 10.8, with the PRC providing a vastly higher percentage of needed goods than the United States to all the countries under assessment, and the United States contributing less than 10% of any country's requirements.[82] This is another source of potential dependency and potential access that the United States should consider as a gap in its strategic portfolio, and one that could be ameliorated by the use of free trade agreements (FTAs) or the reduction or elimination of tariffs.[83] As of early 2022, the United States was not in any FTAs with ASEAN, nor did it have any bilateral free trade agreements with any ASEAN nations except Singapore.[84] The PRC, on the other hand, had bilateral FTAs with Thailand, Singapore, and Vietnam, as well as with ASEAN.[85] The PRC is also current member of the Regional Comprehensive Economic Partnership (RCEP) and has applied for membership in the Comprehensive and Progressive Agreement for Trans-Pacific Partnership (CPTPP), though acceptance to the latter is far from assured.[86]

OVERARCHING ASSESSMENT OF ACCESS TO ASEAN NATIONS

In a Center for Strategic and International Studies (CSIS) survey of strategic elites, the number one concern of every nation surveyed was U.S.–China strategic competition, excepting only the Philippines, whose first concern was conflict in the South China Sea, and Indonesia, whose primary concern was the PRC's economic influence.[87] Across Southeast Asia, there was high regional support (by a vast majority of survey participants) for democratic values, which were seen as beneficial for regional stability and prosperity.[88] With the intent of nations to hedge economically between the two major powers, the expectation would be that while some avenues of access would be limited by MSR investments, there would be few completely constrained opportunities for the United States. Table 10.1 couples the information gathered and analyzed in previous sections on MSR investments and access measures across instruments of national power to assess U.S. access with respect to U.S.–Chinese competition.[89]

Overall, diplomatic access for the United States does not appear to be limited by any of the MSR initiatives, and in some cases the regional backlash against the PRC due to issues with MSR projects has worked against the PRC in terms of trust and public opinion. Since the United States has had similar issues with overseas infrastructure projects in the past, it would do well to learn from the PRC's lessons if it expands overseas investment initiatives. If, as expected, ASEAN nations continue to hedge between major powers, this entry point is likely to continue as countries seek a balance of political power.

In contrast, informational access for the United States has been curtailed by cyber initiatives associated with the MSR's Digital Silk Road. Though the projects themselves were not included in the financial investments evaluated, their impacts on the informational access are critical to include in this assessment. The PRC's access to the global digital commons in ASEAN is growing as regional nations can expand first their cell phone and then data networks. The U.S. exclusion from these regional platforms is a self-imposed aspect of the Silk Road, and one that is financially costly to China rather than an intentional aspect of PRC strategy. Regardless of origin, this access issue should be considered of strategic importance to American whole-of-government operations in the Indo-Pacific.

MSR projects are working to improve the PRC's military access, both for maritime (port) and air, through ASEAN. Dual-use rail, port, and road projects also exist in neutral countries. In general, these projects will slightly expand transit access for the PRC and provide additional early warning capability for sensors operating in the South China Sea. Other construction initiatives (separate from the MSR) are neutral to the United States, even those (such as military construction on reefs in the South China Sea) which are intended to be military or political deterrents. As legal nonentities, they pose no threat to maritime transit through the UN Convention on the Law of the Sea (UNCLOS). And as a military threat, they are not likely to survive combat operations for long.

Economically, the United States is choosing to limit its access even as the PRC seeks to increase its access at great national expense. Beijing's loans, investments, and agreements are not contingent on excluding other nations, and some Western nations have taken advantage of this opportunity to compete with the PRC. The MSR has increased access for the PRC, but opportunities still exist for the United States should it want to invest in ASEAN either bilaterally or multilaterally.

As described in Table 10.1, the PRC and the United States have varying degrees of access to each nation, and these are independent of MSR projects and project investment.[90] In addition, the deficits in access do have some commonality—they are not intractable. The nations with whom the United States has lower levels of diplomatic access still have significant trust in the United States as displayed by need and willingness to work with either formal or second track diplomatic efforts, especially those focused on environmental issues such as climate change and riparian rights.[91] Despite limited alternatives to Huawei, many nations that are currently seeking Chinese support for cyber infrastructure have stated their willingness to pay more for other providers and seek uncorruptible 5G capabilities while preserving diplomatic independence.[92] The difference between extremes in government enthusiasm for the MSR, such as between Laos and Indonesia, is at least partly a matter of the extent to which each nation can afford to negotiate loan terms offered by the PRC and whether they can find alternate bidders for development projects in their countries.[93] These nations would likely be willing to take investment from the United States or other allies if offered.

Humanitarian aid and second track diplomacy are a method for developing access in countries where opportunities are otherwise limited. In 2021, USAID spent almost $100 million in Myanmar in humanitarian aid (not including amounts spent in Bangladesh for refugees) and $20 million in the Philippines for typhoon response.[94] That same year, five ASEAN countries and the United States signed the Mekong–U.S. Partnership, an initiative that builds on a 10-year, $3.5 billion USAID program focused on the five countries of the Lower Mekong River (Cambodia, Laos, Myanmar, Thailand, and Vietnam).[95] Though not specifically a direct competitor to the Beijing-sponsored Lancang-Mekong Corporation or the two MSR corridors, this partnership is a means for investments that could certainly rival those planned by the PRC in the region, and it mirrors one of ASEAN's possible objectives based on survey data.[96]

Free-trade agreements (along with other sources of low- or eliminated-tariff trade) are emerging as a source of economic access whose true cost will only become apparent in the next few decades. Here, the reticence of the United States may prove strategically costly in the military realm as trade improves among the PRC and ASEAN nations. U.S. bilateral trade agreements with ASEAN nations, though less focused on "centrality," are more targeted toward specific needs and are more palatable to the U.S. public. Therefore, they can yield a degree of economic access to the country in question.

Table 10.1 Access Analysis Synopsis

State	MSR Investment Rank	U.S. Diplomatic Access	China Diplomatic Access	Comparative Trust	Information Access	Military Access	Alignment with China/U.S.	Key Operational Access	Ends from U.S.–China Competition
Indonesia	1	Higher	Medium	Much more trust in U.S. than in China	Opportunities for infrastructure support	Security cooperation exercises and transit/ logistics access permitted for U.S.	Suspicious of China, some ties to U.S.	Malacca Strait access	Facilitated development strategies and independence
Malaysia	2	Medium	Medium	Much more trust in U.S. than in China	Opportunities for infrastructure support		Close with China but also with the U.S.	South China Sea disputes with China	Freedom of navigaton/ economic development, Hedging nation
Vietnam	3	Medium	Medium	Much more trust in U.S. than in China	Open access	Logistics access permitted recently for U.S. access	Defense ties with U.S.; maintain commercial/ diplomatic ties to China	Maritime disputes with China	FON and economic development support
Laos	4	Lower	Medium	Slightly more trust in China	Limited	Limited	Reluctant but few options	Rail line through capital/ developmental rather than strategic	Economic development without compromising sovreignty

(continued)

Table 10.1 (continued)

State	MSR Investment Rank	U.S. Diplomatic Access	China Diplomatic Access	Comparative Trust	Information Access	Military Access	Alignment with China/ U.S.	Key Operational Access	Ends from U.S.–China Competition
Philippines	5	Higher	Lower	Much more trust in U.S. than in China	Opportunities for infrastructure support	Basing and military exercises with U.S.	Alliance with U.S., disputes with China	South China Sea disputes with China; key U.S. log basing	FON and economic development support, independence
Cambodia	6	Very Low	Medium-high	More trust in U.S. than in China	Limited	Equivalent security cooperation exercises	Strong; blocked ASEAN condemnation for territorial claims in South China Sea	Koh Kong base (Malacca Strait access)	Economic development without compromising sovreignty
Singapore	7	Higher	Medium	Much more trust in U.S. than in China	Open Access	Limited from a capabilities stand point	Strong defense ties with U.S., diplomatic and commercial ties to China	Malacca Strait access; information and economic powerhouse in SE Asia	Economic leadership without alliance to China
Thailand	8	Medium	Lower	More trust in U.S. than in China	Opportunities for infrastructure support	FMS, alliance, and military exercises with U.S.; security cooperation with China as well	Aligned economically with China, militarily aligned with U.S.	High speed rail through Laos to China; canal bypass possible (not a proposed project)	Greater independence of action, hedging nation

Myanmar	9	Lower	Medium-high	More trust in U.S. than in China	Limited	Limited	Diplomatically dependent on China; economically isolated due to human rights abuses	Pipeline alternative to Malacca Strait Deep-sea Kyaukphyu Port/Log hub (Malacca Strait)	Economic development without compromising sovreignty
Brunei	10	Lower	Medium	More trust in U.S. than in China	Limited	Limited access for both U.S. and China due to limited capabilities	Close with U.S., also with China	Maritime disputes with China/Not significant	Continued economic development and minor security coopreration hedging

Sources: Mobley, Terry. "The Belt and Road Initiative: Insights from China's Backyard." *Strategic Studies Quarterly* 13, no. 3 (2019): 52–72; Shambaugh, David. "U.S.-China Rivalry in Southeast Asia: Power Shift or Competitive Coexistence?" *International Security* 42, no. 4 (2018): 85–127; Ujvari, Balazs. "The Belt and Road Initiative—the ASEAN Perspective." Egmont Institute, 2019.

Commonalities among ASEAN Countries

Though the ASEAN nations are individual in their culture, outlook, and alignment (see Figure 10.9), the ASEAN charter and community focuses on three complementary facets of connectedness: physical, institutional, and people-to-people.[97] Though generalizing across these very diverse countries is not a basis for generating country-specific policy, some commonalities do emerge that can help direct America's ability to maintain a strategic advantage in the region:

- ASEAN nations have endorsed the consensus-based "ASEAN way" rather than majority rules or bilateral treaty-based methods.
- From a regional and later a global standpoint, the nations value connectivity from a physical, institutional, and interpersonal perspective.
- ASEAN does not see the People's Republic of China as an immediate threat from a military perspective.[98]
- ASEAN nations have a varying degree of trust in regional and international partners, and a willingness to engage those partners from a position of ASEAN centrality.[99]
- In a 2021 survey, Japan remained the most trusted international partner, which 61% of ASEAN consider as a reliable champion on issues such as rule of law, governance, and sustainability. The United States rose significantly in trust perception as favored strategic partners, from 30% to 48% from 2020 to 2021, which still leaves the country as third in

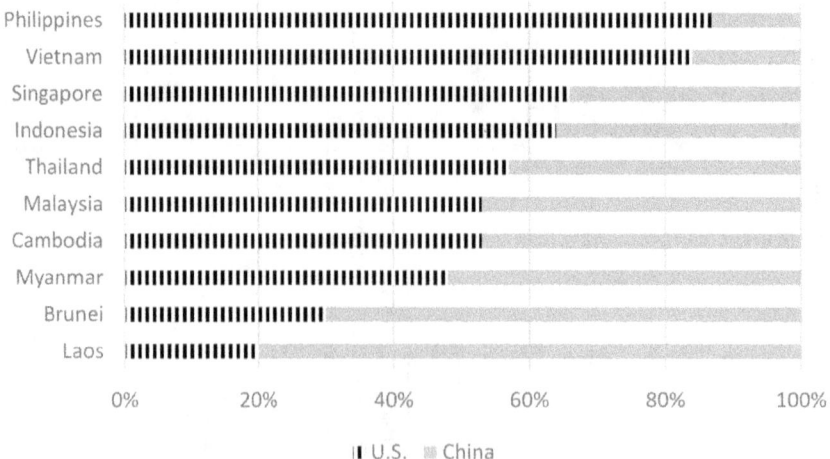

Figure 10.9   Which Strategic Rival Should ASEAN Choose?

Source: Iseas–Yusof Ishak Institute, "Survey of Southeast Asia Report 2021," Q31, 33. Accessed January 31, 2022. https://www.iseas.edu.sg/articles-commentaries /state-of-southeast-asia-survey/the-state-of-southeast-asia-2021-survey-report/

the rankings, slightly behind the EU, but well ahead of the PRC, which is only trusted by 16% of respondents.[100] This finding is important in terms of recommendations from a strategic perspective.

Recommendations for Maintaining and Developing Access

The majority of ASEAN nations are neither aligning themselves formally with the United States (balancing) the way Australia has, nor bandwagoning with the PRC. Rather, they are pursuing a risk-mitigation strategy of hedging. Indonesia, Vietnam, and the Philippines are heavy hedgers against the PRC, as they have more vocally criticized Chinese maritime claims and economic practices. The remainder of ASEAN countries are classified as light hedgers, maintaining deference to the PRC and establishing diplomatic, economic, and military partnerships with both major powers.[101] Understanding and working within this dynamic is essential in crafting a deliberate strategy to maintain the U.S. strategic advantages in the global commons across the domains of land, airspace, sea, and cyber. Using multilateral agreements with trusted allies and partners to create multichannel options across instruments of national power has two concrete advantages—first, it enables host nations to participate while minimizing concerns of Chinese retaliation, and second, it reduces pressure on U.S. domestic leaders for funding, personnel, and material support overseas.

Targeting specific countries with access opportunities, and including redundant options, ensures that the United States is prepared with contingency plans to support joint operational access as necessary in both competition and in case of conflict. Nesting these U.S. objectives with ASEAN and country-specific goals increases the chances for long-term project success, gains legitimacy for the "outsider" United States in Southeast Asia, and facilitates the goals of regional partners for collaborative success.

Objectives by Country

Low-access countries (Cambodia, Myanmar, Laos, Brunei): Based on the analysis in this study, the United States has the lowest overall access to these four countries. These countries tend toward a stronger alignment with the PRC and lower access for the United States and should be a lower strategic priority for U.S. engagement. Belying this alignment, public opinion surveys still demonstrate a strong support for democratic values in these nations, and probably due to humanitarian aid and interaction, the people do report having more trust for the United States than would be expected by their leaders' actions.[102] The United States can exploit this seam by continuing to build diplomatic access and public opinion through aid projects such as the Mekong River initiatives and other environmentally based multilateral initiatives. Continuing security cooperation exercises, especially in established partnerships, will maintain the existing military access, especially with long-term partner Cambodia. Finally, the United States should seek low-risk methods for expanding economic access without significant requirements for capital investment, synchronized with the second track diplomatic projects described previously. These measures will offer low-threat access gains to countries that are likely to align with the PRC due to significant cultural and geopolitical ties.[103]

Medium-access countries (Malaysia and Indonesia): These two countries exist with a similar mid-range access profile and present tremendous opportunities for gaining influence. Malaysia and Indonesia are also ripe for access competition between the United States and the PRC. Together they receive the two highest amounts in MSR investment from the PRC. Yet, both demonstrate a strong degree of trust in the United States. This trust should be leveraged both economically and politically to expand partnerships and agreements including cultural and institutional exchanges, bilateral trade agreements, and infrastructure evaluations and investments. Both Malaysia and Indonesia benefit from subject-matter expert exchanges, including analysis of engineering work and project loans, which have helped the United States demystify MSR terms in past projects.[104] Both Malaysia ($13 billion in 2020) and Indonesia ($18.7 billion in 2020) receive foreign direct investment from the United States, but there is capacity for growth, especially in targeted industries.[105] In Indonesia, Japan's ability to compete against the PRC for rail projects is helping this emerging power remain fully independent. Providing 5G cyber infrastructure alternatives to the PRC-sponsored options is also likely to increase access in these the PRC-skeptical, growth-hungry nations. Finally, maintaining security cooperation and maritime exercises with both Malaysia and Indonesia, and expanding transit and logistics opportunities, will increase military access options. Indonesia has a U.S.-funded coast guard facility on Batam Island, and there is a U.S.-funded portion of Ambon Navy Base on Maluku that could potentially be used for transit access.[106] And both Indonesia and Malaysia have maritime disputes with the PRC that can be subtly exploited to support U.S. freedom of navigation operations. Australia as a key U.S. ally (and Indonesian trusted partner) can facilitate U.S. relationships with these two countries, especially from a security cooperation perspective.[107]

Allies with fluctuating access (Philippines and Thailand): Both the Philippines and Thailand distanced themselves somewhat from the United States due to domestic politics in the last decade. However, both show signs of renewed openness to U.S. access in their countries. Duterte's political maneuvering regarding the U.S. VFA has been resolved, and the United States resumed operations on the five military bases located in the Philippines.[108] South China Sea maritime issues continue to provide a source for continued diplomatic access in the region, as the Philippines is especially vocal concerning the PRC's continued violation of the 2016 UN Arbitration.[109] The United States can continue to use naval and diplomatic resources in the region to support the international rule of law and gain strategic advantages in the area. Similar maritime disputes are particular issues for Malaysia and Vietnam; these pose potential sources of additional diplomatic and security partnerships with those nations as the United States continues its focus on upholding the UNCLOS.

Though Thailand has a cultural and geographical connection with the PRC, it has a historical alliance and long connection with the United States as well, at least until relations began souring with military coups in 2006 and especially in 2014. The Thai continue to be skeptical of Chinese infrastructure lending and, as a result, have internally funded the bulk of the project separately from proffered Chinese loans. In addition, the Thai government has shifted the project's priority (perhaps due to its location in opposition territory) in favor of the Eastern Economic Corridor, despite Chinese pressure for the former project.[110] Despite the potential for a high-value canal across the Isthmus of Kra, the Thai government has long resisted Chinese pressure to construct such a project. Similarly, they have resisted pressure to widen the Mekong River, allowing the passage of wider vessels through to southern China.[111] This demonstrates that opportunities exist for competition to Chinese projects in Thailand, even today. USAID no longer operates a bilateral program in Thailand, though it does include Thailand in the Mekong and other regional programs.[112] This creates vacancies within Thailand for investment opportunities. Attempting to create economic partnerships between Thailand and democratic nations would help create access for U.S. future requirements and support Thai development. Militarily, Thailand vacillates between courting submarine purchases and base expansions from the PRC and continuing annual exercises and military sales with the United States.[113] Understanding this hedging behavior and supporting Thai military growth through foreign military sales and training would enhance U.S. access and potential long-term strategy objectives.[114]

Emerging partnership (Vietnam): One of the highest priorities for the United States should be expanding the Comprehensive Partnership with Vietnam. Vietnam has publicly repudiated some of the PRC's positions in the South China Sea. They are potentially ready to provide U.S. transit access to ports or even off-mainland airstrips, of which there are two on islands in the South China Sea.[115] Their refusal to accept foreign basing or alliances may be a challenge for the United States, but the South China Sea conflict with the PRC (as with the Philippines) makes an opportunity for a stronger posture against the PRC.[116] One of the challenges to increasing diplomatic access hinges on the 2019 Vietnam Defense White Paper's discussion of internal and external threats. Though concerned slightly about the PRC's power, Vietnam is even more concerned about a U.S. threat to Vietnamese socialism.[117] Limiting U.S. narrative around promoting democracy in Asia will aid access objectives in Vietnam and limit concern that other nations have about U.S. intentions as well.

High-value partner (Singapore): Continuing to strengthen the U.S.–Singapore strategic partnership is also a method for increasing U.S. access throughout ASEAN. This relationship is already diplomatically and economically strong, and Singapore sees an opportunity in strengthening U.S. commitment to the region by serving as a hub for U.S. economic engagement in Southeast Asia.[118] Singapore's status as a trusted agent within ASEAN (and its expansion of a network of allies with Australia, India, and Japan) make it an invaluable partner as the United States looks to expand regional economic and diplomatic initiatives.[119] As a particularly rules-focused state, Singapore is likely to support UNCLOS-based proclamations, even if they cause friction with the PRC, which may help leverage ASEAN or other multilateral initiatives supporting maritime transit access for the U.S. and other allies.[120] Singapore is particularly interested in cyber security initiatives. This could generate a partnership in creating cyber infrastructure in competition with the PRC's Huawei 5G as they continue to expand cell phone and Internet technology.[121]

RECOMMENDATIONS FOR GAINING STRATEGIC ADVANTAGES IN SOUTHEAST ASIA

To gain and maintain existing strategic advantages in Southeast Asia, the United States should begin by re-evaluating its strategic narratives from the perspective of the proposed partner nations. Two particularly problematic narratives in the region are the denunciation of Chinese lending as debt-trap diplomacy, which places participating countries in the status of foolish victims, and the heavy promotion of democratic values, which may alienate potentially critical partners. Centering narratives around financial transparency and human rights, for example, might be more palatable without gainsaying traditional U.S. values.

Leveraging international allies and partners in investment opportunities has several distinct advantages. The European Union is seen as a well-trusted agency even in ASEAN countries.[122] The United States can also continue to expand agreements with Indo-Pacific allies and partners for economic and regional initiatives (Australia, Japan, India, South Korea, New Zealand) such as the trilateral Blue Dot Network between Australia, the United States, and Japan, a multistakeholder initiative designed to evaluate and certify nominated infrastructure projects based on high quality standards and principles.[123] Diversifying global investment opportunities reduces their risk as well, and international partnerships mitigate American domestic political and economic risk of overseas investments as well.

In addition to allies and international organizations, working with ASEAN partners (Singapore, Indonesia, Malaysia, etc.) provides internal credibility as well as regional expertise in addition to the advantages described previously. This also has the added benefit of supporting ASEAN as a regional organization, which may prove advantageous for the United States in the future.

To expand the scope of secure cyber infrastructure, the United States should incentivize U.S. or other trusted cyber and cell infrastructure companies to operate in Southeast Asia. This initiative could be combined with investment coalitions as described previously. Singapore, especially, has demonstrated a willingness to partner on cyber infrastructure investment initiatives.

From a military perspective, the United States should continue and expand security cooperation exercises, whether naval, littoral maritime, or other military operation. Counterdrug and counterterrorism exercises have also been popular regional options. Emulating the Maritime Silk Road projects, the U.S. military can continue capacity-building operations, including fixed site, resupply location, and airstrip construction.

Diplomatically and militarily, the United States should evaluate additional South China Seas locations for forward posturing or rotational forces (targeting Vietnam and Philippines as potential partners for staging on island locations).

CONCLUSION

Though the PRC's geopolitical strategy in implementing the Maritime Silk Road initiatives is clear, it has not yet proven particularly successful in either significantly increasing the PRC's access to participating countries, nor in limiting U.S. access to those nations. This analysis demonstrated that the PRC has not yet achieved its desired results from MSR investments, from factors including mixed project outcomes, the hedging instincts of ASEAN countries, the natural protective instincts of sovereign nations, and continued U.S. activities in the region. This suggests that the United States can continue to maintain and gain strategic advantages in the region. In a corollary note, the United States tends to focus on military and counterterrorism focused security cooperation exercises and training rather than economic initiatives.[124] Unfortunately, these activities do not tend to promote long-term alliances or partnerships, as Patricia Sullivan discussed in a series of meta-analyses evaluating the success of security cooperation exercises.[125]

Overall, ASEAN countries are likely to continue a hedging strategy between the two major powers until and unless forced by conflict to choose between them. Building resiliency and capacity in these nations will enhance their ability to balance with the United States if supported with investment across the instruments of national power described as DIME (diplomatic, informational, military, and economic). Diplomatically, the United States can maximize agreements with regional leaders like Singapore, Indonesia, and Malaysia to protect ASEAN centrality and focus on regional goals. From an informational perspective, incentivizing Western companies to build 5G infrastructure in competition with Huawei is essential for maintaining open access in Southeast Asia. Militarily, the United States should continue to build partnerships with key nations like Vietnam, Thailand, and the Philippines (among others) to counter the PRC's challenges to freedom of navigation within the parameters of international law. Most importantly, the United States needs to bolster the economic components of the Indo-Pacific strategy to counter the PRC's MSR and economic influence, prioritizing the use of trusted allies (Japan, Australia, South Korea, and India) to maximize the use of resources and mitigate domestic constraints while helping meet ASEAN needs for growth. Without these changes, the United States risks losing strategic access to Southeast Asia, especially in the cyber and economic realms.

NOTES

1. Joseph R. Biden Jr., *Interim National Security Strategic Guidance* (Executive Office of the President, Washington DC, 2021).

2. Terence Wesley-Smith, "A New Cold War? Implications for the Pacific Islands," in *The China Alternative: Changing Regional Order in the Pacific Islands*, eds. Graeme Smith and Terence Wesley-Smith (Canberra: ANU Press, 2021), 71–106.

3. Andrew Novo, "Strategic Failure: America Is (Literally) Missing the Boat Competing with China," *Strategy Bridge*, May 18, 2020, https://thestrategybridge.org/the-bridge/2020/5/18/strategic-failure-america-is-literally-missing-the-boat-competing-with-china

4. Michael S. Chase and Arthur Chan, *China's Evolving Approach to "Integrated Strategic Deterrence*," Rand Corporation, 2016, https://www.rand.org/pubs/research_reports/RR1366.html

5. Organization for Economic Co-operation and Development, "China's Belt and Road Initiative in the Global Trade Investment and Finance Landscape," in *OECD Business and Finance Outlook* (Paris: OECD Publishing, 2018), https://www.oecd.org/finance/Chinas-Belt-and-Road-Initiative-in-the-global-trade-investment-and-finance-landscape.pdf

6. *Ibid.*

7. *Ibid.*

8. John D. Ciorciari and Jürgen Haacke, "Hedging in International Relations: An Introduction," *International Relations of the Asia-Pacific* 19, 3 (September 2019), 367–374.

9. Luke Hunt, "Cambodia Bans US Visits to Ream Naval Base," *The Diplomat*, December 6, 2021, https://thediplomat.com/2021/12/cambodia-bans-us-visits -to-ream-naval-base-after-sanctions

10. Ellen Nakashima and Cate Cadell, "China Secretly Building Naval Facility in Cambodia, Western Officials Say," *Washington Post*, June 6, 2022, https:// www.washingtonpost.com/national-security/2022/06/06/cambodia-china -navy-base-ream/

11. Hunt, "Cambodia Bans US Visits to Ream Naval Base."

12. Ciorciari and Haacke, "Hedging in International Relations: An Introduction."

13. Van Jackson, "Power, Trust, and Network Complexity: Three Logics of Hedging in Asian Security," *International Relations of the Asia-Pacific* 14, 3 (2014), 331–356.

14. Kuik and Rozman, "Light or Heavy Hedging."

15. Bonny Lin, Michael S. Chase, Jonah Blank, Cortez Cooper III, Derek Grossman, Scott W. Harold, Jennifer D. Moroney, et al., *Regional Responses to US-China Competition in the Indo-Pacific* (Santa Monica, CA: RAND Corp., 2020), https:// www.rand.org/pubs/research_reports/RR4412.html

16. Iseas–Yusof Ishak Institute, "Survey of Southeast Asia Report 2021," accessed January 31, 2022, 32–34, https://www.iseas.edu.sg/articles-commentaries/state-of -southeast-asia-survey/the-state-of-southeast-asia-2021-survey-report/

17. Lloyd Austin, *National Defense Strategy* (Washington, DC: Office of the Secretary of Defense, 2022).

18. Jane Hardy, "Integrated Deterrence in the Indo-Pacific: Advancing the Australia-United States Alliance," *United States Studies Centre,* 2021, https://www .ussc.edu.au/analysis/integrated-deterrence-in-the-indo-pacific-advancing-the -australia-united-states-alliance

19. OECD, "Belt and Road Initiative," 4.

20. Hoang Ha, "Understanding China's Proposal for an ASEAN-China Community of Common Destiny and ASEAN's Ambivalent Response." *Contemporary Southeast Asia* 41, 2 (2019), 223–254.

21. *Ibid.*

22. Yose Rizal Damuri, Vidhyandika Perkasa, Raymond Atje, and Fajar Hirawan, "Perceptions and Readiness of Indonesia towards the Belt and Road Initiative: Understanding Local Perspectives, Capacity, and Governance," Centre for Strategic and International Studies, 2019, 2–6.

23. Koya Jibiki, "Indonesia Turns to State Coffers," *Nikkei Asia*, October 14, 2021.

24. Agatha Kratz and Dragan Pavlićević, "Norm-Making, Norm-Taking or Norm-Shifting? A Case Study of Sino–Japanese Competition in the Jakarta–Bandung High-Speed Rail Project," *Third World Quarterly*, 40, 6 (2019), 1107–1126.

25. Anna Gelpern, Sam Horn, S. Morris, B. Parks, and C. Trebesch, *How China Lends: A Rare Look into 100 Debt Contracts with Foreign Governments* (Peterson Institute for International Economics, Kiel Institute for the World Economy, Center for Global Development, and AidData at William & Mary, 2021).

26. OECD, "Belt and Road Initiative," 15–16, 22.

27. Lee Jones and Shahar Hameiri, "Debunking the Myth of Debt Trap Diplomacy," *Chatham House, The Royal Institute of International Affairs*, August 19, 2020, https://www.chathamhouse.org/2020/08/debunking-myth-debt-trap-diplomacy

28. Brahma Chellaney, "China's Debt Trap Diplomacy," *The Hill*, May 2, 2021.

29. OECD, "Belt and Road Initiative," 15–16.

30. Abdhi Latif Dahir, "China Is Pushing Africa into Debt, Says America's Top Diplomat," *Quartz Africa*, March 7, 2018.

31. Gelpern et al., *How China Lends*, 36–44.

32. Jones and Hameiri, "Debunking the Myth of Debt Trap Diplomacy."

33. Jones and Hameiri, "Debunking the Myth of Debt Trap Diplomacy."

34. Bhavan Jaiparagas, "Chinese 'Projects Will Not Go On': Mahathir Blasts Najib's 'Stupidity'," *South China Morning Post*, August 21, 2018.

35. Deborah Brautigam and Yinxuan Wang, "Global Debt Relief Dashboard: Tracking Chinese Debt Relief in the COVID-19 Era," China Africa Research Initiative (CARI), Johns Hopkins University School of Advanced International Studies, Version 1.6, January 2021, https://www.worldbank.org/en/topic/debt/brief/covid-19-debt-service-suspension-initiative

36. Pradumna Rana, Wai-Mun Chia, and Xianbai Ji, "China's Belt and Road Initiative: A Perception Survey of Asian Opinion Leaders," 2019, 71–92. https://think-asia.org/handle/11540/11536

37. *Ibid.*

38. *Ibid.*

39. *Ibid.*

40. *Ibid.*

41. "2021 Edelman Trust Barometer," accessed 1 Feb. 2022, https://www.edelman.com/trust/2021-trust-barometer.

42. Rana, Chia, and Ji, "China's Belt and Road Initiative," 71–92.

43. Deborah Brautigam and Yinxuan Wang, "Global Debt Relief Dashboard: Tracking Chinese Debt Relief in the COVID-19 Era," China Africa Research Initiative (CARI), Johns Hopkins University School of Advanced International Studies, Version 1.6, January 2021.

44. Organisation for Economic Co-operation and Development, "China's Belt and Road Initiative in the Global Trade Investment and Finance Landscape," in *OECD Business and Finance Outlook* (Paris: OECD Publishing, 2018), 18–19.

45. *Ibid.*, 22

46. Rana, Chia, and Ji, "China's Belt and Road Initiative," 71–92.

47. Office of the Inspector General, "Semiannual Reports to Congress," U.S. Agency for International Development, accessed March 22, 2022, https://oig.usaid.gov/sites/default/files/2021-12/USAID%20OIG%20SARC_April%201-September%2030%202021.pdf

48. Joint Operational Access Concept (JOAC), Version 1.0 (Washington, DC: Department of Defense, January 17, 2012).

49. Jonathan Moyer, Sara D. Turner, and Collin J. Meisel, "What Are the Drivers of Diplomacy? Introducing and Testing New Annual Dyadic Data Measuring Diplomatic Exchange," *Journal of Peace Research* 58, 6 (November 2021), 1300–1310.

50. From Joseph S. Nye, "Public Diplomacy and Soft Power," *The Annals of the American Academy of Political and Social Science* 616 (2008), 94–109; and James Pamment, *New Public Diplomacy in the 21st Century: A Comparative Study of Policy and Practice,* 1st ed. (Routledge, 2012).

51. Bonny Lin, Michael S. Chase, Jonah Blank, Cortez A. Cooper III, Derek Grossman, Scott W. Harold, Jennifer D. P. Moroney, Lyle J. Morris, Logan Ma, Paul Orner, Alice Shih, and Soo Kim, *Regional Responses to U.S.-China Competition in the Indo-Pacific* (Santa Monica, CA: RAND Corporation, 2020), 32–38.

52. Full details are available in Appendix B of the RAND report by Lin et al., which included both regional and PRC assessments of diplomatic influence. This chapter has used the RAND regional assessments when available and approximated their methodology for evaluating diplomatic ties and shared commitment/alignment for Laos, Brunei, Cambodia, and Myanmar, which were not included in the original RAND study.

53. Lin et al., *Regional Responses to U.S.-China Competition in the Indo-Pacific,* 34–35, Appendix B.

54. Iseas–Yusof Ishak Institute, "Survey of Southeast Asia Report 2021," 26, https://www.iseas.edu.sg/articles-commentaries/state-of-southeast-asia-survey/the-state-of-southeast-asia-2021-survey-report/

55. Data from Bonny Lin, Michael S. Chase, Jonah Blank, Cortez A. Cooper III, Derek Grossman, Scott W. Harold, Jennifer D. P. Moroney, Lyle J. Morris, Logan Ma, Paul Orner, Alice Shih, and Soo Kim, *Regional Responses to U.S.-China Competition in the Indo-Pacific* (Santa Monica, CA: RAND Corporation, 2020), Appendix B, https://www.rand.org/pubs/research_briefs/RB10137.html.

56. David Shambaugh, "U.S.-China Rivalry in Southeast Asia: Power Shift or Competitive Coexistence?" *International Security* 42, 4 (2018), 85–127.

57. Iseas–Yusof Ishak Institute, "Survey of Southeast Asia Report 2021," 52.

58. Iseas–Yusof Ishak Institute, "Survey of Southeast Asia Report 2021," 42–43.

59. Shahbaz, Funk, Slipowitz, Vesteinsson, Baker, Grothe, Vepa, and Weal, eds., *Freedom on the Net 2021* (Freedom House, 2021), 30–33. freedomonthenet.org.

60. Nicol Turner Lee, "Navigating the US-China 5G Competition," in *Global China: Assessing China's Growing Role in the World* (Washington, DC: Brooking Institution, 2020).

61. Shahbaz et al., *Freedom on the Net 2021*; H. R. McMaster, "What China Wants," *The Atlantic*, May 19, 2020, https://www.theatlantic.com/magazine/archive/2020/05/mcmaster-china-strategy/609088/

62. Lee, "Navigating the US-China 5G Competition."

63. Michael Goodier, "The Definitive List of Where Every Country Stands on Huawei," *New Statesman*, July 29, 2020.

64. Bruno Mascitelli and Mona Chung, "Hue and Cry over Huawei: Cold War Tensions, Security Threats or Anti-Competitive Behaviour?" *Research in Globalization* (2019).

65. Shahbaz et al., *Freedom on the Net 2021*.

66. Kentaro Iwamoto, "Huawei 5G Dominance Threatened in Southeast Asia," *Nikkei Asia*, July 20, 2020, https://asia.nikkei.com/Spotlight/Huawei-crackdown/Huawei-5G-dominance-threatened-in-Southeast-Asia.

67. "Transmissions and Broadcasting," U.S. Agency for Global Media, March 23, 2022, https://www.usagm.gov/our-work/transmissions-and-broadcasting/

68. Mascitelli and Chung, "Hue and Cry over Huawei."

69. Iwamoto, "Huawei 5G Dominance Threatened."

70. U.S. Joint Chiefs of Staff, *Joint Operations, Joint Publication 3-0* (Washington, DC: U.S. Joint Chiefs of Staff, 2017).

71. Amy Searight, "'Revitalizing U.S.–ASEAN Relations,'" Center for Strategic and International Studies (CSIS), 2017.

72. Lin et al., *Regional Responses to US-China Competition*; Joseph S. Nye, Jr., "Has Economic Power Replaced Military Might?," Project Syndicate, June 6, 2011.

73. Prashanth Parameswaran, "What's Behind the New China-Philippines Coast Guard Exercises," *The Diplomat*, March 15, 2017.

74. Rene Acosta,"Philippines Reverses Course and Commits to a U.S. Visiting Forces Agreement," *USNI News*, July 30, 2021.

75. Cheng-Chwee Kuik and Gilbert Rozman, "Light or Heavy Hedging: Positioning between China and the United States," in Gilbert Rozman, ed., *Joint U.S.-Korea Academic Studies 2015* (Washington, DC: Korea Economic Institute of America, 2015).

76. Kuik and Rozman, "Light or Heavy Hedging."

77. Data from https://oec.world/ as of December 19, 2021.

78. Amy Searight and Michael Green, eds., *Powers, Norms, and Institutions: The Future of the Indo-Pacific from a Southeast Asia Perspective: Results of a CSIS Survey of Strategic Elites* (Center for Strategic & International Studies, 2020).

79. U.S. Department of the Treasury, "Sanctions Programs and Country Information," accessed February 12, 2022, https://home.treasury.gov/policy-issues/financial-sanctions/sanctions-programs-and-country-information

80. James Reilly, "China's Unilateral Sanctions," *The Washington Quarterly* 35, 4 (2012), 121–133.

81. Ketian Zhang, "Cautious Bully: Reputation, Resolve, and Beijings Use of Coercion in the South China Sea," *International Security* 44, 1 (Summer 2019), 117–159.

82. Data from https://oec.world/ as of December 19, 2021.

83. Brian Harding and Kim Mai Tran, "U.S.–Southeast Asia Trade Relations in an Age of Disruption," Center for Strategic and International Studies, June 27, 2019.

84. Harding and Tran, "U.S.–Southeast Asia Trade Relations."

85. *Ibid.*

86. "China Country Guide," International Trade Administration, accessed February 14, 2022, https://www.trade.gov/country-commercial-guides/china-trade-agreements

87. Searight and Green, eds. *Powers, Norms, and Institutions.*

88. Searight and Green, eds. *Powers, Norms, and Institutions.*

89. Yue Yang, and Fujian Li, "ASEAN–China Cooperation under the Framework of the Belt and Road Initiative: A Comparative Study on the Perspectives of China and ASEAN," in *The Belt and Road Initiative: ASEAN Countries' Perspectives* (2019), 1–58.

90. *Ibid.*

91. Iseas-Yusof Ishak Institute, "Survey of Southeast Asia Report 2021."

92. *Ibid.*

93. Yose Rizal Damuri, Vidhyandika Perkasa, Raymond Atje, and Fajar Hirawan, "Perceptions and Readiness of Indonesia towards the Belt and Road Initiative:

Understanding Local Perspectives, Capacity, and Governance," Centre for Strategic and International Studies, 2019.

94. "Mekong-U.S. Partnership," U.S. Agency for International Development, November 29, 2021, https://www.usaid.gov/asia-regional/lower-mekong-initiative-lmi

95. *Ibid.*

96. *Ibid.*

97. Tan Sri Rastam Mohd Isa, "Cooperation and Competition in the Asia-Pacific: ASEAN and the Superpower Dynamics Dilemma," *Horizons: Journal of International Relations and Sustainable Development*, 11 (2018).

98. Lin et al., *Regional Perspectives in U.S.-China Competition,* 32–38.

99. Iseas–Yusof Ishak Institute, "Survey of Southeast Asia Report 2021," 42–55.

100. *Ibid.*

101. Kuik and Rozman, "Light or Heavy Hedging."

102. Iseas–Yusof Ishak Institute, "Survey of Southeast Asia Report 2021," 52.

103. Lin et al., *Regional Responses to US-China Competition,* 32–38.

104. "Mekong-U.S. Partnership"

105. Bureau of Economic Analysis, "Direct Investment by Country and Industry," 2020, https://www.bea.gov/data/intl-trade-investment/direct-investment-country-and-industry

106. "U.S., Indonesia Building Maritime Training Base in Batam," *Indo Pacific Defense Forum,* July 24, 2021.

107. Sebastian Strangio, "Australia, Indonesia Agree to Ramp Up Defense Relationship," *The Diplomat,* September 10, 2019.

108. Joseph Hammond, "Philippine, U.S. Forces Improve Defense Cooperation," *Indo Pacific Defense Forum,* 9 November 2021.

109. Office of the Department of State, "Joint Vision for a 21st Century United States-Philippines Partnership," November 16, 2021.

110. Council of Foreign Relations, "Of Questionable Connectivity: China's BRI and Thai Civil Society," June 7, 2021.

111. *Ibid.*

112. U.S. Agency for International Development, "USAID Thailand Country Programs," accessed February 20, 2022, https://www.usaid.gov/thailand.

113. Thomas Parks, "Can U.S. Assistance Reinvigorate the U.S.-Thai Alliance?," *The Asia Foundation,* December 8, 2021.

114. Kuik and Rozman, "Light or Heavy Hedging."

115. "Airpower in the South China Sea," Asia Maritime Transparence Initiative, July 29, 2015.

116. Socialist Republic of Viet Nam Ministry of National Defense, *2019 Viet Nam National Defense* (Vietnam: National Political Publishing House, 2019).

117. *Ibid.*

118. Lin et al., *Regional Responses to US-China Competition.*

119. Iseas–Yusof Ishak Institute, "Survey of Southeast Asia Report 2021."

120. Government of Singapore, Ministry of Foreign Affairs, "ASEAN," webpage, undated. https://www.mfa.gov.sg/SINGAPORES-FOREIGN-POLICY/International-Organisations/ASEAN

121. Sumathi Bala, "Singapore Wants to Be an e-Commerce Hub as Asia's Digital Economy Grows," *CNBC,* July 27, 2021.

122. Iseas–Yusof Ishak Institute, "Survey of Southeast Asia Report 2021," 36.

123. U.S. International Development Finance Corporation (DFC), "The Launch of Multi-Stakeholder Blue Dot Network," November 4, 2019, http://www.dfc .gov/

124. Allen et al., *Chinese Military Diplomacy.*

125. Patricia L. Sullivan, Brock F. Tessman, and Xiaojun Li, "US Military Aid and Recipient State Cooperation," *Foreign Policy Analysis* 7, 3 (2011), 275–294.

CHAPTER 11

Advancing the U.S.–Vietnam Security Partnership from Obstacles to Opportunities

Thomas J. Bouchillon

Sustaining America's competitive advantage requires an expanded network of allies and partners who have the capability and will to meaningfully contribute to the rules-based international order. In the Indo-Pacific, the paradigm of decades past was whether or under what conditions the United States would provide aid and come to the defense of those countries under its influence and leadership. Today, however, many more middle and upper-tier Asian states enjoy economic wealth, credible national defense capabilities, and domestic political stability. With persistent destabilizing threats coming from multiple directions, the region will require an organizing principle of collective economic and security contributions based on states' resources, interests, and values. The United States is still best positioned to define and shape this order, but it must now share the burden of action with others who have reaped its benefits.

Vietnam provides a unique and important case study for how a state can join, contribute to, and benefit from this new regional order. Strategists looking to expand and strengthen America's influence can look to the U.S.–Vietnam partnership to draw lessons on historical mindedness, strategic patience, optimistic persistence, and the frictions that sometimes arise when interests and values do not neatly intersect. Equally compelling studies can be made from other rising Indo-Pacific states such as Indonesia, Malaysia, or Bangladesh, but the authors and editors of this book chose Vietnam primarily for the paradox it represents. An enemy less than a half-century ago, Vietnam is now poised to become one of America's top-tier strategic partners—and a responsible and respected international actor—for the foreseeable future.

Throughout the course of its "4,000 years of cultural development" iterations of Vietnamese dynasties have frustrated the successive plans of Han, Mongol, Champa, Khmer, Japanese, French, and American invaders.[1] But a short half-century removed from a strategic blunder in Southeast Asia's graveyard of empires, the United States is building a new relationship with Vietnam that, by every metric, is on a meteoric rise. What is driving this relationship? What is its potential? What are its limitations?

Strengthening the U.S.–Vietnam bilateral security relationship is a critical component to advancing national interests in the Indo-Pacific. A host of strategic documents spanning three presidential administrations consistently identify Vietnam as a priority partner in the region, with a growing emphasis on this relationship.[2] The United States' Indo-Pacific Strategy, published in March 2022, outlines at least four objectives that rely on a partnership with Vietnam: advance a free and open Indo-Pacific; build connections within and beyond the region; drive regional prosperity; and bolster Indo-Pacific security.[3] The U.S. military enterprise spanning from Washington to Hawaii has rewritten strategic plans and reallocated resources to answer this call. Growth in the bilateral defense relationship has been matched across other instruments of national power and increasingly includes broad economic partnerships, education exchanges, health sector capacity building, environmental security, and cooperation on emerging domains such as space and cyber.[4]

Other countries have taken notice and are expanding security relationships with Vietnam as well, with like-minded partners such as Japan, Australia, India, South Korea, and the European Union (EU) joining the United States to compete for market share with Russia and China. But Vietnam's position in the geographic heart of the Association of Southeast Asian Nations (ASEAN) is a metaphor for much of its contemporary foreign policy. Its strategic location and fierce independence drive it forward with caution, especially in its military engagements with the United States. By understanding Vietnam's geostrategic position, its core national interests, and its history of conflict with great powers, U.S. policymakers and defense officials can craft an engagement strategy that satisfies both countries' objectives—albeit at a pace and scope that may at times frustrate impatient and overly optimistic American military leaders.

HISTORY MATTERS: LOOKING BACK BEFORE LOOKING FORWARD

Vietnam's Modernization and Reform

The early part of the twentieth century saw Vietnam under French colonial rule. The rise of communism in Eurasia sparked minor civil disruptions in Vietnam, but none that threatened France's hold on power. This lasted until World War II, when Japan took control of Indochina in 1941. France's attempt to reestablish colonial rule over Indochina after the war was met with opposition from the Viet Minh, led by Ho Chi Minh and General Vo Nguyen Giap. The Viet Minh fought France in the First Indochina War from 1946 to 1954, ultimately securing victory at Dien Bien Phu. The country was divided along the seventeenth parallel with the Soviet- and Chinese-backed communist regime controlling the north and France supporting the south. But this division did not stop the fighting in Vietnam as the country plunged into civil war.

The United States first intervened militarily in Vietnam as early as 1955 to conduct operations against communist guerillas in the south. U.S. involvement in the Vietnam War raged on through three presidencies until the signing of the Paris Agreement on January 27, 1973.[5] Fighting between the North and South, however, continued for two more years. With President Ford unwilling to reengage militarily in Vietnam, the communist North launched its final assault into South Vietnam in early March 1975 and by late April had control over the entire country. The newly unified Socialist Republic of Vietnam had little reason to celebrate, however. Casualties were estimated at over 3 million, much of the industrial North lay in waste, and the land was littered with unexploded munitions and remnants of toxic herbicides.[6]

The ensuing decade saw economic turmoil caused by the collectivization of private property and forced agrarian output in the South. Southern citizens experienced human rights abuses leveled against former soldiers, property owners, capitalists, and virtually anyone approaching what could be perceived as the "bourgeois class."[7] Further exacerbating an already dire situation, in 1978 Vietnam attacked neighboring Cambodia, prompting a short but sharp People's Republic of China (PRC) incursion along Vietnam's northern border in 1979 and beginning the Third Indochina War.[8] The result was destitute poverty, international outrage and condemnation, and hundreds of thousands of fleeing refugees.

In 1985, Vietnam pivoted and launched its most recent reform—which carries on today—with the initiation of the *Doi Moi* (economic reform) policy moving the country toward a market-based economy. Since the beginning of *Doi Moi*, Vietnam has been among the world's fastest growing economies, reducing poverty rates from over 40% to under 7%.[9] Along with economic reform, in 1985 Hanoi and Washington began bilateral cooperation on the recovery and repatriation of missing service members from the Vietnam War. The United States and Vietnam finally normalized relations in 1995—the same year it joined ASEAN.

Since joining ASEAN, Vietnam's ascension has been meteoric across almost all sectors of government and society. In the past decade alone, Vietnam has modernized its military by increasing its importation of foreign defense articles by almost 700%.[10] Likewise, its overall defense budget increased from 2.23% to 2.36% of GDP while its economy grew an average of five to seven percent per year.[11] Vietnam's bilateral trade with the United States has exploded over the past two decades, growing from $1.5 billion to $77.6 billion annually. Simultaneously, its exports to the United States transitioned from textiles and low-end products to furniture, machinery, and electronic components.[12] Vietnam has 190,000 students studying abroad, with over 22,000 attending university in the United States—the fifth most of any country in the world and far more than any other country in Southeast Asia.[13] In 2020, Vietnam was ASEAN chair, a nonpermanent member of the UN Security Council, and deployed UN Peacekeepers to South Sudan.

Despite the COVID-19 pandemic, Vietnam's 2020 ASEAN Chair year was not an opportunity lost. As ASEAN chair, Vietnam successfully shifted ASEAN's focus, agenda, and efforts toward pandemic response and economic recovery while emerging from the pandemic postured stronger than before. While many countries struggled to balance economic health with restrictions on movement and gatherings, Vietnam managed to do both, achieving GDP growth rates of 2.9% in 2020 and 2.6% in 2021.[14] Vietnam's public health measures may have been considered authoritarian by American standards, but it maintained relatively low fatality rates, and its developing health care system did not experience overload.[15] Equally important, Vietnam used the pandemic to strengthen its position as an emerging regional powerhouse, even donating personal protective equipment to the United States in the early stages of the outbreak.[16] In 2021, Vice President Kamala Harris officially opened the U.S. Center for Disease Control's new Southeast Asia regional headquarters in Hanoi.[17]

America's Legacy of War

The mental and physical scars left from the Vietnam War cast a looming shadow over the bilateral relationship and have in the past limited the scope and scale of security partnership. With careful management and proper resourcing, however, programs aimed at resolving a legacy of war issues have formed a foundation of mutual respect and growing trust on which the bilateral relationship can be rebuilt. In particular, three manpower and resource-intensive legacy of war issues are being resolved in a manner considered fair to both Vietnam and the United States. These programs are freeing many chains binding the relationship, allowing both sides to invest valuable resources devoted to advancing the security partnership into new and emerging areas.

The first and primary issue for the U.S. population is achieving the fullest possible accounting for service member left behind during the Vietnam War. Out of almost 2,000 missing U.S. service members lost during the war, 1,244 remain unaccounted for.[18] The Defense POW/MIA

Accounting Agency (DPAA) maintains a full-time presence in Hanoi and conducts joint recovery operations with Vietnamese counterparts to account for missing U.S. service members. This mission began in 1985 and was the foundation upon which all other aspects of the bilateral relationship were built. Unfortunately, time is not an endless resource in the completion of this mission. Eyewitnesses are becoming harder to find, and Vietnam's notorious jungle canopy and acidic soil decomposes remains at a rapid rate. Furthermore, the Vietnamese government applies tremendous resources to this mission that, upon completion, could be shifted to advancing other areas of the bilateral relationship. It is therefore imperative that the United States prioritizes achieving the fullest possible accounting and brings the mission to an honorable conclusion as quickly as practically possible.

Second, clearing unexploded ordnance (UXO) from the war is a priority for both the U.S. and Vietnam governments. Since 1993 the United States has funded over $153 million in UXO remediation in Vietnam and now averages over $15 million per year in assistance. Included in this partnership is U.S. Indo-Pacific Command's support to the Vietnam National Mine Action Center and the Level I and Level II International Mine Action Standard certification of Vietnamese de-miners.[19] Relatedly, the United States has pledged over $400 million in dioxin remediation to include a project at Da Nang Airport completed in 2018 and a second project at Bien Hoa Airbase in southern Vietnam that began in 2019.[20] The dioxin remediation projects are dual funded by the U.S. Agency for International Development (USAID) and the Department of Defense, and fully executed under USAID programming. The United States has also contributed over $126 million in assistance through USAID programs to persons with disabilities living in areas heavily sprayed by dioxin during the war.[21]

The final issue hindering the bilateral relationship is the persistent and outsized message of Vietnamese prodemocracy groups inside the United States. Originally descended from former soldiers, political dissidents, refugees, and ethnic minorities, these groups have garnered large legislative attention and support from both U.S. political parties and even earned a briefing at the White House in 2012. The aims of these groups have changed over recent years, with all but the most fringe elements advocating for U.S. policy that encourages positive changes in Vietnam rather than positions that undermine Vietnam's sovereignty.[22] Nevertheless, the government in Hanoi senses interference and even state sponsorship of these groups and issues veiled language asserting as much in its strategy documents.[23] Two successive U.S. administrations have emphasized a rules-based international order and respect for other nations' sovereignty, which has helped alleviate some—but not all—angst from Communist Party officials.

China's Troubled History with Vietnam

The United States is not the only country with a complicated history with Vietnam. China and Vietnam have coexisted along the competition continuum for over a thousand years, with China playing the role of Vietnam's most pernicious and enduring antagonist.[24] Today's China—the PRC—seeks hegemony in the South China Sea and is rapidly expanding down the Mekong River, threatening Vietnam's fisheries, farmlands, and natural environment. The PRC also seeks to exert influence over Cambodia and Laos, which when combined with its dominance of the South China Sea would effectively encircle Vietnam. Despite Communist Party ties and much-needed assistance during and after the war against America, Vietnam has largely rejected the PRC's touted "community of common destiny."

Relations with Vietnam have never fully recovered since the PRC's most recent land border incursion in 1979—ironically referred to as a "preemptive defense" in Chinese national security circles. Since then, the PRC's strategy of choice against Vietnam has been to leverage its military and economic overmatch to compel acquiescence in the international arena. As an example, in 2014 the China National Offshore Oil Company floated an exploratory oil rig 120 nautical miles off the coast of Vietnam—well within Vietnam's exclusive economic zone—where it remained despite uproar from international onlookers. Beijing correctly calculated the risk of escalation was greatest for Hanoi and its neighbors, which managed to garner only notional support from ASEAN counterparts. Subsequent PRC harassment of Vietnamese fishermen and exploitation of minerals in Vietnam's exclusive economic zone has cost Hanoi an estimated loss of $1 billion.[25] Perceived favorable economic concessions to Chinese investors triggered nationwide protests in 2014 and 2018, leading the Vietnamese prime minister to scale back terms of the deal after receiving "enthusiastic feedback" from the public.[26] Short of withdrawing its claims in the South China Sea, there is not much Beijing can do in the near to mid-term to improve its favorability and trust with either the Vietnamese public or the government in Hanoi. The PRC's looming shadow and its provocative behavior will cause Hanoi to look for economic and security partnerships elsewhere.

PERCEPTIONS ACROSS THE REGION

Vietnam within ASEAN at the Heart of the Indo-Pacific

With the PRC rising as a major military competitor, the Indo-Pacific is the U.S. military's priority theater and likely will be for the foreseeable future.[27] Straddling the Indian and Pacific Oceans, ASEAN lies at the geographic and economic center of the Indo-Pacific and is key to advancing national defense strategy objectives. With this in mind, it is instructive to briefly examine ASEAN as an institution, Vietnam's role within it, and what this portends for both U.S. and Vietnamese security interests.

Official statements affirming ASEAN centrality and consensus decision-making point to a region uncomfortable in the middle of great power competition. ASEAN's slow and deliberate processes can frustrate U.S. policymakers, but in some instances they work for U.S. benefit. For example, on contentious issues such as South China Sea Code of Conduct negotiations, sovereign member states are speaking as a unified voice behind the ASEAN collective.[28] The theme of ASEAN centrality also bodes well for the U.S. military, whose official statements, policies, and defense engagements all demonstrate high levels of respect for ASEAN as an institution and multilateralism as a mechanism.[29] Conversely, the PRC's attempts to undermine ASEAN centrality and resolve disputes bilaterally limit its success advancing some of its core interests.[30]

U.S. policy toward ASEAN is evolving to meet the current security environment. Beginning in July 2020, Washington began issuing statements shifting away from its policy of taking no position on territorial disputes to one in which openly repudiates the PRC's claims in the South China Sea.[31] While most in the region—including Vietnam—welcome these positions, others allege the United States is using provocative rhetoric against PRC claims in order to escalate tensions with Beijing rather than stand on principles of international law.[32] Moreover, some Vietnamese interlocutors have privately expressed concern over the longevity of U.S. policy due to frequent turnover of presidential administrations and global commitments that could divert America's attention. However, three successive administrations have placed the region—and by extension ASEAN and Vietnam—at or near the top of their foreign policy priorities. This continuity of policy points to the formation of a national grand strategy and should give assurances to the region of the United States' commitment into the future.

The Mekong subregion, which includes Burma, Laos, Thailand, Cambodia, and Vietnam, is an oft-overlooked area of PRC economic and military expansion. The United States has a multitude of interagency programs aimed at security cooperation, economic development, countertrafficking, health sector capacity building, and environmental resilience throughout mainland Southeast Asia. As the nation at the bottom of the Mekong Delta, environmental changes to the river affect Vietnam more than any other. Vietnamese fishermen and farmers are harmed by salinization and pollution of Mekong wetlands caused by headwater dams, overdevelopment, and water manipulation upstream.[33] Productive cooperation among riparian nations has been elusive in recent years. Although both countries now see benefits of closer cooperation, Vietnam has historical enmity with Thailand, it feels threatened by Cambodia's perceived capitulation to the PRC, and it is desperately attempting to keep Laos out of Beijing's orbit.[34] As closer partnership among lower Mekong countries benefits Vietnam, this is an area in which the United States, Japan, and other like-minded partners can make significant progress with focused investments in resources and relationships.

How Vietnam and the United States See Each Other

U.S. optimism toward the Vietnam relationship is primarily driven by improvements in diplomatic relations, growth in the depth and breadth of security cooperation, and a shared recognition of common threats to the rules-based international order. Since 2016, Vietnam has hosted three U.S. presidential visits, five Secretary of Defense and Secretary of State visits, and two U.S. aircraft carrier visits. Additional increases in equipment sales, professional military education exchanges, and trainings contribute to the optimism. Diplomatically, the United States sees Vietnam as a nation contributing to international institutions and promoting regional stability and prosperity. Leaders of the two nations affirmed as much during a 2017 meeting, producing a joint statement that addressed a range of issues including trade, defense cooperation, sovereignty, and respect for international law.[35] Perhaps most importantly, the United States believes Vietnam can serve as a bulwark against an expansionist China in both the South China Sea and the Mekong subregion. These two key terrain features drive virtually every geopolitical, geoeconomic, and geosecurity issue in Southeast Asia, and Vietnam is the country best positioned to advance U.S. interests there.

Vietnam sees utility in growing its partnership with the United States for several reasons, not least of which is because it benefits when the PRC's hegemonic ambitions are checked. With ASEAN stubbornly neutral, Vietnam is also encouraged by the stances of like-minded U.S. partners such as India, Japan, and Australia. Second, Vietnam enjoys the benefits of military modernization resulting from a U.S. partnership that includes peacekeeping training, disaster response, military medical capacity building, professional military education, English language training, fighter trainer aircraft, unmanned aerial systems, and Coast Guard training. In addition to better equipment and training, Vietnam is increasingly interoperable with its neighbors in U.S.-led security mechanisms. Since 2016, Vietnam has benefitted from U.S.-led programs such as the Global Peace Operations Initiative, Maritime Security Initiative, Southeast Asia Maritime Law Enforcement Initiative, ASEAN-U.S. Maritime Exercise, and Rim of the Pacific Exercise.[36] Participating in these activities provides the Vietnam military critical operational experience and exposure to other nations' best military practices. Finally, partnering with the United States and other like-minded countries increases Vietnam's stature and legitimacy in the international community. Once the scourge of ASEAN, the past three years have seen Vietnam, as ASEAN chair, hold a nonpermanent seat on the UN Security Council and host the second summit between President Trump and North Korean Supreme Leader Kim Jong-un in 2019.

The U.S.–Vietnam bilateral security relationship will continue to grow in depth and breadth based on shared interests. However, the fifth section of this chapter will demonstrate how deep-seated positions on both sides will, if not reconciled, limit the size, scope, and pace of bilateral cooperation.

Vietnam in the Greater International Security Architecture

The coming decade will likely see Vietnam increasing its participation in multinational security organizations to better integrate its interests into those of the rules-based international order. This includes participation in UN Peacekeeping Operations and associated trainings; ASEAN Defense Ministers Meeting (ADMM)-Plus meetings, expert working groups, and exercises; and activities that enforce UN Security Council Resolutions.[37] Vietnam will also increase security cooperation with Japan, India, and Australia and will draw further away from pariah states such as North Korea. Vietnam will likely not sever military ties with Russia in the near to mid-term future, although subsequent sections will illustrate fault lines in the relationship that may portend long-term malalignment between Hanoi and Moscow.

WHERE NATIONAL SECURITY INTERESTS ALIGN

With a few important exceptions, the U.S. proclivity of values-based framing of its foreign policy—as the next section will show—can be problematic with Communist Party leaders. Therefore, U.S. defense and security leaders should carefully identify and clearly communicate key areas of mutual interests when crafting an engagement strategy with Vietnam. For it is the alignment with the following U.S. objectives that has and will continue to be the primary driver behind the evolution of the U.S.–Vietnam bilateral defense relationship.

Resisting PRC Aggression in the South China Sea and the Mekong Delta

Not all South China Sea claimant states consistently assert their sovereignty in the face of PRC pressure. In recent years and to varying degrees, the Philippines, Brunei, and Malaysia have had uneven and unpredictable responses to PRC incursions. Vietnam has been the most vocal, consistent claimant state willing to stand against PRC aggression in the South China Sea. This trend continues when looking at the Mekong subregion. Among the five riparian countries in the Mekong Delta, only Thailand and Vietnam have effectively resisted the pressure of PRC expansionism. The

United States and Vietnam need other ASEAN states to push back against PRC extraterritorial expansion in both the South China Sea and southward down the Mekong River.

Building Vietnam's Military Deterrence Capability

Vietnam seeks to modernize its military to deter foreign invasion on land and breaches of sovereignty at sea. Vietnam has military modernization programs with a range of diverse partners including Russia, India, Japan, Australia, and the United States. With the exception of Russia (which will be addressed later), the United States should welcome like-minded partners filling critical gaps in Vietnam's military modernization goals. Nevertheless, the United States should aim to carve out as much market share as practically possible in helping Vietnam achieve its national defense objectives. This is especially true in areas which the United States enjoys a clear comparative advantage over its allies—for example, manned and unmanned aircraft.

Security cooperation programs offer another way to build Vietnam's deterrence capability. U.S. security cooperation programs "build defense and security relationships; promote specific U.S. security interests; develop allied and friendly military capabilities for self-defense and multinational operations; and provide U.S. forces with peacetime and contingency access to host nations."[38] Security cooperation builds long-term relationships between leaders and units and is a long-term investment in relationships. U.S. security cooperation goals for Vietnam aim to build trust and overcome legacy of war issues, as well as help Vietnam build a more professional, resilient, and modern military. Despite the growing security cooperation portfolio with Vietnam, tactical-level activities are limited, especially between the nations' armies. Before 2022, military cooperation between the United States and Vietnam was limited to the areas of maritime security, search and rescue, peacekeeping, disaster response, and high-level dialogues. In the summer of 2022, however, the two countries signed an updated defense agreement that allows for new cooperation in the air and cyber domains, defense trade and technology, and information sharing. Delayed due to the pandemic, fear of PRC retaliation, and domestic political inertia, this agreement now provides the Vietnamese military needed political cover to measurably expand bilateral defense cooperation with the United States for years into the future.

Enhancing U.S. Posture and Presence in the First Island Chain

Routine U.S. presence in Vietnam would frustrate China's strategy in both the Mekong subregion and the South China Sea. It is unlikely Vietnam would provide the same level of access, basing, or overflight as other

ASEAN partners such as Thailand, Singapore, or the Philippines in the near to mid-term future. However, the U.S. military can achieve similar effects with routine ship visits and expanded bilateral and multilateral security cooperation activities. Japan, India, and Australia can increase military activities with Vietnam as well, advancing U.S. interests through a federated approach to burden sharing. Vietnam has proven most willing to expand its defense activities to those that adhere to principles of multilateralism and strengthening of a rules-based international order.

Resolving Legacy of War Issues

The United Sates is committed to fulfilling its obligation of fullest possible accounting of U.S. service members missing from the Vietnam War. The League of POW/MIA Families represents those missing service members and their families, while DPAA is the DoD organization with responsibility for the mission. With a large support base in Congress and active support from Gold Star Families, DPAA enjoys steady funding that will likely continue into the near and mid-term future.

Vietnam, for its part, will continue to facilitate the U.S. POW/MIA recovery mission for the foreseeable future, as it sees tangible benefits from the joint cooperation. In addition to financial and economic incentives, Vietnam welcomes U.S. investigative and archival research and capacity building assistance in the recovery of its missing, as well. Vietnam also welcomes U.S. expansion of de-mining efforts from Quong Tri Province to Quang Binh and Tuah Thien Hue Provinces. U.S. dioxin remediation projects at Da Nang Airport and, most recently, Bien Hoa Airfield received laudatory praise from the highest levels of the Vietnamese government. USAID is reassessing its disability support programs in Vietnam with an eye toward expansion. A senior U.S. military officer familiar with these programs noted that continued cooperation with Vietnam to address legacy of war issues builds trust and confidence and opens doors to increased security cooperation in other areas.

Strengthening People-to-People Ties

The U.S. military desires increased training and relationship building opportunities for younger, impressionable Vietnamese officers and soldiers. Strong people-to-people ties build resilience in the bilateral relationship. Many older, entrenched officers distrust the United States and temper the pace and scope of the relationship. But younger Vietnamese officials do not hold the same views and generally have a higher favorability of the United States.[39] Increasing the scope and scale of interaction between the two militaries will set conditions for generational growth in the bilateral relationship.

Respecting Vietnam's Sovereignty

Regime survival is the Communist Party of Vietnam's (CPV's) number one priority. Vietnam's most recent Defense White Paper in 2019 highlights the dangers of forces "inside the country who have not given up their plots against the revolution."[40] These forces, according to Party leadership, focus on "eliminating the leading role of the Party and the socialist regime in Vietnam."[41] Unfounded as it may be, this almost certainly is suggestive of prodemocracy groups suspected of receiving support from inside the United States. While the United States has zero designs or desires to see regime change in Vietnam, its interlocutors will nevertheless need to emphasize respect for Vietnam's sovereignty and system of government until such time as communist leaders in Hanoi are assuaged of this fear.

Increasing Cooperation among Mekong Subregion Countries

In September 2020, Vietnam cohosted a virtual summit with the United States and Mekong River nations to launch the Mekong–U.S. Partnership—an expansion of the existing Lower Mekong Initiative.[42] While China's competing Lancang–Mekong Initiative has been accused of acting as a vessel to control the flow of water, funds, and illicit activity down the river, the United States and like-minded countries such as Japan, Australia, and South Korea offer sustainable solutions for the region's environmental, infrastructure, and security challenges.[43] It is too early to effectively measure results of these various programs, but it is likely Vietnam and the region welcome alternatives to the PRC's largesse. Adding cooperative security measures to an already full Mekong multilateral initiative will be challenging for a host of reasons, not least of which is resource competition among the countries themselves. However, China is furiously exploiting voids in transboundary river governance, and the entire region would benefit from a transparent, rules-based, and comprehensive security cooperation mechanism focused on the river and its associated tributaries and deltas.

AND WHERE THEY DIVERGE

Despite positive momentum in the bilateral relationship, both the United States and Vietnam have self-imposed obstacles that will limit the growth of cooperative activities in the near to mid-term future. These obstacles result from political policies, long-held perceptions, and relationships with internal constituencies and external actors that interfere with bilateral progress.

Trade and Monetary Policy

Both the United States and Vietnam have demonstrated tendencies toward economic protectionism that, if not properly managed, may create friction and limit growth in other areas of the bilateral relationship. Trade with Vietnam has ballooned since relations were normalized in 1995, with the greatest growth coming since Vietnam's ascension into the World Trade Organization in 2006. Since 2017 many companies exporting products to the United States have shifted production from the PRC to Vietnam as result of trade tariffs on certain PRC goods. As a result, as of 2022 the U.S. trade deficit with Vietnam is the third largest in the world and growing.[44] In December 2020, the U.S. Treasury labeled Vietnam a currency manipulator, accusing Hanoi of devaluing the dong for trade advantage.[45] The Biden administration's Treasury Department has taken a more strategic approach than its predecessor, reaching an agreement with Vietnamese counterparts and issuing a joint statement in July 2021 removing the designation.[46] Still, U.S. domestic political forces from both parties—like the ones responsible for abandoning the Trans Pacific Partnership (TPP) in 2016—will keep pressure on any presidential administration to reduce trade deficits and advance protectionist trade policies. Like the Trump administration, the Biden administration appears averse to multilateral free trade agreements. While successive U.S. administrations have ceded America's opportunity to establish international rules and norms on trade, Vietnam has joined the region and moved on to the Comprehensive and Progressive Trans Pacific Partnership (CPTPP) without the United States. As of 2022, Vietnam still has progress to make in meeting CPTPP standards on intellectual property, state owned enterprises, and cross-border data flow.[47] Despite decades of market reform, this struggle hints at Hanoi's proclivity toward nationalism and protectionism and may carry over into the defense and security environments.

Removals

Vietnam was previously one of nine countries determined by the Department of Homeland Security as "recalcitrant" for failing to accept the return of its citizens residing in the United States with final orders of deportation.[48] President Biden's Department of Homeland Security has backed off sanctions on some of these countries, including neighboring Laos, and this has reduced tensions over this issue with Vietnam. In response, Vietnam and the United States recently signed an agreement to expedite the removals process and stave off sanctions. As a result, the Department of Homeland Security continues to quietly repatriate Vietnamese nationals residing in the United States, although at a scope and scale smaller than in previous years.[49] Stoked by fears of U.S. efforts to overthrow the Communist Party from within, Vietnam previously refused to accept a large number of these individuals, many of whom fled the country in the years immediately following the Vietnam War.[50] If another U.S. administration takes a hardline approach to removals, or if this mistrust reemerges in Vietnam, the bilateral relationship could cool over fears of undermining party sovereignty.

Democracy and Human Rights

The Biden administration has taken a decidedly values-based approach to foreign policy, at least in name if not in practice. The Interim National Security Strategic Guidance, issued in March 2021, implies democracy as the organizing principle of U.S. foreign policy and places it at the antithesis of autocracy.[51] While this likely does not register well among the inner circles of the Communist Party, Vietnam has thus far enjoyed inconsistencies between the administration's words and its deeds. The December 2021 virtual Summit for Democracy, for example, received much initial media attention but quickly faded into memory as another round of COVID-19 swept the region and the United States doubled down on new aid and agreements with ASEAN. Still, the United States continues to raise human rights issues with Vietnam over religious and Internet freedom and its treatment of political activists. This will continue to be a thorn in the side of the bilateral relationship and will cause Vietnam's Ministry of National Defense to temper its exuberance for expanded partnerships with the United States.

In 2022 Vietnam landed on the State Department's Trafficking in Persons (TIP) Tier 3 List. In a bit of diplomatic maneuvering, State Department and National Security Council leaders successfully advocated for the president to sign a waiver allowing Vietnam to avoid sanctions which would have suspended select Title 10 and Title 22 security assistance and security cooperation activities.[52] It is unlikely the president would waive sanctions in future years if Vietnam fails to make notable improvements in this area. This would significantly set back ongoing security cooperation programs, and more importantly damage years of progress building trust and confidence in the bilateral relationship. It is therefore critical to direct some DoD-funded activities to help Vietnam address human trafficking in particular, and transnational crime in general.

Vietnam's Relationship with Russia

Vietnamese leaders are likely unappreciative of perceived U.S. interference in their foreign and defense policymaking, particularly when it comes to choosing friends. Hanoi has strong historic ties with Moscow, and Vietnam purchases around 80% of its military equipment from Russia.[53] Its continued reliance on Russian military equipment risks U.S.

sanctions based on Section 231 of the Countering America's Adversaries Through Sanctions Act (CAATSA).[54] CAATSA was designed to punish Russian defense and intelligence industries and those nations that support them. However, with a large number of U.S. priority partners—most notably Vietnam, India, and Indonesia—reliant on Russian military equipment, CAATSA sanctions could have an unintended consequence of driving these countries' defense establishments even closer to Russia. To date, the U.S. government has not imposed CAATSA sanctions on Vietnam (or India or Indonesia), but did so to its ally Turkey. Even a heightened threat or rhetoric of doing so to Vietnam could prove harmful to the bilateral relationship.

While Vietnam is slowly trending in the direction of diversifying its sources of military equipment—to include some from the United States— its military intelligence relationship remains firmly ensconced with Russia. Publicly available details of this relationship are sparse, but it is widely understood that the two countries' military intelligence services maintain an open and highly cooperative relationship.[55] U.S. defense leaders must carefully consider this when sharing sensitive information or providing advanced equipment to the Vietnamese military.

Despite its historical ties to Russia, the 2022 invasion of Ukraine has shaken Vietnam's confidence in Moscow as a long-term strategic partner. With Russia's status as a global pariah state transcending the most recent war in Europe, Vietnam will look elsewhere to balance its foreign and defense policy options. Aside from the United States, Vietnam is looking to India, Japan, Australia, South Korea, and even Israel and the Czech Republic to fill this future gap.

Mistrust

Many Communist Party hardliners suspect the U.S. government of undermining the CPV's legitimacy through the use of nongovernmental organizations, aid groups, and U.S.-based Vietnamese prodemocracy advocates.[56] Further, some in the party maintain that suggested reforms from international institutions such as the World Bank are, in actuality, a U.S. conspiracy to overthrow the party using "friendly gestures."[57] Hanoi openly denounces outside attempts of "peaceful evolution" and considers U.S. statements over human rights as meddling in its internal affairs. This mistrust is largely generational and most prevalent in military and security circles.

"The Four No's"

Vietnam in 2019 updated its longstanding foreign policy of "The Three Nos" (interpreted as no alliances, no foreign basing of troops, and no working with one country against another[58]) to add a fourth—no use of force or threatening the use of force in international relations.[59] By definition, these principles prevent Vietnam from entanglement in the conflict of other nations. In practice, however, they allow Vietnam the flexibility to temper defense engagements based on domestic political considerations or fears that those activities may be provocative to others.

Absorptive and Language Capacity

The Vietnamese military has long focused on internal security and maintains a heavy political role. Its top priority is regime and party protection—a concept alien to U.S. defense policymakers as well as generally an impediment to building a Western-style merit-based officer and noncommissioned officer corps.[60] This concept holds especially true for the People's Army, whose operating concept throughout the country—from the villages to the highest levels of the political system—is an antithesis to U.S. precepts of civilian–military relations. The Vietnam People's Army does not devote significant resources to external engagements with the West. Those units that do are limited and largely compartmented under the Ministry of Defense's Foreign Relations Department, not the more powerful and influential General Staff Department. The center of gravity in Vietnam is the People's Army, and U.S. defense planners may be frustrated by the difficulty accessing its rank-and-file soldiers as we seek to strengthen interoperability and build capacity.

The Vietnamese military has very few officers—and even fewer enlisted personnel—with working proficiency of the English language. The majority of English-speaking officers work in the Foreign Relations Department which, as previously mentioned, does not hold the same influence as the General Staff Department. This limits the pool of Vietnamese soldiers with which the United States can train. Australia and the United States have modest English language training programs with the Vietnamese military, but the throughput of soldiers trained remains insufficient to meet Vietnam's modernization demands.

OPPORTUNITIES TO ADVANCE MUTUAL INTERESTS

Reaffirm Vietnam's Sovereignty

U.S. defense leaders should reaffirm respect for Vietnam's sovereignty and affirm its system of government at every available opportunity. U.S. defense interlocutors should also clearly explain that the U.S. government does not support U.S.-based organizations that seek to overthrow the party. While this may seem pedantic to U.S. leaders, its significance to Vietnam cannot be overstated. Vietnam's Prime Minister Pham Minh Chinh, in his historic speech at the Center for Strategic and International Studies in Washington, DC on May 11, 2022, mentioned "sovereignty" and respect for each other's "political system" six times.[61] This was not a coincidence, and his remarks were not impromptu—they lie at the core for how Vietnam defines its relationship with the United States. Relations with Communist Party officials in neighboring Laos have markedly improved since 2016, due in large part to then-President Obama's overt expressions recognizing Lao sovereignty and system of government.[62]

Promote the Status of the Bilateral Relationship

The United States and Vietnam should work toward advancing the bilateral relationship from a comprehensive partnership to a strategic partnership. While the semantics may seem more symbolic than practical, this step would be extremely significant for Vietnam and allow it to devote additional resources to the bilateral relationship. Furthermore, promoting the status of the relationship would signal to the international community strengthened U.S. commitments to the region. Vietnam currently has sixteen strategic partners, many of which have far fewer diplomatic, economic, and defense ties than does the United States.[63] A strategic partnership, in Vietnamese diplomatic vernacular, gives high attention to another nation's strategic interests when formulating policy.[64] Vietnam has sent mixed signals on its willingness to promote the relationship, with concerns over U.S. domestic political stability, ranklings over human rights statements, and a lasting hangover from the failed TPP. Most importantly are concerns over backlash from the PRC in the economic and security spheres if Vietnam upgrades its partnership status with the United States.[65] However, persistent outreach from senior U.S. officials, aid during the COVID-19 pandemic, and continued PRC provocations bode well for the United States' desire for upgrade. The United States should capitalize on this momentum and continue to press with patience.[66]

Increase Military People-to-People Ties

The United States has an overwhelmingly high favorability rating among the Vietnamese population. In a 2020 survey across multiple sectors of civil society, almost 86% of Vietnamese people surveyed would prefer their county to align itself with the United States to the PRC.[67] American universities host over 22,000 students from Vietnam—more than any other ASEAN country by a factor of almost three.[68] However, these trends have not carried over to the military—distrust among senior officers is widespread, and the number of military exchange students is paltry. The U.S. military should make a concerted effort to better align perceptions of the Vietnamese military with those of the public. To begin, the U.S. Army,

U.S. Navy, and U.S. Coast Guard should seek agreements to begin sending Vietnamese cadets to the U.S. Military Academy, U.S. Naval Academy, and U.S. Coast Guard Academy. This would build on progress made by the U.S. Air Force, which saw its first Vietnamese exchange cadet in 2020. Second, the U.S. and Vietnam militaries should expedite progress on the U.S.-funded ASEAN English Language Training Center located at the Vietnam Military Science Academy. Stalled in recent years—first by Title 10 Section 333 limitations and most recently by COVID-19—this relatively inexpensive, high-payoff project should receive the highest attention by defense policy makers. Finally, the U.S. Army should invite the Vietnam Army to more fully participate in training exercises such as Cobra Gold and attend tactical courses such as those offered by the 25th Infantry Division Lightning Academy (Jungle Training Course, Air Assault School, etc.). This would serve as a launch point to build junior leader relationships that will pay dividends for decades to come.

Prioritize Security Cooperation

Despite eagerness on the part of the U.S. defense establishment, growth of security cooperation should not outpace the political, economic, or other sectors of the bilateral relationship. Likewise, the United States would send mixed signals with the transfer of advanced military hardware while simultaneously condemning the country for human rights, trade imbalances, and its relationship with Russia. As the pace and scope of the bilateral relationship allows, the United States should prioritize the following security cooperation programs in addition to those currently ongoing.

- Cyber defense security and critical infrastructure security. Vietnam has indicated a desire for capacity building from the United States in these two related areas since the establishment of its Cyber Command in August 2017. However, the mechanism to conduct such activities has not been established.
- Air Defense forces. Vietnam has also requested U.S. assistance to develop its Air Defense forces. It is likely Vietnam lacks the capability to identify, track, and target foreign aircraft operating in or near its territorial airspace, most notably over the South China Sea. Furthermore, it lacks integrated air defense for some of its newly acquired capabilities.
- Vietnam has agreed to purchase T-6 trainer jets through foreign military sales. Once this program nears life-cycle completion, the U.S. military should consider transferring or selling more advanced fighter jets to Vietnam. This would have the dual benefit of building interoperability with the United States and reducing Vietnam's military dependence on Russia.

- Peacekeeping. The United States, Australia, Canada, Korea, Japan, and France assist Vietnam in forming, training, and certifying rotational medical peacekeepers deployed to South Sudan. The United States funded a Level 2 hospital and built a Peace Keeping Operations training center and barracks as part of this effort. Continued U.S. assistance to build Vietnam's peacekeeping forces strengthens the military relationship and helps Vietnam contribute to a rules-based international order.
- Bolster Vietnam's deterrence capability in the maritime gray zone. The U.S. Coast Guard has ongoing maritime law enforcement programs with Vietnam that can be used as a bridge to add complexity and depth to training and capacity building. The most likely crisis escalation event in the South China Sea will not be from regional navies but will come from maritime law enforcement agencies, nationalized fishing fleets, or deputized maritime militias. Like most states in the region, Vietnam maintains a maritime militia as part of its enduring concept of "people's war," but its capabilities are more akin to that of a national fishing force and rudimentary compared to that of the PRC.
- Build Vietnam's resilience to withstand both environmental degradation and transnational crime emanating from the Mekong River. The U.S. Army Corps of Engineers has active programs with the Mekong River Commission to model and monitor water and sedimentation flows, but this partnership could be expanded to highlight the risks emanating from upstream Chinese development projects. To help the Vietnam military and police tackle transnational crime on the Mekong, the U.S. military can offer cost-effective, nonstandard training and equipment solutions from the U.S. Coast Guard, Special Operations Command Pacific, or Joint Interagency Task Force—West.

Leverage Allies and Partners to Fill Engagement Gaps

Given the challenges facing the bilateral relationship, the U.S. military must be comfortable with a federated approach to partnership with Vietnam. Quad partners Japan, Australia, and India have rapidly growing defense cooperation programs with Vietnam that complement—not compete with—those of the United States. A federated approach has the additional benefit to Vietnam of lowering its strategic risk of exposure.

- Japan's "free and open Indo-Pacific" concept prioritizes security cooperation and capacity building in Southeast Asia.[69] Since its inception in 2016, Japan has substantially accelerated defense cooperative programs with Vietnam, particularly in the areas of maritime security and peacekeeping operations.[70] Moreover, a plurality of Vietnamese citizens consider Japan an ideal alternative to a binary choice of the United States or China.[71]

- Despite Australia's sizable contributions to U.S. efforts in the Vietnam War, party leaders in Vietnam do not harbor the same distrust toward America's closest Indo-Pacific ally. Similar to Japan, Australia's defense partnerships with Vietnam have grown substantially in recent years. Australia is the primary source of English language training for the Vietnamese military and provides training, equipment, and materiel support for its peacekeeping mission. Australia could fill a critical gap in areas where the United States faces roadblocks engaging the Vietnamese military, such as army tactical-level training.

- India has shown increasing interest in its military posture in Southeast Asia. An active member of the ASEAN Defense Ministers Meeting (ADMM)-Plus and previous co-chair of several working groups, India seeks stronger defense partnerships with Vietnam as part of its Indo-Pacific Oceans Initiative and in search of potential hedges to expanding PRC influence.[72] A more active and engaged India compliments U.S. objectives for the region and, as an extension, Vietnam.

- Engagement in multilateral mechanisms reduces friction and builds legitimacy for U.S. defense policy positions. The ADMM-Plus mechanism hosts a series of meetings and exercises on priority issues such as maritime security, disaster response, cyber security, and counterterrorism. ADMM-Plus incorporates ASEAN's eight dialogue partners, including the United States as well as China and Russia, so it is imperative for the United States to actively contribute to this forum.[73] The ASEAN Regional Forum, ASEAN Coast Guard Forum, and other multilateral security mechanisms provide opportunities to advance U.S. interests without the commitment of expensive military resources. In these limited instances, however, U.S. defense leaders may need to take an appetite suppressant on the incessant need to build readiness, as these activities are focused partnerships and process development.

Reduce Vietnam's Reliance on Russia

The United States should build a strategy to compete with Vietnam's relationship with Russia by exploiting fault lines and creating friction in the relationship and offering better alternatives. First, Russia's recent overtures toward China undermine Vietnam's strategic interest. An information campaign highlighting the risks of a Sino-Russian axis could inject a dose of much-needed realism into Vietnam's foreign policy planning. Second, the United States should conditionalize some key security cooperation programs (air defense and cyber security, for example) on disaggregating from Russian systems but stop short of applying CAATSA sanctions. Third, despite Vietnam's public waffling and ultimate refusal to condemn Russia's invasion of Ukraine, party leaders in Hanoi have reason to be dismayed by their old friend. Publicly, Vietnamese leaders have expressed disapproval for "war in Europe," and analysts have noted private musings comparing Vietnam's position to that of Ukraine—threatened by a larger neighbor and with no military allies.[74,75] American messaging should remind Vietnam of Russia's flagrant violation of the rules-based international order on which it relies. Finally, the United States should exercise strategic patience. The older generation of Vietnam's military and political leadership has strong historic ties to their former Soviet benefactors. The younger generation of leaders is tying Vietnam's future to ASEAN, the United States, Japan, Australia, India, and even the European Union.

CONCLUSION

Achieving U.S. policy objectives in the Indo-Pacific will require a stronger partnership with Vietnam. With a rapidly rising economy, favorable demographics, and strong political and military will, Vietnam is the ASEAN state with perhaps the greatest opportunity to resist PRC extraterritorial expansionism. A hegemonic China with de facto control of the South China Sea and the Mekong subregion is a worst-case scenario for Vietnam. Hanoi will look to ASEAN first, then to the United States, Japan, Australia, and India for reassurances to uphold the rules-based international order from which it has benefited for the past three decades. If Russia's warmongering in Europe and its alignment with the PRC persists, a new generation of Vietnamese leaders will further distance the country from its former patron.

The United States is postured to increase diplomatic, economic, and military engagements with Vietnam, but enduring obstacles emanating from both Hanoi and Washington will limit the pace and scope of this growth. Security cooperation is a key component to advancing the bilateral relationship but must not outpace the diplomatic or economic sectors of the relationship. As such, the United States and Vietnam should upgrade their relationship status to signal to the region the realities of this growing partnership. The U.S. military should prioritize bilateral programs that help Vietnam build a credible deterrence capability and strengthen people-to-people ties. Finally, U.S. policymakers should support efforts from regional powers Japan, Australia, and India to diversify Vietnam's portfolio of partners and to fill critical capability gaps. Most important, however, is for the United States to exercise strategic patience with Vietnam. Scarred by war yet supremely confident in its strategic position, Vietnam has the potential to be a long-term, top-tier U.S. security partner for decades to come.

NOTES

1. Vu Hong Lien and Peter D. Sharrock, *Descending Dragon, Rising Tiger: A History of Vietnam* (London: Reaktion Books Ltd, 2014), 7–11.

2. Compare, for example, the *National Security Strategy* from 2015, 2017, and the *Interim National Security Strategic Guidance* in 2021.

3. White House, *Indo-Pacific Strategy of the United States* (Washington, DC: White House, 2022), https://www.whitehouse.gov/wp-content/uploads/2022/02/U.S.-Indo-Pacific-Strategy.pdf.

4. White House Briefing Room, "FACT SHEET: Strengthening the U.S.-Vietnam Comprehensive Partnership," August 25, 2021, https://www.whitehouse.gov/briefing-room/statements-releases/2021/08/25/fact-sheet-strengthening-the-u-s-vietnam-comprehensive-partnership.

5. Lien and Sharrock, *Descending Dragon, Rising Tiger*, 233.

6. Lien and Sharrock, *Descending Dragon, Rising Tiger*, 240–241.

7. Asia Pacific Foundation of Canada, "Vietnam after the War," https://asiapacificcurriculum.ca/learning-module/vietnam-after-war.

8. Frank Frost, *Engaging the Neighbors: Australia and ASEAN since 1974* (Acton ACT, Australia: Australia National University Press, 2016), 34.

9. Huong Le, "Economic Reforms, External Liberalization, and Macroeconomic Performance in Vietnam," *International Research Journal of Finance and Economics* 176 (November 2019), 130.

10. Austin Wyatt and Jai Galliott, "Closing the Capability Gap: ASEAN Military Modernization during the Dawn of Autonomous Weapon Systems," *Asian Security* 16, 1 (2020), 54–55, https://doi.org/10.1080/14799855.2018.1516639.

11. Thoi Nguyen, "The Trouble with Vietnam's Defense Strategy," *The Diplomat*, January 17, 2020, https://thediplomat.com/2020/01/the-trouble-with-vietnams-defense-strategy.

12. Michael F. Martin, "U.S.-Vietnam Economic and Trade Relations: Issues in 2020," *Congressional Research Service*, February 13, 2020.

13. Erin Duffin, "Number of International Students in the U.S. by Origin, 2019/2020," *Statistica*, November 23, 2020, https://www.statista.com/statistics/233880/international-students-in-the-us-by-country-of-origin.

14. Linh Do and Dang Khoa, "From 'Zero Covid' to Adaptation, Vietnam's Response to Global Pandemic," *VN Express International*, March 20, 2022, https://e.vnexpress.net/news/trend/from-zero-covid-to-adaptation-vietnams-response-to-global-pandemic-4440623.html.

15. Minh Vu and Bich T. Tran, "The Secret to Vietnam's COVID-19 Success: A Review of Vietnam's Response to COVID-19 and Its Implications," *The Diplomat*, April 18, 2020.

16. "Vietnam to Ship 450,000 Protective Suits to the U.S.," *Reuters*, April 8, 2020, https://www.reuters.com/article/us-health-coronavirus-vietnam/vietnam-to-ship-450000-protective-suits-to-united-states-idUSKCN21Q2BK.

17. Centers for Disease Control and Prevention, "Vice President Kamala Harris Opens New CDC Southeast Asia Regional Office in Vietnam," August 25, 2021, https://www.cdc.gov/media/releases/2021/p0825-new-cdc-office.html#:~:text=Vice%20President%20Kamala%20D.,Health%20Ministers%20from%20eleven%20countries.

18. Defense POW/MIA Accounting Agency, "Progress in Vietnam" fact sheet as of April 13, 2022, https://www.dpaa.mil/Resources/Fact-Sheets/Article-View/Article/569613/progress-in-vietnam.

19. U.S. Department of State, Bureau of Political-Military Affairs, "To Walk the Earth in Safety: Documenting the United States' Commitment to Conventional Weapons Destruction (2020)," https://www.state.gov/reports/to-walk-the-earth-in-safety-2020, accessed November 3, 2020.

20. U.S. Agency for International Development, "United States and Vietnam Launch Dioxin Remediation Project at Largest Hotspot in Vietnam," April 20, 2019, https://www.usaid.gov/vietnam/press-releases/apr-20-2019-united-states-and-vietnam-launch-dioxin-remediation-project.

21. U.S. Agency for International Development, "Fact Sheet: Improving the Quality of Life of Persons with Disabilities," September 2020, https://www.usaid.gov/sites/default/files/documents/FS_DisabilitySupport_Sept2020_Eng.pdf.

22. Cindy M. Dinh and Bao Nguyen, "The Rise of the Vietnamese American Political Consciousness Advocacy on Capitol Hill," *Asian American Policy Review* 25 (2014–2015), December 16, 2015, https://aapr.hkspublications.org/2015/12/16/the-rise-of-the-vietnamese-american-political-consciousness-advocacy-on-capitol-hill.

23. See, for example, references to "color revolutions" in Vietnam's latest defense white paper. Vietnam Ministry of National Defence, *2019 Viet Nam National Defence* (Hanoi: National Political Publishing House, 2019).

24. Carlyle A. Thayer, "Vietnam and the Challenge of Political Civil Society," *Contemporary Southeast Asia* 31, 1 (2009), 1–23.

25. Huang Le Thu, "Vietnam's Response to the United States' Changing Approach to the South China Sea," *Asia Unbound*, blog post from *Council on Foreign Relations*, August 3, 2020, https://www.cfr.org/blog/vietnams-response-united-states-changing-approach-south-china-sea.

26. Gary Sands, "In Vietnam, Protests Highlight Anti-Chinese Sentiments," *The Diplomat*, June 12, 2018, https://search-proquest-com.usawc.idm.oclc.org/docview/2053269070?accountid=4444.

27. White House, *Indo-Pacific Strategy of the United States* (Washington, DC: White House, 2022), https://www.whitehouse.gov/wp-content/uploads/2022/02/U.S.-Indo-Pacific-Strategy.pdf.

28. See, for example, "ASEAN Outlook on the Indo-Pacific," signed June 23, 2019, https://asean.org/asean-outlook-indo-pacific.

29. Department of Defense, *Department of Defense Indo-Pacific Strategy Report: Preparedness, Partnerships, and Promoting a Networked Region* (Washington, DC: Department of Defense, 2019), 46–48, https://media.defense.gov/2019/Jul/01/2002152311/-1/-1/1/DEPARTMENT-OF-DEFENSE-INDO-PACIFIC-STRATEGY-REPORT-2019.PDF.

30. Huang Le Thu, "China's Dual Strategy of Coercion and Inducement towards ASEAN," *The Pacific Review* 32, 1 (2019), 21, https://www.cfr.org/blog/vietnams-response-united-states-changing-approach-south-china-sea.

31. Canberra, too, has issued similar statements, most recently in the form of a Note Verbale to the United Nations.

32. Huang Le Thu, "Vietnam's Response to the United States' Changing Approach to the South China Sea," *Asia Unbound*, blog post, Council on

Foreign Relations, August 3, 2020, https://www.cfr.org/blog/vietnams-response
-united-states-changing-approach-south-china-sea.

33. Su Yean Teh and Hock Lye Koh, "Climate Change and Soil Salinization: Impact on Agriculture, Water, and Food Security," *International Journal of Agriculture, Forestry, and Plantation* 2 (February 2016), 7, https://ijafp.com/wp-content/uploads/2016/03/KLIAFP2_11.pdf.

34. Voice of Vietnam, "Vietnam, Thailand Pledge Stronger Cooperation," November 7, 2020, https://vovworld.vn/en-US/news/vietnam-thailand-pledge-stronger-cooperation-919868.vov.

35. Donald J. Trump and Tran Dai Quang, "Joint Statement: Between the United States and the Socialist Republic of Vietnam," November 13, 2017, https://vn.usembassy.gov/20171112-joint-statement-united-states-america-socialist-republic-viet-nam.

36. Department of Defense, *Department of Defense Indo-Pacific Strategy Report: Preparedness, Partnerships, and Promoting a Networked Region* (Washington, DC: Department of Defense, 2019), 48–51, https://media.defense.gov/2019/Jul/01/2002152311/-1/-1/1/DEPARTMENT-OF-DEFENSE-INDO-PACIFIC-STRATEGY-REPORT-2019.PDF.

37. Artem Sherbinin, "Enforcing Sanctions on North Korea is an Opportunity for Cooperation at Sea, *War on the Rocks*, March 16, 2021, https://warontherocks.com/2021/03/enforcing-sanctions-on-north-korea-presents-an-opportunity-for-cooperation-at-sea.

38. Defense Security Cooperation Agency, "Electronic Security Assistance Management Manual," see Chapter 1, section 1.1, https://samm.dsca.mil/chapter/chapter-1.

39. Kat Devlin, "40 Years after Fall of Saigon, Vietnamese See U.S. as Key Ally," *Pew Research Center*, April 30, 2015, https://www.pewresearch.org/fact-tank/2015/04/30/vietnamese-see-u-s-as-key-ally, accessed December 5, 2020.

40. Vietnam Ministry of National Defence, *2019 Viet Nam National Defence* (Hanoi: National Political Publishing House, 2019).

41. *Ibid.*

42. Department of State, "The Mekong-U.S. Partnership: The Mekong Region Deserves Good Partners," September 14, 2020, https://www.state.gov/the-mekong-u-s-partnership-the-mekong-region-deserves-good-partners.

43. Mervyn Piesse, "U.S. Launches Mekong Partnership as Chinese Debt Trap Closes on Laos," *Future Directions International*, September 16, 2020, https://www.futuredirections.org.au/publication/us-launches-mekong-partnership-as-chinese-debt-trap-closes-on-laos.

44. Mark E. Manyin and Michael F. Martin, "U.S.-Vietnam Relations," *Congressional Research Service*, February 16, 2021, https://crsreports.congress.gov/product/pdf/IF/IF10209/12.

45. Tal Axelrod, "Treasury Labels Switzerland, Vietnam Currency Manipulators," *The Hill*, December 16, 2020, https://thehill.com/policy/finance/trade/530487-treasury-department-labels-switzerland-vietnam-currency-manipulators.

46. Sebastian Strangio, "Vietnam, US Reach Accord on Alleged Currency Manipulation," *The Diplomat*, July 20, 2021, https://thediplomat.com/2021/07/vietnam-us-reach-accord-on-alleged-currency-manipulation.

47. Nguyen Anh Duong, "Time for Vietnam to Get Cracking on CPTPP Reforms," *East Asia Forum*, January 13, 2022, https://www.eastasiaforum.org/2022/01/13/time-for-vietnam-to-get-cracking-on-cptpp-reforms.

48. U.S. Department of Homeland Security, Immigration and Customs Enforcement, "Visa Sanctions against Two Countries Pursuant to Section 243(d) of the Immigration and Nationality Act," August 13, 2020, https://www.ice.gov/visa sanctions, accessed December 9, 2020.

49. Brittany Valentine, "The Recent Deportation of 33 Vietnamese Refugees Shows How Far the U.S. Must Still Go to Progress Its Immigration System," *Al Dia*, March 16, 2021, https://aldianews.com/politics/policy/groundtheplane.

50. Michael Tatarski, "Why Is the U.S. Deporting Protected Vietnamese Immigrants," *The Diplomat*, June 5, 2018, https://thediplomat.com/2018/06/why-is-the-us-deporting-protected-vietnamese-immigrants.

51. White House, *Interim National Security Strategic Guidance* (Washington, DC: White House, 2021), 3.

52. U.S. Department of State, "Trafficking in Persons Report: July 2022," July 2022, full report available at https://www.state.gov/reports/2022-trafficking-in-persons-report/.

53. David Hutt, "Why China Is Picking a Fight with Vietnam," *Asia Times*, September 5, 2019, https://asiatimes.com/2019/09/why-china-is-picking-a-fight-with-vietnam.

54. Section 231 of the Countering America's Adversaries Through Sanctions Act can be found at https://www.state.gov/section-231-of-the-countering-americas-adversaries-through-sanctions-act-of-2017.

55. See, for example, "Vietnam/Russia: Vietnam, Russia Intensify Cooperation in Military Security," *Asia News Monitor*, March 23, 2018. https://search-proquest-com.usawc.idm.oclc.org/newspapers/vietnam-russia-intensify-cooperation-military/docview/2016509582/se-2?accountid=4444.

56. Bill Hayton, "The United States and Vietnam: An Emerging Security Partnership," United States Studies Centre at the University of Sydney, November 2015, 22.

57. Tsuboi, Yoshiharu, "Corruption in Vietnam," *Center of Excellence, Contemporary Asian Studies*, Waseda University, 2005, https://core.ac.uk/download/pdf/286945466.pdf.

58. Vietnam Ministry of National Defence, *2019 Viet Nam National Defence* (Hanoi: National Political Publishing House, 2019), 23–24.

59. Derek Grossman and Christopher Sharman, "How to Read Vietnam's Latest Defense White Paper: A Message to Great Powers," *War on the Rocks*, December 31, 2019, https://warontherocks.com/2019/12/how-to-read-vietnams-latest-defense-white-paper-a-message-to-great-powers.

60. Vietnam Ministry of National Defence, *2019 Viet Nam National Defence* (Hanoi: National Political Publishing House, 2019), 21.

61. Pham Minh Chinh, "Remarks Ahead of the ASEAN-U.S. Summit," given at the Center for Strategic and International Studies, May 11, 2022, https://www.csis.org/analysis/prime-minister-pham-minh-chinh-ahead-us-asean-summit.

62. This assessment is based on the author's personal experience while assigned to the U.S. Embassy in Laos from 2017 to 2019.

63. Xuan Loc Doan, "U.S., Vietnam Strategic Partners in All But Name," *Asia Times*, April 10, 2019, https://asiatimes.com/2019/04/us-vietnam-strategic-partners -in-all-but-name.

64. Nguyen Thi Thuy Hang, "U.S.-Vietnam Comprehensive Partnership Presents an [*sic*] Possible," *Journal of International Relations, Peace Studies, and Development* 4, 1 (2018), https://scholarworks.arcadia.edu/cgi/viewcontent.cgi?article =1046&context=agsjournal.

65. Phuong Vu, "What's in a Name: The Promise and Peril of a US-Vietnam 'Strategic Partnership,'" *The Diplomat*, March 2, 2022, https://thediplomat.com/2022/03 /whats-in-a-name-the-promise-and-peril-of-a-u-s-vietnam-strategic-partnership.

66. See, for example, remarks from Ambassador Le Cong Phung on January 28, 2020, in Washington, DC marking the beginning of Lunar New Year celebration, http://vietnamembassy-usa.org/relations/vn-us-vow-build-strategic -partnership.

67. Siew Mun Tang et al., *The State of Southeast Asia: 2020* (Singapore: ISEAS Yusof Ishak Institute, 2020), 29.

68. Erin Duffin, "Number of International Students in the U.S. by Origin, 2019/2020," *Statistica*, November 23, 2020, https://www.statista.com/statistics /233880/international-students-in-the-us-by-country-of-origin.

69. Japan Ministry of Defense, "Achieving the Free and Open Indo-Pacific Vision: Japan Ministry of Defense's Approach," July 2020, https://www.mod.go.jp/e /publ/pamphlets/pdf/indo_pacific/indo_pacific_e.pdf.

70. A comprehensive list of Japan-Vietnam defense engagements since 2012 can be viewed at https://www.mod.go.jp/e/d_act/exc/cap_b/vietnam/index.html.

71. Siew, *State of Southeast Asia: 2020*, 30

72. Alexander L. Vuving, ed., and Srini Sitaraman, "Are India and China Destined for War? Three Future Scenarios," *Hindsight, Insight, Foresight: Thinking About Security in the Indo-Pacific* (Honolulu: Asia Pacific Center for Security Studies, 2020), 296–97.

73. The full list of ADMM and ADMM-Plus activities can be found at https:// admm.asean.org.

74. Pham Minh Chinh, "Remarks Ahead of the ASEAN-U.S. Summit," given at the Center for Strategic and International Studies, May 11, 2022, https://www.csis .org/analysis/prime-minister-pham-minh-chinh-ahead-us-asean-summit.

75. Derek Grossman, "Ukraine Conflict Echoes Loudest in Vietnam, not Taiwan," *Nikkei Asia*, March 21, 2022, https://asia.nikkei.com/Opinion/Ukraine -conflict-echoes-loudest-in-Vietnam-not-Taiwan.

CHAPTER 12

Competing in the Middle East

Christopher J. Bolan

INTRODUCTION

There has been a long debate in both academia and the policymaking world about the role played by national security interests and values in the formulation of American foreign policy and decision-making. Foreign policy decision-making is perhaps easiest when both national interests and values align. United States (U.S.) military support to democratic nations in Europe through regional security structures such as the North Atlantic Treaty Organization (NATO) has enjoyed bipartisan support for decades. In this situation, there is little tension between American national interests arguing for the provision of security guarantees to Europe and American values promoting free-market economies and democratic governance.

Unfortunately, the Middle East has presented American policymakers with difficult choices and trade-offs to be made in terms of securing national security interests at the expense of promoting American values. U.S. national security interests in regional stability, nonproliferation, terrorism, and the defense of Israel have often driven U.S. decision-makers to provide military and financial support to authoritarian and repressive governments in the region at odds with traditional American democratic values.

This chapter argues that both changing American national security interests and evolving internal regional political and societal transformations demand a new American approach to the Middle East. Charting a new American strategy in the midst of such significant change will not be easy or straightforward. Nonetheless, this chapter seeks to make the case that such change is necessary to effectively achieve declining American national interests in a region undergoing significant transformation itself.

A new strategic approach must meet several new criteria. American strategy will need to strike a new balance between competing and cooperating with outside actors to accomplish American interests spanning energy security, nonproliferation, terrorism, and the defense of Israel. It must move away from an approach overemphasizing American military dominance and toward broader considerations aimed at maintaining American influence through diplomacy to reduce internal regional divisions and tensions and through economic engagement that assists the region in moving away from centralized government controls and heavy reliance on the export of fossil fuels. Finally, America should seek to expand beyond an approach focused narrowly on government-to-government contacts and programs to more fully engage civic societies and the private economic sector both in the United States and in the region itself.

RECALIBRATING U.S. INTERESTS

In a speech at the U.S. State Department in 2011, President Obama delivered a crisp distillation of America's historical "core interests" in the Middle East.[1] Although priorities have varied across the decades, these include the free flow of energy, preventing the proliferation of nuclear weapons, countering terrorism, the defense of Israel, and the pursuit of an Arab–Israeli peace.

Energy Security and Building a Permanent U.S. Military Presence

It was the growing strategic significance of oil during the latter half of the twentieth century that propelled the Middle East from a relative backwater to center stage in American foreign policy. The insatiable demand for oil to fuel extensive allied military operations during World War II convinced American strategists of the importance of the massive oil reserves located in Saudi Arabia and the Gulf. As Daniel Yergin writes in his Pulitzer prize-winning history of oil and its impact on international politics:

The lessons of World War II, the growing economic significance of oil, and the magnitude of Middle Eastern resources all served, in the context of the developing Cold War with the Soviet Union, to define the preservation of access to that oil as a prime element in American . . . security.[2]

Recognition of American interest in access to the region's substantial energy reserves propelled President Roosevelt to become the first sitting president to visit the region and declare in 1943 that "the defense of Saudi Arabia is vital to the defense of the United States."[3] Throughout much of the Cold War, numerous regional challenges—such as Iranian nationalization of its own extensive oil reserves in 1951 and the 1973 Saudi-led

OPEC oil embargo—threatened to disrupt America's uninterrupted access to Middle East oil. However, lacking any realistic military options, these disruptions were countered through a combination of American diplomacy, covert intelligence operations, and domestic calls for conservation, increased domestic production of fossil fuels, and expansion of renewable energy sources.

It was the Soviet invasion of Afghanistan in 1979 that prompted then President Carter to openly declare Middle Eastern oil a vital national security interest[4] and to take the initial steps to create an independent American military force capable of deploying to protect those distant energy resources. President Carter moved to create the Rapid Deployment Joint Task Force (RDJTF). The RDJTF was the predecessor to present-day U.S. Central Command headquartered in Florida charged with providing an "over-the-horizon" military capability to intervene forcefully and quickly to defend American interests in the Middle East.

However, the first test of this newly constructed military capability would not come until decades later in response to Saddam Hussein's August 1990 invasion of Kuwait. President George H. W. Bush and his talented Secretary of State James Baker marshalled an international coalition of thirty-nine countries[5] and deployed over 500,000 American troops to Saudi Arabia as part of Operation Desert Storm to reverse this Iraqi aggression.[6] In addressing a joint session of Congress, President Bush emphasized the danger this invasion posed to the free flow of energy resources to the world.

Vital economic interests are at risk as well. Iraq itself controls some 10 percent of the world's proven oil reserves. Iraq plus Kuwait controls twice that. An Iraq permitted to swallow Kuwait would have the economic and military power, as well as the arrogance, to intimidate and coerce its neighbors—neighbors who control the lion's share of the world's remaining oil reserves. We cannot permit a resource so vital to be dominated by one so ruthless.[7]

Although the number of U.S. troops would ebb and flow, the physical presence of American troops became the centerpiece of America's strategy in the region. Today, the American military footprint remains substantial at somewhere between 60,000 and 70,000 troops spread across a vast network of bases throughout the region.[8]

U.S. Interests No Longer Require a Large Permanent U.S. Military Footprint

The foundational strategic logic for America's commitment to safeguarding the region's energy supplies is eroding. In this arrangement the U.S. role was to provide regional security while America's oil exporting partners would surge to pump as much oil and natural gas as required to fuel global economic growth at a sustainable cost for energy consuming countries. However, this bargain has broken down on both sides.

For one, the United States is simply no longer as dependent on Middle Eastern oil as it was throughout much of the Cold War. Throughout the 1970s, for instance, the United States imported as much as 30–50% of the oil it consumed—much of it from the Middle East.[9] Today, the Persian Gulf accounts for 8–9% of total American imports of petroleum and crude oil.[10] According to the BP Statistical Review of World Energy, the United States became the world's top oil producer in 2014 and the top energy producer in 2017.[11] As such, the relationship between the United States and the oil producing countries in the Middle East has been fundamentally transformed from one of consumer (U.S.) and producer to one of global competitors in energy production. As a result of this shift, the two countries with the vast majority of the world's spare production capacity (Saudi Arabia and the United Arab Emirates, UAE) are simply no longer as responsive to U.S. requests to increase production to ease oil market volatility.[12]

Of course, the Saudis and other Arab Gulf leaders have their own complaints about Americans unwillingness to hold up their end of this bargain. In particular, they see President Obama's (ultimately successful) efforts at forging a nuclear deal with Iran as an implicit willingness to turn a blind eye to Tehran's other threatening behaviors including its advanced ballistic missiles and its extensive political, economic, and military support to Sh'ia militia in Lebanon (Hizbollah), Yemen (Houthis), and Iraq (Popular Mobilization Forces).[13] Their confidence was further shaken when despite his tough rhetoric against Iran, President Trump did not respond more forcefully to aggressive Iranian actions that included downing a U.S. Global Hawk drone,[14] attacks on international shipping in the Gulf,[15] and drone and missile strikes on Saudi oil facilities[16] and U.S. occupied military bases in Iraq.[17]

Additionally, the region faces no comparable external or regional conventional military threat capable of dominating the region's vast energy resources that have long justified maintaining such an expansive and expensive on-the-ground U.S. military presence. The Soviet Union that invaded Afghanistan in 1979 and exposed the region's vulnerability to outside aggressors is no longer. Apart from its substantial nuclear arsenal, Russia today is a mere shadow of the former USSR. In particular, Russia's conventional military forces in Ukraine have proved themselves incompetent across the strategic, operational, and tactical levels of war. Arab leader confidence in Russia as a strong military partner has undoubtedly been shaken and is not likely to recover any time soon.

Similarly, there is no single country in the region capable of exercising regional hegemony in a way that would threaten the energy reserves concentrated in the Persian Gulf. Iraq's military was largely destroyed during

Operation Iraqi Freedom in 2003 and is today racked by civil war and primarily concerned by internal threats posed by the Islamic State of Iraq and Syria (ISIS) and Iranian-backed militia forces. Meanwhile, Iran poses no realistic threat or capacity to occupy territory outside of the country itself. Iran's conventional military forces have been severely degraded by decades of Western sanctions and starved of resources by clerics in Tehran who prefer to bolster the capabilities of the Islamic Republican Guard Corps (IRGC) as the true defenders of the regime.

Furthermore, America's core interests today can simply be secured without this substantial American military footprint. The absence of physical threat to the region's energy resources and vastly reduced American dependence on Middle Eastern oil allow U.S. policymakers to assume more risk in the Middle East. U.S. policymakers can restructure and reduce America's military footprint in the region and shift these resources to higher priority theaters in Asia (competing with a rising China) and Europe (confronting a resurgent if stumbling Russia).

Meanwhile, the terrorist threat emanating from groups like Al Qaeda and ISIS has been significantly reduced and is best handled by a strategy that emphasizes building the capability of our partners in the region to deal with these increasingly localized threats. Daniel Byman, a terrorism expert at Georgetown University and the Brookings Institute, argues that Al Qaeda is a much-diminished threat as it has failed to accomplish any of its major strategic goals since 9/11 and currently suffers from significant organizational weaknesses including financial troubles, infighting, and the lack of a physical safe haven.[18] Meanwhile, the focused but limited application of U.S. military force in support of allied ground forces operating in Iraq and Syria has effectively destroyed the physical caliphate created by the Islamic State.[19]

The most immediate threat to America's nonproliferation objectives for the region is clearly Iran. However, the potential for preventing Iran from emerging as a nuclear weapon state is best handled through negotiations. It was the skillful combination of economic pressure and successful diplomacy that ultimately imposed the most significant constraints on Iranian civilian nuclear activities. The Joint Comprehensive Plan of Action (JCPOA) or Iran nuclear deal concluded in 2015 accomplished the following: (1) reduced Iran's stockpile of enriched uranium by 98%; (2) eliminated two-thirds of its centrifuges; (3) extended its "breakout" time required to assemble enough enriched uranium for a single nuclear bomb from two to three months to at least one year; and (4) and imposed the world's most comprehensive and intrusive arms inspection regime over Iran's nuclear program.[20] America's withdrawal from the JCPOA in 2018 has only marshalled further historical evidence bolstering the case for negotiations and a return to this internationally backed nuclear deal. In response to President Trump's "maximum pressure" strategy, Tehran adopted a counterstrategy of "maximum resistance" that has brought Tehran to within striking range of becoming a nuclear threshold state. Iran today has stockpiled enough enriched uranium to build a single nuclear bomb[21] (bringing its break-out time effectively to zero) and is using advanced centrifuges to enrich uranium at higher levels of 60%[22] (JCPOA capped enrichment levels to 3.67%). In the absence of a negotiated deal, Iran has only further developed its civil nuclear capabilities, placing it closer to effectively becoming another nuclear threshold state.

As for defending Israel, Israel today is arguably more secure than it has been its entire history and is quite capable of defending its own interests independent of U.S. military capabilities. It has signed peace agreements with Egypt, Jordan, the United Arab Emirates, Bahrain, Morocco, and Sudan. Meanwhile, the bordering states of Lebanon and Syria are subsumed by domestic civil wars and economic collapse—rending neither capable of seriously threatening Israel militarily. Additionally, Israel is the region's only nuclear weapons state,[23] and decades of massive U.S. assistance totaling over $140 billion have enabled Israel to build the region's most capable military forces.[24] Israel has repeatedly used its military advantages to defeat opponents in four Arab–Israeli wars, to launch multiple incursions into Lebanon, and today conducts regular unilateral air strikes against Iranian targets in Syria while conducting sophisticated covert operations against Iranian nuclear facilities and scientists. In short, Israel is quite capable of militarily defending its own vital interests in the region.

CHARTING U.S. STRATEGY IN A CHANGING MIDDLE EAST

Internal Sources of Regional Instability[25]

The real but limited security threat posed by Iran has all too often been presented as the primary cause of instability in the Middle East. This allegation has been repeated often and loudly enough both inside and outside of the region that a casual observer could be forgiven for succumbing to the misleading belief that if not for the malign activities of Iran, the Middle East would be a stable, prosperous, and peaceful region. Iranian leaders themselves exaggerate their role and influence in regional affairs, claiming that Tehran now controls four Arab capitals in Lebanon, Iraq, Syria, Yemen.[26]

U.S. senior military leaders largely echo these assessments leaving the impression that Iran is the primary instigator of violence and instability in the region. In testimony to the Senate Armed Services Committee in 2017 as then commander of all U.S. forces in the Middle East, General Joseph Votel identified Iran as the "greatest long-term threat to stability for this part of the world."[27] His successor General Kenneth McKenzie, Jr., in one of his first public speeches at a Washington-based policy institute, similarly characterized Iran as the "long-term, enduring, most significant threat to stability in the Central Command AOR."[28]

These assessments are not entirely misconceived from a narrow military viewpoint as Iran is a significant state actor seeking to undermine U.S. influence in the region. Iran's civilian nuclear program has justifiably raised concerns about the risks that leaders in Tehran might at some point decide to actively pursue the development of a nuclear weapon. Iranian support to a wide range of Shi'a militias (estimated by some analysts to include a network of as many as 200,000 fighters) poses a direct threat to U.S. allies.[29] Additionally, Iran also possesses a potent ballistic missile force that exposes U.S. and allied forces and infrastructure to attack.

However, a singular and obsessive emphasis on Iranian activities (even if understandable given rising U.S.–Iran tensions) obscures a much larger truth. The fundamental source of instability in the region stems not from Tehran, but rather from the failure of Arab leaders themselves to satisfy the basic political, economic, and societal needs of their own people.

The instability that has roiled the region since the Arab uprisings in 2011 is a direct product of failed domestic policies in these countries that include widespread political repression, pervasive corruption by the ruling elites, poor education systems that have not done enough to develop Arab human capital, and economies that have proved incapable of generating meaningful employment for much of the region's population. These are the enduring domestic challenges in the Arab world that must be more effectively addressed if the region is to enjoy any semblance of stability over the longer term.

While Iran has certainly sought to take advantage of the instability in the wake of the Arab uprisings for its own purposes, it is not a proximate cause of that instability. For instance, Iran played no role whatsoever in ousting long-time pro-Western Arab leaders in Tunisia, Egypt, or Yemen. In Libya, it was a NATO-led military operation that was responsible for ousting Muammar Qaddafi, creating a leadership vacuum that set the stage for today's chaos enveloping the county. The Syrian civil war gathered momentum only after Assad's brutal overreaction to domestic protests in the provincial capital of Deraa that were led by Syrian teenagers inspired by the Arab Spring revolution taking place in other Arab countries. In no Arab capitals were protesters calling for the emulation of Iran's revolutionary call for leadership by Shi'a clerics (velayat-e-faqih). Indeed, public opinion polls in the Arab world indicate quite the reverse—as public attitudes toward Iran are increasingly negative. Instead, the protests that roiled the region were by all accounts driven by the failure of Arab leaders themselves to satisfy the desires of people for political expression, economic prosperity, and personal dignity.

Indeed, the slogans repeated throughout Arab capitals echoed the protest themes of "bread, freedom, and social dignity" sounded in Egypt's

Tahrir Square. In this sense, a myopic focus on the secondary role played by Iran in aggravating these problems only obscures the more salient need for Arab leaders to address these more foundational political, economic, and social challenges themselves. Encouraging and assisting Arab leaders in making these necessary reforms should be a central diplomatic and economic task for U.S. policymakers.

Changing Intra-Regional Power Dynamics[30]

At same time as U.S. regional interests are evolving, and domestic social, political, and economic pressures are building on Arab leaders, the region itself is undergoing significant internal power shifts that will both confound and present opportunities for U.S. policymakers and strategists.

One of these most significant internal shifts of power is the relative rise of the non-Arab states of the region, namely, Iran, Turkey, and Israel. For varied reasons, these countries have emerged from the relative chaos and uncertainties of the Arab uprisings relatively unscathed and in many ways have seen their influence expand—often at the expense of the region's historical regional power centers in Cairo, Baghdad, and Damascus. The perception of U.S. disengagement from the region has further spurred leaders in Tehran, Ankara, and Jerusalem to increasingly assert themselves and pursue their own interests—independently and with increasing frequency in direct defiance of U.S. interests and considerations.

U.S. regional interests today are most directly and aggressively opposed by Iran. Particularly since the Iranian revolution in 1979, leaders in Tehran have actively sought to undermine the American interests in the region. They have done so in no small measure by seeking to undermine the legitimacy of America's partners and allies through direct and indirect means. One of primary means of expanding Iranian regional influence has been the IRGC—the vanguard of the clerical regime in Tehran. This force has been instrumental in financing, training, and equipping a region-wide network of more than one hundred Shi'a partners and militias extending from Lebanon (Hizbollah), to Syria (the Assad regime and assorted militia groups), Iraq (Popular Mobilization Forces), and Yemen (Houthis).[31] Iran has also developed advanced ballistic missile and drone forces that have successfully attacked international shipping in the Arabian Sea,[32] oil refinery facilities in Saudi Arabia,[33] and U.S. forces operating in Iraq.[34] Of particular concern to the broader international community have been advances in Iran's (thus far) civilian nuclear program that could one day provide the foundation for development of a nuclear weapons capability. The internationally backed JCPOA, commonly referred to as the Iran nuclear deal, had effectively constrained Iran's civil nuclear program through a series of mutually agreed upon limitations (e.g., caps on uranium stocks and limiting the number and types of centrifuges) coupled with the world's most intrusive arms control inspection regime.[35] Unfortunately, since President Trump's withdrawal from this deal in May 2018 and announcement of his strategy of maximum pressure, Iran has doubled down on its own countervailing strategy of maximum resistance and exceeded many of the JCPOA constraints, having amassed enough enriched uranium for a single nuclear bomb, moving Tehran ever closer to becoming a nuclear threshold state.[36]

While Turkey has been a member of the U.S.-backed North Atlantic Treaty Organization for decades, it has increasingly taken strong positions at odds with U.S. interests and values. For instance, in the run up to the American invasion of Iraq in 2003, the Turkish Parliament refused American requests to deploy U.S. forces on Turkish soil.[37] Meanwhile, under President Erdogan's leadership, Turkey has grown more anti-democratic, authoritarian, and repressive.[38] Despite vehement U.S. objections, he has moved Turkey closer to Russia, going so far as to purchase the S-400 advanced Russian air defense missile, posing a direct threat to NATO aircraft.[39] Erdogan has openly threatened U.S. forces operating in Syria who are supporting Kurdish militias considered terrorists by Ankara.[40] He has also defied U.S. sanctions aimed at isolating Iran and instead has announced plans to significantly expand Ankara's bilateral trade with America's arch enemy in Tehran.[41] Meanwhile, at the regional level Turkey has backed Islamic political movements and movements, including the Muslim Brotherhood, in direct opposition to American-backed allies in Saudi Arabia, the UAE, and Egypt. Similarly, Turkey's intervention into Libya backing Islamist elements is opposed by Egypt and several of the Arab Gulf states. Turkey also forged a strategic political, economic, and military alliance with Qatar despite a three-year-long Saudi–UAE-led boycott of this small Arab Gulf country that only recently ended in 2021.[42]

Israel too is increasingly adopting policy positions openly at odds with U.S. preferences and interests. For decades, Israel has defied U.S. pressure to abandon settlements in the West Bank. Nonetheless, leaders in Tel Aviv and Washington have strived to present a united front on other regional security issues. During U.S. military operations in 1991 to liberate Kuwait from Iraqi occupation, U.S. political leaders convinced a reluctant U.S. General Schwarzkopf to devote scarce military resources to hunting for Iraqi SCUD missile units to ease Israeli concerns and preempt the potential for unilateral Israeli military action.[43] However, as discussed previously, Israel today is arguably more secure than ever in its history. Coupled with the perception of an American disengagement from the region, Israel is increasingly willing to challenge U.S. policies and take independent policy positions. Nowhere is this more apparent than in Israeli public opposition to American efforts to conclude a nuclear agreement with Iran. Israeli Prime Minister Benjamin Netanyahu went so far as to deliver an unprecedented address to a joint session of the U.S. Congress in 2015, declaring his public opposition to President Obama's seminal foreign policy accomplishment concluding the JCPOA.[44] Such a provocation risked creating a U.S. partisan divide over decades of American military support totaling over $140 billion.[45] More recently, as the Biden administration was struggling to conclude negotiations to reinstate the JCPOA, Israel undertook covert operations killing Iranian IRGC officials that (whatever their justification) would inevitably derail these delicate discussions.[46] In criticizing efforts to revise the JCPOA, Israel Prime Minister Neftali Bennett has also reminded international and American officials alike that Israel would not hesitate to take independent military action to block Iran's ability to develop nuclear weapons.[47]

The other major development in shifting regional power structures has to do with the devolution of power from the historic regional centers of Baghdad, Egypt, and Damascus to the Arab Gulf countries. This shift is the result of social, economic, and political factors. Egypt and Damascus were centers of the Arab uprisings beginning in late 2010 demanding new leadership and more effective governance. Meanwhile, Iraq and Syria are suffering from internal societal divisions fueling civil wars. Despite possessing the region's largest population, Egypt's traditional role as regional power broker has suffered because of its departure from the Arab League consensus by signing a peace treaty with Israel and by its own internal economic and political struggles. Although not entirely immune from the Arab uprisings, the oil-rich Arab Gulf states had sufficient economic resources to stave off large-scale public discontent at least temporarily. By some estimates, Saudi Arabia and other Gulf states devoted some $150 billion to increased social welfare programs in response to this potential unrest.[48] The monarchies of the region also seem to have emerged from the Arab uprising relatively unscathed—perhaps because of a deeper sense of cultural, political and religious legitimacy as contrasted with the regional republics and because of the financial resources they can throw at social services.[49]

This focus on internal threats posed by the Arab uprisings coupled with the growing significance of Asian economies as the primary consumers of Middle Eastern energy, the rise of a more risk-acceptant younger generation of leaders in the Gulf, and the impression of an intensified threat posed by Iran at a time of U.S. disengagement from the region have led the Arab Gulf leaders to increasingly assert their own independence from Washington. Leaders in Riyadh and Abu Dhabi have most visibly charted a new, more aggressive foreign policy that departs from more traditional quietist policies that took their cue from leaders in the United States. This was especially evident in the brash decision-making of then Saudi Defense Minister Mohammed bin Salman (now the crown prince and effective ruler of the kingdom) to launch his disastrous military campaign in Yemen beginning in 2015. This campaign has cost the kingdom $100 billion[50] and has failed in its primary strategic objective of eliminating the threat posed by Iranian-backed Houthis on Riyadh's southwestern border. Moreover, the campaign has further sullied the kingdom's already shaky international reputation,[51] created momentum in the U.S. Congress to end military assistance,[52] and produced what the United Nations in 2019 declared the world's worst humanitarian disaster.[53]

The UAE has been Saudi Arabia's major Gulf partner in the disastrous military campaign in Yemen.[54] Aside from Yemen, other visible signs of strain in the U.S.–UAE bilateral relationship were the UAE's initial abstention on a UN vote to condemn Russia for its invasion of Ukraine[55] and secretly allowing China to build port facilities near Abu Dhabi.[56] Moreover, U.S. officials have publicly reprimanded the UAE for its continuing trade and financial dealings with Iran in contravention of U.S. sanctions.[57] All this has led the UAE ambassador to the United States, Yousef Al Otaiba, to admit that this important bilateral relationship is currently undergoing a significant "stress test."[58]

U.S. POLICY RECOMMENDATIONS

Adopt a Strategy of Partnership

Globally, American strategies will need to prioritize the more critical and urgent theaters of Asia and Europe—meaning that the Middle East can simply no longer occupy center stage in U.S. calculations. Charting an American regional strategy for the Middle East will require policymakers to make an accurate assessment of America's declining interests while recognizing the opportunities and challenges presented by shifting intraregional power dynamics. U.S. policymakers will have to be creative, flexible, and agile in reacting to these changing currents. U.S. strategy will also need to account for the reality that Arab Gulf countries are increasingly tied economically to Asia, which will result in growing Chinese influence. Some of this influence will necessarily come at the expense of America's historical dominance of the region—particularly in terms of economics and politics. This trend will be reinforced by the rising trend of political authoritarianism in the region. Moreover, the widespread perception of American disengagement will fuel current trends of regional leaders taking unilateral actions that are at times at odds with American calculations of interests and values. The sum of these trends suggest that American policymakers will no longer be able to meaningfully dictate many regional outcomes. With fewer resources at hand, American strategy must be smarter and laser-liked focused on prioritizing those issues that matter most. In short, instead of approaching the region from a position of assumed dominance, U.S. policymakers will need to adopt the mindset of building genuine partnerships with regional leaders.

Acknowledge That the Military Competition Has Been Won

Nowhere is this needed shift in overall strategic approach more evident than in the military sphere. The Middle East was sensibly America's strategic priority when counterterrorism was the most immediate threat in the wake of the 9/11 attacks. However, consecutive U.S. national security strategies now have demanded a strategic focus on great power competition to confront a rising China and resurgent Russia. Fortunately for American policymakers, the decades-long focus on threats emerging from the Middle East have left the United States in a position of unchallenged military dominance. Although estimates vary, approximately 40,000 to 60,000 American troops are stationed throughout the region across an extensive array of military bases in twenty-one countries within the U.S. Central Command Area of Operations. This network of bases provides the United States a unique capability to surge and facilitate large deployments of military forces and equipment as threats emerge.[59]

Additionally, the United States has been the largest single arms supplier to the region for decades.[60] As a result, the most capable of these regional military forces are today heavily reliant on U.S. military equipment and training, creating a dependence on Washington that will be difficult if not impossible to replicate or reverse. In the military realm at least, U.S. policymakers should be generally comfortable with the status quo. China has not yet demonstrated either the desire or capability to replace America as the provider of regional security. Meanwhile, beyond the ability to provide second-rate military equipment, Russia is unlikely to significantly chip away at America's military dominance as Moscow will be consumed by its campaign in Ukraine for the foreseeable future.

Reshape the U.S. Military Footprint in the Middle East[61]

Although U.S. interests are diminished, there is still a role to be played by American military forces. The terrorist threat that justified the deployment of over 200,000 troops to invade and then occupy Afghanistan and Iraq is much more diffuse and diminished. Meanwhile a tailored and intelligent reduction of the U.S. military forces in the region will expose fewer U.S. troops to ballistic missile and drone strikes by Iran and Iranian-backed militia operating in Syria, Lebanon, and Iraq.

As demonstrated by the transition of America's military force posture in Iraq from one of occupier to that of enabler, the implementation of a "by-with-and-through" strategy[62] of building local capacity to deal with localized terrorist threats means the U.S. military mission will primarily focus on training and equipping. The focus of U.S. ground forces present in the region should be this preparation of local military forces to carry the burden of counterterrorism operations. Bolstering regional missile defenses against the Iranian missile and drone threats will serve as another mission for U.S. forces and as an important deterrent against Iranian strikes. Meanwhile U.S. ground, naval, and air forces can take advantage of America's region-wide access to basing to periodically surge in and out of the region to ensure freedom of navigation, bolster regional deterrence efforts targeting the multifaceted threat posed by Iran, conduct joint coalition exercises, and undertake limited combat operations when necessary.

A focus on this set of narrowed military missions should allow for a meaningful reduction in the overall U.S. military footprint in the region. Additional steps could include consolidating U.S. base facilities at fewer locations; relocating major portions of the U.S. forward military headquarters back to the continental United States; and shifting some of the intelligence, surveillance, and reconnaissance assets out of the region while bolstering local capabilities to perform at least some of these functions.

Move Toward a Regional Security Architecture

The strategic imperative to shift resources and attention to Asia and Europe implies a parallel reduction in America's investment in the Middle East. U.S. policymakers should encourage both outside and regional leaders to take constructive steps to bear a heavier responsibility for stability and security by promoting cooperation over competition. Encouraging Saudi leaders to end their counterproductive boycott of Qatar was an important step in this regard. Additionally, the Biden administration can seek to exploit the potential openings for Arab–Israeli economic and military cooperation created by the Abraham Accords. The United States should also lend support to emerging Arab dialogues with Tehran that have the potential to lower the temperature of sectarian competition and thus reduce the prospect of open military conflict. Whether or not the JCPOA can be resurrected, the United States should begin a diplomatic initiative to create an Organization for Security and Cooperation in Europe (OSCE)-like regional security and consultation structure that eventually includes Iran. Such an organization could provide a framework for improving transparency; encouraging regional cooperation in areas of mutual concern such as freedom of navigation in the Gulf, climate change, and infectious disease; and providing an established mechanism for communication in the event of crisis. Leaders from Middle Eastern states themselves including the two key rivals—Saudi Arabia and Iran— have expressed the need and formally proposed regional security initiatives along these lines.[63] A number of models for such an architecture exist across the globe, and each offers insights that might be useful in informing one best tailored to the conditions and challenges of the Middle East.[64] In particular, the OSCE model could be used to initially focus on arms control and crisis management issues aimed at reducing the prospect for armed confrontation. This could build off the existing Gulf Cooperation Council

(GCC) Arab dialogues with Iran. Other major efforts could examine prospects for improving regional economic integration and building civil societies that are tolerant and resilient.

On an international level, the United States should abandon a Cold War approach to Chinese and Russian engagement in the region that frames everything in terms of a zero-sum competition. U.S. leaders should actively seek to channel this engagement in constructive ways when interests overlap. This is not a particular novel idea and in fact is a regular feature of great power *cooperation* in the Middle East. For instance, the Obama administration worked tirelessly with Russia and China to secure Iran's agreement to the nuclear deal in 2015. Both President Obama and President Trump maintained U.S.–Russian military deconfliction communications to prevent unintended escalation in Syria. Meanwhile, Chinese combat engineers support UN demining operations in Lebanon, and Chinese naval ships are active participants in antipiracy operations in the Gulf of Aden.

Reenergize the Nonmilitary Instruments of Power

The main challenge for American policymakers will to be reorient Middle East strategy away from one dominated by military means and instead rely on creative and aggressive use of the instruments of diplomacy, economics, finance, and information. Deeper U.S. military engagement in the Middle East in the aftermath of the 9/11 terrorist attacks has not led to greater regional stability. The internal sources of the region's instability—political repression, societal divisions, corruption, ineffective governance, economic inequalities—must be addressed primarily through actions taken by regional leaders and through a broader U.S. approach that better leverages contributions from the nonmilitary instruments of power. U.S. policymakers can be most effective at promoting these efforts by conditioning and scaling U.S. financial assistance to specific government reforms that improve transparency and bolster civic society, promote small-scale startups, and otherwise build the private sector. A parallel effort is needed to engage U.S. business leaders to help identify specific steps that could be taken to foster a business environment more conducive to direct foreign investment. Additionally, the U.S. government should be willing to provide technical expertise in critical areas such as education, water conservation, health services, and green energy production that would contribute to regional stability over the longer term.

U.S. leadership will also be required to coordinate a larger multilateral effort along these same lines. This will necessarily involve discussions with international actors with the potential to make substantial investments in the region, including traditional Western allies. For instance, the G7 in June 2022 announced its intent to raise $600 billion to counter China's Belt and Road Initiative.[65] However, channeling investments from China and

Russia into projects that also contribute to regional stability is also smart policy. China's Belt and Road Initiative offers attractive opportunities for foreign investment to many countries in the region. Opposing all such projects as a reflexive U.S. policy is not sustainable. The reality is that U.S. private sector companies will be reluctant to invest in areas or countries that are at high risk of civil war (e.g., Lebanon, Syria, Yemen) or lack the necessary transparency and rule of law that will guarantee a fair return on investment. As a result, U.S. policymakers will need to be selective about where and in which sectors to resist or constructively engage Chinese or Russian activities. In this context, encouraging Chinese and Russian projects in countries like Syria, Libya, and Yemen where the U.S. private sector is unlikely to invest makes some degree of strategic sense. Meanwhile, U.S. policymakers should continue to strongly resist the sale of advanced Russian missile systems that compromise the ability of the GCC countries to develop an integrated air defense and missile system. Similarly, U.S. policymakers should continue to oppose Chinese projects to build port or other major infrastructure facilities in partner nations (e.g., Israel and the UAE) when they create unacceptable vulnerabilities for visiting U.S. naval and other military forces.

But there are also important opportunities where the United States and regional leaders can find mutual advantage in economic, financial, and technological cooperation. One important area ripe for potential coordination and investment is energy. In a recent report, Karen Young, the founding director of the Program on Economics and Energy at the Middle East Institute, writes:

There is an increasing global demand for climate finance, new green investments, and the creation of circular economies of scale. The Gulf and the wider Middle East will be hard hit by climate change. The Gulf states' ability to lead solutions will enable their own survival and ability to thrive in the challenges ahead. U.S. policy coordination through blended finance of international financial institutions, U.S. government development finance arms and Gulf state development funds might seek to provide financing for projects like renewable power plants, as well as adaptive infrastructure in transport, including electric fleets and buses in public transport, across the Middle East.[66]

Finally, while China and Russia will continue to seek advantage by highlighting affinities between authoritarian models and elites, the U.S. can emphasis its advantages with both regional leaders and the broader Arab publics by serving as a competing model for political and economic openness. Indeed, Arab leaders themselves have an interest over the long term in building resilient and tolerant civil societies more resistant to extremism. One recent opinion poll taken involving nearly 30,000 respondents from thirteen Arab countries found over two-thirds of Arabs support a pluralistic democratic system of government for their country.[67]

Another survey of over 200 million Arab youth says that more are increasingly turning to business entrepreneurship as a means to guarantee their own prosperity and that they fully expect their government to promote reforms needed to foster these opportunities.[68] The United States can play a constructive role in encouraging and fostering these attitudes among the future leaders of this region. Even in a region dominated by authoritarian leaders, U.S. policymakers still have important ideological cards to play in the great game of geopolitics.

CONCLUSION

The immediate security challenges associated with Iran will continue to occupy a priority spot on the list of urgent American foreign policy priorities in the Middle East. There will also be the understandable if ill-considered temptation to view any inroads made by Russia or China in the region as something to be reflexively resisted rather than constructively channeled. However, the challenges confronting a region wrecked by a series of civil wars, sectarian conflict, political repression, and economic inequalities cannot be addressed by American military might alone. U.S. diplomatic, political, and economic leadership will be needed to forge genuine and effective international and regional partnerships that bridge government, private sector, and civil societies. Competing and winning in this space will require a broad U.S. strategic approach that emphasizes investments in the nonmilitary instruments of power. Fortunately, a reduced U.S. dependence on Middle Eastern oil and the absence of any competing power capable of dominating the region provides a window of opportunity to shrink and restructure America's military footprint to enable this needed strategic shift.

NOTES

1. Barack H. Obama, "Remarks by the President on the Middle East and North Africa," U.S. State Department, Washington, DC, May 19, 2011, https:// obamawhitehouse.archives.gov/the-press-office/2011/05/19/remarks -president-middle-east-and-north-africa.

2. Daniel Yergin, *The Prize: The Epic Quest for Oil, Money & Power* (New York: Free Press, 1992), 410.

3. P.T. Hart, *Saudi Arabia and the United States: Birth of a Security Partnership* (Bloomington: Indiana University Press, 1999) as quoted in Jack Covarrubias and Tom Lansford, eds., *Strategic Interests in the Middle East: Opposition and Support for US Foreign Policy* (Burlington, VT: Ashgate Publishing Company, 2007), p. xvii.

4. President Jimmy Carter, "State of the Union Address," January 23, 1980, https:// www.jimmycarterlibrary.gov/assets/documents/speeches/su80jec.phtml.

5. "Gulf War Fast Facts," CNN.com, August 2, 2021, https://www.cnn.com /2013/09/15/world/meast/gulf-war-fast-facts/index.html.

6. Shannon Collins, "Desert Storm: A Look Back," U.S. Department of Defense, Washington, DC, January 11, 2019, https://www.defense.gov/News/Feature-Stories/story/Article/1728715/desert-storm-a-look-back/.

7. George H. W. Bush, "Address Before a Joint Session of Congress on the Persian Gulf Crisis and the Federal Budget Deficit," Washington, DC, August 11, 1990, https://bush41library.tamu.edu/archives/public-papers/2217.

8. "How Many Troops Are Deployed in the Middle East and Surrounding Region?" *USA Facts*, January 8, 2020, https://usafacts.org/articles/how-many-troops-are-deployed-middle-east/.

9. "Oil Dependence and U.S. Foreign Policy, 1850–2022," Council on Foreign Relations, Washington, DC, accessed May 27, 2022, https://www.cfr.org/timeline/oil-dependence-and-us-foreign-policy.

10. "Oil and Petroleum Products Explained," U.S. Energy Information Agency, Washington, DC, April 21, 2022, https://www.eia.gov/energyexplained/oil-and-petroleum-products/imports-and-exports.php.

11. Robert Rapier, "Yes, the U.S. Is the World's Top Energy Producer," *Forbes*, August 8, 2018, https://www.forbes.com/sites/rrapier/2018/08/08/yes-the-u-s-is-the-worlds-top-energy-producer?

12. Noah Browing, "Saudi Arabia and UAE Could Ease Oil Market Volatility, IEA Says," *Reuters*, February 11, 2022, https://www.reuters.com/business/energy/saudi-arabia-uae-could-ease-oil-market-volatility-iea-says-2022-02-11/; Edward Helmore, "Saudi Arabia and UAE Leaders 'Decline Calls with Biden' amid Fears of Oil Price Spike," *The Guardian*, March 8, 2022, https://www.theguardian.com/us-news/2022/mar/09/saudi-arabia-and-uae-leaders-decline-calls-with-biden-amid-fears-of-oil-price-spike.

13. Stephen Kalin and Sarah Dadouch, "Gulf Arab Allies Hail Triumph after U.S. Quits Iran Deal," *Reuters*, May 8, 2018, https://www.reuters.com/article/us-iran-nuclear-gulf-reaction/gulf-arab-allies-hail-triumph-after-u-s-quits-iran-deal-idUSKBN1I93CU.

14. Joshua Berlinger et al., "Iran Shoots Down US Drone Aircraft, Raising Tensions Further in Strait of Hormuz," CNN, June 20, 2019, https://www.cnn.com/2019/06/20/middleeast/iran-drone-claim-hnk-intl/index.html.

15. H. I. Sutton, "Spate of Attacks on Ships in the Middle East Points to Iran-Backed Group," *USNI News*, January 6, 2021, https://news.usni.org/2021/01/06/spate-of-attacks-on-ships-in-middle-east-points-to-iran-backed-group.

16. Geoff Brumfiel, "Outside Experts See Iran's Hand in Attack on Saudi Oil Facility," *NPR WITF*, September 16, 2019, https://www.npr.org/2019/09/16/761378683/outside-experts-see-irans-hand-in-attack-on-saudi-oil-facility.

17. Amina Ismail and John Davison, "Iran Attacks Iraq's Erbil with Missiles in Warning to U.S., Allies," *Reuters*, March 13, 2022, https://www.reuters.com/world/middle-east/multiple-rockets-fall-erbil-northern-iraq-state-media-2022-03-12/.

18. Daniel Byman and Asfandyar Mir, "How Strong sI Al-Qaeda? A Debate," *War on the Rocks*, May 20, 2022, https://warontherocks.com/2022/05/how-strong-is-al-qaeda-a-debate/.

19. J.M. Berger and Amarnath Amarasingam, "With the Destruction of the Caliphate, the Islamic State Has Lost Far More Than Territory," *The Washington Post*, October 31, 2017, https://www.washingtonpost.com/news/monkey-cage/wp/2017/10/31/the-caliphate-that-was/.

20. "The Historic Deal that Will Prevent Iran from Acquiring a Nuclear Weapon," The White House, Washington, DC, January 6, 2016, https://obamawhitehouse .archives.gov/issues/foreign-policy/iran-deal.

21. Dan De Luce, "Iran Has Enough Uranium to Build an Atomic Bomb, U.N. Agency Says," *NBC News*, May 31, 2022, https://www.nbcnews.com/politics /national-security/iran-enough-uranium-build-atomic-bomb-un-says-rcna31246.

22. "Explainer: Iran's Centrifuges," *The Iran Primer*, November 22, 2021, https://iranprimer.usip.org/blog/2021/nov/22/explainer-controversy-over -iran%E2%80%99s-centrifuges.

23. Hans M. Kristensen and Matt Korda, "Nuclear Notebook: Israeli Nuclear Weapons, 2022," *Bulletin of the Atomic Scientists*, January 17, 2022, https://thebulletin .org/premium/2022-01/nuclear-notebook-israeli-nuclear-weapons-2022/.

24. Anthony H. Cordesman, "The Arab-Israeli Military Balance," Center for Strategic & International Studies, Washington, DC, June 2010, https://csis -website-prod.s3.amazonaws.com/s3fs-public/legacy_files/files/publication /100629_Arab-IsraeliMilBal.pdf.

25. Much of this section is extracted directly from previously outlined arguments in Christopher J. Bolan, "Dealing with Iran Will Not Be Enough to Restore Regional Stability," Foreign Policy Research Institute, June 14, 2019, https://www.fpri.org /article/2019/06/dealing-with-iran-will-not-be-enough-to-restore-regional -stability/ and is reproduced here with the permission of the original publisher.

26. Mamoon Alabbasi, "Iran Continues to Boast of Its Regional Reach," *Middle East Eye*, March 13, 2015, https://www.middleeasteye.net/news /iran-continues-boast-its-regional-reach.

27. Shawn Snow, "CENTCOM Commander Calls Iran Greatest Threat to the Region," *Military Times*, March 13, 2017, https://www.militarytimes.com/news /pentagon-congress/2017/03/13/centcom-commander-calls-iran-greatest -threat-to-the-region/.

28. Remarks by General Kenneth F. McKenzie, Jr, Foundation for Defense of Democracies, May 8, 2019, https://www.fdd.org/wp-content/uploads/2019/05 /Transcript-CMPP-McKenzie.pdf.

29. Seth G. Jones, "War by Proxy: Iran's Growing Footprint in the Middle East," Center for Strategic & International Studies, Washington, DC, March 11, 2019, https://www.csis.org/war-by-proxy.

30. Expanded treatment of this analysis can be found in Christopher J. Bolan, Jerad I. Harper, and Joel R. Hillison, "Diverging Interests: US Strategy in the Middle East," *The US Army War College Quarterly: Parameters* 50, 4 (Winter 2020), https://press.armywarcollege.edu/cgi/viewcontent.cgi?article=2691&context =parameters.

31. Phillip Smyth, "The Shia Militia Mapping Project," The Washington Institute for Near East Policy, Washington, DC, May 20, 2019, https://www .washingtoninstitute.org/policy-analysis/shia-militia-mapping-project.

32. Jonny Hallam et al., "Deadly Drone Attack on Tanker Escalates Iran-Israel Maritime Tensions," *CNN*, August 2, 2021, https://www.cnn.com/2021/07/31 /middleeast/iran-israel-tanker-attack-drone-oman-intl/index.html.

33. Martha Raddatz, "Iran Fired Cruise Missiles in Attack on Saudi Oil Facil-ity: Senior US Official," *ABC News*, September 15, 2019, https://abcnews.go.com /Politics/iran-fired-cruise-missiles-attack-saudi-oil-facility/story?id=65632653.

34. "Iran Attack: US Troops Targeted with Ballistic Missiles," *BBC News*, January 8, 2020, https://www.bbc.com/news/world-middle-east-51028954.

35. The White House, "The Iran Nuclear Deal: What You Need to Know about the JCPOA," Washington, DC, July 2015, https://obamawhitehouse.archives.gov/sites/default/files/docs/jcpoa_what_you_need_to_know.pdf.

36. Dan De Luce, "Iran Has Enough Uranium to Build an Atomic Bomb, U.N. Agency Says," *NBC News*, May 31, 2022, https://www.nbcnews.com/politics/national-security/iran-enough-uranium-build-atomic-bomb-un-says-rcna31246.

37. Richard Boudreaux and Amberin Zaman, "Turkey Rejects U.S. Troop Deployment," *Los Angeles Times*, March 2, 2003, https://www.latimes.com/archives/la-xpm-2003-mar-02-fg-iraq2-story.html.

38. "Freedom in the World 2022: Turkey," Freedom House, https://freedomhouse.org/country/turkey/freedom-world/2022.

39. David Gauthier-Villars and Ann M. Simmons, "Turkey Receives Russian Missile System, Risking U.S. Sanctions," *The Wall Street Journal*, July 12, 2019, https://www.wsj.com/articles/turkey-starts-taking-delivery-of-russian-air-defense-missile-system-risking-u-s-sanctions-and-testing-its-position-in-nato-11562920372.

40. Joel Gehrke, "Recep Tayyip Erdogan Threatens US Forces with 'Ottoman Slap'," *Washington Examiner*, February 13, 2018, https://www.washingtonexaminer.com/recep-tayyip-erdogan-threatens-us-forces-with-ottoman-slap.

41. "Turkey's President Criticizes US Sanctions on Iran," The Associated Press, December 20, 2018, https://abcnews.go.com/International/wireStory/turkeys-president-criticizes-us-sanctions-iran-59928349.

42. Giorgio Cafiero and Daniel Wagner, "Turkey and Qatar's Burgeoning Strategic Alliance," Middle East Institute, June 8, 2016, https://www.mei.edu/publications/turkey-and-qatars-burgeoning-strategic-alliance.

43. Mark Thompson, "The Great SCUD Hunt," *Time*, December 15, 2002, https://content.time.com/time/magazine/article/0,9171,400021,00.html.

44. Dan Williams and Matt Spetalnick, "Israel's Netanyahu Draws Rebuke from Obama over Iran Speech to Congress," *Reuters*, March 3, 2015, https://news.yahoo.com/israels-netanyahu-address-congress-speech-strained-ties-obama-060926075.html.

45. Jeremy M. Sharp, "U.S. Foreign Aid to Israel," Congressional Research Service, RL33222, November 16, 2020, https://crsreports.congress.gov/product/pdf/RL/RL33222/40.

46. Farnaz Fassihi and Ronen Bergman, "Israel Tells U.S. It Killed Iranian Officer, Official Says," *New York Times*, May 25, 2022, https://www.nytimes.com/2022/05/25/world/middleeast/iran-israel-killing-khodayee.html.

47. "Bennett to IAEA Chief: Israel Reserves the Right to Act Against Iran's Nuclear Program," *The Times of Israel*, June 3, 2022, https://www.timesofisrael.com/bennett-to-iaea-chief-israel-reserves-right-to-act-against-irans-nuclear-program/.

48. "To Stave Off Arab Spring Revolts, Saudi Arabia and Fellow Gulf Countries Spend #150 Billion," Knowledge at Wharton, University of Pennsylvania, September 21, 2011, https://knowledge.wharton.upenn.edu/article/to-stave-off-arab-spring-revolts-saudi-arabia-and-fellow-gulf-countries-spend-150-billion/.

49. F. Gregory Gause, III, "Kings for All Seasons: How the Middle East's Monarchies Survived the Arab Spring," Brookings Doha Center, September 2013,

https://www.brookings.edu/wp-content/uploads/2016/06/Resilience-Arab
-Monarchies_English.pdf.

50. Martin Jay, "Saudi Arabia Looks for Peace Deal in Yemen, But at What Price?" *Inside Arabia*, April 2, 2020, https://insidearabia.com/saudi-arabia-looks -for-a-peace-deal-in-yemen-but-at-what-price/.

51. Joyce Sohyun Lee et al., "Saudi-led Airstrikes in Yemen Have Been Called War Crimes. Many Relied on U.S. Support," *The Washington Post*, June 4, 2022, https://www.washingtonpost.com/investigations/interactive/2022/saudi-war -crimes-yemen/.

52. Joe Gould, "US House Bill Would Close Door on Saudi Arms Sales," *Defense News*, October 24, 2018, https://www.defensenews.com/congress/2018/10/24 /us-house-bill-would-close-door-on-saudi-arms-sales/.

53. "Humanitarian Crisis in Yemen Remains the Worst in the World, Warns UN," *UN News*, February 14, 2019, https://news.un.org/en/story/2019/02/1032811.

54. Akbar Shahid Admed, "UAE Faces Growing Scrutiny for Role in Bloody U.S.-backed Campaign in Yemen," *HuffPost*, September 30, 2017, https://www .huffpost.com/entry/uae-yemen-growing-scrutiny_n_59ce9b1ae4b05f005d34396c.

55. Con Coughlin, "The UAE's Vote on Ukraine Signals a Strategic Shift," *The National News*, February 26, 2022, https://www.thenationalnews.com/opinion /comment/2022/02/26/the-uaes-vote-on-ukraine-signals-a-strategic-shift/.

56. Katie Bo Lillis et al., "Construction Halted on Secret Project at Chinese Port in UAE After Pressure from US, Officials Say," *CNN*, November 19, 2021, https:// www.cnn.com/2021/11/19/politics/china-uae-us-construction-port/index .html.

57. Benoit Faucon and Sune Engel Rasmussen, "As U.S. Cracks Down on Iran's Oil Sales, It Calls Out an Ally," *The Wall Street Journal*, September 8, 2019, https:// www.wsj.com/articles/as-u-s-cracks-down-on-irans-oil-sales-it-calls-out-an -ally-11567944003

58. "Ties Between UAE, US Going through 'a Stress Test,' Says Ambassador," *The Hill*, March 3, 2022, https://news.yahoo.com/ties-between-uae-us-going -164422769.html.

59. J. P. Lawrence, "US Troop Level Reduction in Middle East Likely as Focus Shifts Elsewhere," *Stars and Stripes*, January 14, 2022, https://www.stripes.com /theaters/middle_east/2022-01-14/centcom-central-command-drawdown-iraq -afghanistan-kuwait-saudi-arabia-4289137.html.

60. "Arms Sales in the Middle East: Trends and Analytical Perspectives for U.S. Policy," Congressional Research Service, November 23, 2020, https://crsreports .congress.gov/product/pdf/R/R44984.

61. A number of defense policy analysts have published thoughtful work on how the United States should adjust its military force posture in the Middle East, including Becca Wasser and Elisa Ewers, "Rightsizing in the Middle East," *Foreign Affairs*, December 16, 2021, https://www.foreignaffairs.com/articles /middle-east/2021-12-16/rightsizing-middle-east; Joshua Rovner, "For America's Military, Less Is More in the Middle East," *The National Interest*, September 10, 2014, https://nationalinterest.org/print/feature/americas-military-less-more-the-mid dle-east-11242; Melissa Dalton and Mara Karlin, "Toward a Smaller, Smarter Force Posture in the Middle East," Brookings Institute, August 28, 2018, https://www .brookings.edu/blog/order-from-chaos/2018/08/28/toward-a-smaller-smarter

-force-posture-in-the-middle-east/; Seth G. Jones and Seamus P. Daniels, "U.S. Defense Posture in the Middle East," Center for Strategic & International Studies, Washington, DC, May 19, 2022, https://www.csis.org/analysis/us-defense -posture-middle-east; and Christine McVann, "Reshaping U.S. Force Posture in the Middle East," The Washington Institute for Near East Policy, Washington, DC, March 10, 2021, https://www.washingtoninstitute.org/policy-analysis /reshaping-us-force-posture-middle-east.

62. Diana I. Dalphonse, Chris Townsend, and Matthew W. Weaver, "Shifting Landscape: The Evolution of By, With, and Through," *Strategy Bridge*, August 21, 2018, https://thestrategybridge.org/the-bridge/2018/8/1/shifting-landscape-the -evolution-of-by-with-and-through

63. Lewis Sanders IV, "Iran and Saudi Arabia Need a 'Fresh Security Architecture'," *DW.com*, February 18, 2018, https://www.dw.com/en/iran-and-saudi -arabia-need-a-fresh-security-architecture/a-42636094.

64. See for example analysis done by Ross Harrison, "Toward a Regional Framework for the Middle East: Takeaways from Other Regions," in Ross Harrison and Paul Salem, eds., *From Chaos to Cooperation: Toward Regional Order in the Middle East* (Washington, DC, The Middle East Institute, 2017), 197–217.

65. Andrea Shalal, "G7 Aims to Raise $600 billion to Counter China's Belt and Road," *Reuters*, 27 June 2022, https://www.reuters.com/world/refile-us -aims-raise-200-bln-part-g7-rival-chinas-belt-road-2022-06-26/.

66. Karen E. Young, "Economic and Energy Cooperation for the United States and the Gulf Arab States," in Gerald M. Feierstein, Bilal Y. Saab, and Karen E. Young, eds., *US-Gulf Relations at the Crossroads: Time for a Recalibration* (Washington, DC, The Middle East Institute, April 2022), 20.

67. "Largest Arab Opinion Poll Finds Huge Appetite for Democracy among Arab Citizens," The Arab Center for Research and Policy Studies, Qatar, October 14, 2020, https://www.prnewswire.com/news-releases/largest-arab-opinion-poll-finds -huge-appetite-for-democracy-among-arab-citizens-301151782.html.

68. "Arab Youth Survey: 13th Annual Edition 2021," ASDA'A BCW, United Arab Emirates, 2021, https://arabyouthsurvey.com/en/findings/.

CHAPTER 13

Irregular Warfare and U.S. Landpower

Kevin D. Stringer

INTRODUCTION

Irregular warfare (IW) remains a vexing and contentious subject in U.S. national security discourse with serious implications for American land-power. With the United States facing an era of increasing competition in relation to its adversaries, the U.S. military, particularly the U.S. Army, must improve its irregular warfare competencies to sustain its strategic landpower advantage in the twenty-first century and beyond. Despite the cultural inclination of all the services to prepare for high-end, force-on-force struggle with nation-state adversaries, or what *Joint Publication 1 Doctrine for the Armed Forces* calls "traditional warfare," conflict remains a unified whole.[1] The irregular portion cannot be ignored or avoided due to longstanding organizational preferences.

While the 2020 *Summary of the Irregular Warfare Annex to the National Defense Strategy* purported "that irregular warfare is to be institutionalized as a core competency with sufficient, enduring capabilities to advance national security objectives across the spectrum of competition and conflict,"[2] continuing cultural opposition to irregular warfare within the Department of Defense (DoD), and the Army's lack of focus on this critical landpower competency contribute to a suboptimal national defense situation for this highly prevalent form of warfare. This perspective is not an outlier. In the 2019–2021 period alone, a plethora of credible practitioners, researchers, and academics in venues like *Small Wars Journal*, *War on the Rocks*, *Modern Warfare Institute*, *Irregular Warfare Initiative*, and *Joint Force Quarterly* have voiced their concerns to the inadequate conduct of irregular warfare by the U.S. military, critiqued the current situation, and offered recommendations for adaptation and implementation for the future.[3]

This chapter further contributes to this urgent national security discussion by elaborating the U.S. military's problem with irregular warfare and then proposing three initial and pragmatic areas for change. For the problem, the main issues revolve around the Department of Defense's conceptualization of irregular warfare, its place in the American military culture, and historical U.S. outcomes in the conduct of irregular warfare. For the solutions, the proposition is threefold. First, the U.S. military should change its definition of irregular warfare to simplify the concept and focus only on the components of irregular warfare that are relevant to the military element of national power. This proposal leads to a greater issue beyond this scope of the chapter, which is the necessity to change the overall definition of warfare for the U.S. military. Second, the U.S. Army, as the landpower proponent, needs to alter its professional military education (PME) curriculum beginning at the War College level and continuing downward to the Command and General Staff College and the various Captain's Career courses, to weave irregular warfare into the entire professional military education curriculum just like the subjects of leadership and ethics. Finally, the Army requires dedicated irregular warfare–capable headquarters and units for campaigning at the operational level of warfare. This aim can be reached through a pragmatic degree of force structure specialization and confirming experimentation.

THE PROBLEM

The problem begins with the conceptualization and visualization of irregular warfare in current U.S. Department of Defense culture. Historically and in practice, the Department has articulated directly or indirectly a dichotomous conceptualization where irregular and conventional warfare are separated "into two distinct, mutually exclusive categories,"[4] with an implication that one is for special operations forces (SOF) and the other for conventional formations. In his book *Nonstate Warfare*, political scientist Stephen Biddle underlines this perspective and then rightly rejects the "widespread assumption that conventional and irregular warfare constitute autonomous, exclusive categories of distinct military conduct."[5] In an attempt to overcome this deficiency, the *IW Annex* states, "The Department must institutionalize irregular warfare as a core competency for both conventional and special operations forces, sustaining the ability to impose costs and create dilemmas for our adversaries across the full spectrum of competition and conflict."[6] The achievement of this objective requires actions that change the "conventional warfare" discussion to a "warfare of all types" conversation, which implies an American version of hybrid warfare that seamlessly intertwines both conventional and irregular warfare.

Equally awkward is the endemic cultural resistance to irregular warfare in the U.S. military, particularly the Army. As the late strategist and academic Colin Gray noted, irregular warfare challenges the traditional American notion of warfare.[7] The evidence of this cultural opposition, which commences at the Department of Defense level and extends into the U.S. Army, begins with the placement of irregular warfare in an annex rather than the main text of the *National Defense Strategy*. As David Ucko observed, "there is something curious about the need for the United States to "institutionalize" irregular warfare after a full twenty years of nonstop engagement in precisely such efforts."[8] This institutional cultural antipathy to irregular warfare in the modern era goes back at least to World War II if not earlier. In that war for example, the Office of Strategic Services (OSS) under William Donovan developed a unique irregular warfare capability and structure, but because it threatened the existing American concept of war, and its activities were considered ungentlemanly by the conventional military leaders of the day, the Truman administration dissolved the organization in the aftermath of World War II.[9] This historical expungement of irregular warfare concepts and dialogue from the U.S. Army is a recurring organizational and strategic cultural theme. In the 1990s, irregular warfare, otherwise known as low-intensity conflict, was eliminated from Army thinking by then Chief of Staff General Dennis Reimer with the resultant disappearance of the subject from the U.S. Army War College curriculum.[10]

This decision and similar ones set the conditions for the twenty-first century, where the U.S. Army has struggled to defeat its irregular adversaries. The Army's own cultural and doctrinal rigidity for landpower precluded it from effectively addressing and executing irregular warfare operations because the organization adheres to the mantra that conventional military forces are designed solely for combat against the counterpart forces of other states. In his article "In Era of Small Wars, U.S. Army Must Embrace Training Mission," John Nagl provided damning examples of the haphazard and ad hoc approach to advising in Iraq and Afghanistan that stemmed largely from the Army's institutional (cultural) neglect of irregular warfare campaigning.[11] This cultural disconnect from the reality of conflict takes on new dimensions with the broad marketing of multidomain operations by Army leadership, which are implied to equate to large-scale combat operations (LSCO) against a near-peer adversary. This approach outwardly contradicts the DoD policy directive that irregular warfare is an equal core competence for the Army, is an integral element of most operations, and should be fully integrated within multidomain thinking.

Finally, the question of actual results in irregular warfare looms large. As Lieutenant General Charles Cleveland judged in his book *The American*

Way of Irregular War: An Analytical Memoir, the U.S. Army seems unable to achieve positive strategic outcomes in irregular warfare.[12] He attributes this result to a multitude of factors to include a failure to focus on population-centric conflicts, a failure to "elevate and mature" irregular warfare capabilities, and strategic muddling caused by the inability of the Executive Branch to lead and conduct national level, whole-of-government irregular warfare operations.[13] His credible assessments, based upon careful research and almost forty years of practitioner irregular warfare experience, are echoed by other reliable sources. Hence, for American landpower to succeed in any future conflict and maintain its strategic advantage, the U.S. Army must embrace irregular warfare with an aim of achieving excellence in this form of conflict. The next sections offer some simple proposals to this objective.

DEFINITIONS AND TERMINOLOGY

A simplified definition of irregular warfare would help focus the U.S. military, specifically the Army, on this type of warfare and bring clarity to subroles and missions like unconventional warfare (UW), stabilization operations (SO), foreign internal defense (FID), counterterrorism (CT), and counterinsurgency (COIN). As Kevin Bilms wrote, "words matter in marketing and advertising campaigns" so IW needs to be simply and properly defined.[14] This assessment is buttressed by Australian military research on the topic, which posits that the analysis of irregular warfare has been complicated by its language.[15] This viewpoint is further affirmed by a number of Baltic and Nordic military partners in the aftermath of a European theater-level irregular warfare workshop in 2014. In the after-action report, they highlighted that "Definitions [pertaining to IW as defined in U.S. joint doctrine and non-doctrinal publications] and the understanding of the definitions remained a contentious issue at the seminar. Confusion and conflation reigned, and new terms proliferated when consensus could not be achieved . . . A common terminology and cipher are necessary to achieve intellectual interoperability among the participants . . . and achieve cooperative planning and policy development within the Baltic Sea region."[16] This lack of a universally accepted definition has led to needless debate over irregular warfare meaning and scope.

The current U.S. definition that "Irregular warfare is a struggle among state and non-state actors to influence populations and affect legitimacy"[17] is problematic in two dimensions. First, state and nonstate actors can include a number of stakeholders who are not engaged in warfare and who may not even be relevant for the military element of national power. Second, the fact that there is a struggle for legitimacy and influence over "relevant populations" is a characteristic of irregular warfare, not a definition of the same. Hence a quite simple definition that encompasses the true nature of irregular warfare for the U.S. military would be "Armed conflict, even if it is below the traditional level of what is accepted as war, involving surrogate forces."[18]

This definition is underpinned by the concept of surrogate forces, which encompasses a potentially wide gamut of participants such as insurgents, terrorists, resistance fighters, militias, partisans, mercenaries, private military companies, warlords, religious extremists, and others in irregular warfare, but at the same time does not preclude regular or conventional formations, police units, and territorial defense forces. A surrogate is simply an entity outside of the U.S. military "who takes the place of or acts for another."[19]

This simplified definition serves several purposes. First, it better reflects the realities of warfare with state or nonstate actors and reduces the bifurcation of conventional and irregular warfare, where the latter has been historically associated with insurgency, counterinsurgency, and counterterrorism operations mostly conducted by special operations forces. In other words, conflict between state and nonstate actors was not at one time considered a mission for conventional forces and had limited relevance for overall U.S. national security.[20] This viewpoint does not hold up under closer scrutiny. The American military experience in Vietnam, Iraq, Afghanistan, and the majority of its eighteenth and nineteenth century conflicts demonstrates the relevance of irregular warfare to national security and the necessary co-involvement of both conventional and special operations forces. As Kelly H. Smith noted, "seldom will surrogate operations imply either a pure special operations or conventional force solution."[21] Currently, the international relations actor/military force type boundary is even more amorphous and seamless with private military companies and terrorist groups increasingly used as state proxies (Wagner Group and Hezbollah) and conventional forces deployed in support of irregular operations (Donbass region of Ukraine).

Second, the more simple and restricted definition of irregular warfare as "armed conflict, even if it is below the traditional level of what is accepted as war, involving surrogate forces" prevents the U.S. military from choosing an interpretation of the current definition that best suits its bureaucratic interest or results in mission creep into nonmilitary domains. The "struggle among state and non-state actors" verbiage found in the current Department of Defense definition implies political, economic, social, financial, and informational aspects that the military should probably not address. These realms belong to interagency partners to manage and resource appropriately. The new definition also remains truer to the Clausewitzian nature and character of war which is about "primordial violence, hatred, and enmity," hence armed conflict.[22] By accepting irregular warfare as "Armed conflict, even if it is below the traditional level of what is accepted as war, involving surrogate forces," the Department of

Defense, specifically the U.S. Army, can focus on doctrine, organization, education, and training for those underlying missions that deal with surrogates either as enemies (counterterrorism and counterinsurgency) or as allies (unconventional warfare, stabilization operations, and foreign internal defense) while integrating irregular warfare into all aspects of warfighting and campaigning. With this simple and explicit definition, the next stage of irregular warfare integration into American landpower culture is for the Army to change its PME curriculum, beginning at the War College level, to inject irregular warfare into the curriculum for its next cohort of strategic leaders.

IRREGULAR WARFARE AND PROFESSIONAL MILITARY EDUCATION

Professional military education, specifically at the U.S. Army War College, is the cornerstone of Army senior leader development and reinforces institutional preferences and culture. Hence, excellence in the conduct of irregular warfare by the U.S. Army requires the proper melding of irregular warfare into current War College programs, and all PME programs from the Command and General Staff College to the various Captain's Career courses. The current professional dialogue postulates that there is insufficient professional military education on irregular warfare for Army officers, and this dearth of instruction will limit Army efforts in future, multidomain operations against peer adversaries.[23] The irregular warfare education that is available across PME programs seems to be studied separately from conventional warfare, resulting in further emphasis on the artificial division between conventional and irregular warfare that plagues the U.S. military culture.[24] Additionally, there is "the belief that irregular warfare education is not relevant for conventional forces which creates significant flaws deriving from an increasingly complex and volatile security environment that can adversely impact conventionally educated and trained forces caught in irregular conflicts."[25] These viewpoints are concerning since even as the United States shifts to great power competition and trains for conventional (traditional) war, the Joint Force must continue to study and practice irregular warfare.

Because defense policy determines professional military education topics, contemporary PME inclusion of irregular warfare in the curricula should be based upon existing guidance. For irregular warfare, already in 2012, the Joint Staff's education policy indicated that irregular warfare "is as strategically important as traditional warfare."[26] The 2020 Chairman of the Joint Chiefs of Staff education policy unequivocally stated, "that officer PME programs will . . . maintain a current and relevant curriculum that provides graduates with knowledge, skills, and abilities required to perform successfully across a competition continuum comprising armed conflict, competition below armed conflict, and cooperation in both traditional and irregular warfare (IW) contexts."[27] Correspondingly, the Department of Defense Directive 3000.07 reinforced that "Irregular Warfare (IW) is as strategically important as traditional warfare and DOD must be equally capable in both."[28] These clear policy statements would imply that irregular warfare receives equal and balanced treatment at PME institutions despite the critical academic and practitioner views to the contrary.

In order to gain some insights into the implementation of this guidance, the following paragraphs offer an analysis of the U.S. Army War College curriculum based upon the 2022 resident course directives which are the syllabi for the core blocks of instruction—(1) Introduction to Strategic Studies (ISS), (2) War, Policy, & National Security I (WPNS I), (3) War, Policy, & National Security II (WPNS II), (4) Military Strategy and Campaigning (MSC), (5) Strategic Leadership (SL) and (6) Defense Management (DM). The preliminary analysis suggests only a superficial integration of irregular warfare into the War College curriculum, which would contradict very explicit military educational policy directives mandating the opposite. For scope purposes, this analysis explicitly did not include other War College programs, such as the U.S. Army's Distance Education Program, nor the other senior service colleges. It also did not examine the electives programs, which offer some irregular warfare topics (such as the Special Operations, Strategic Warfare in the 21st Century, and Peace Operations electives) and others that lend themselves to an integrated look at conventional and irregular warfare (such as Insurgency and Civil War and The Vietnam War). The main reason for the latter exclusion was that Joint Staff guidance implies that the study of irregular warfare should be a core PME curriculum topic and not relegated to an electives list.

In the foundational *Introduction to Strategic Studies* (ISS), which aims to prepare senior military leaders to "apply intellectual rigor and adaptive problem solving to multi-domain and joint warfighting" there was no mention of irregular warfare.[29] This entire introductory course centered on the study of the 1990–1991 Persian Gulf War case as a means of understanding warfare at the strategic level.[30] While certainly a meritorious and valuable historical case, the Persian Gulf War, as taught, offers few insights into the conduct of irregular warfare in the policy, strategy, theater, and campaign realms. It is an excellent case study for discussing the lack of irregular warfare planning and options that could have contributed to the overall Desert Shield/Desert Storm campaign. A case in point is the unconventional warfare efforts for occupied Kuwait that never materialized and General Norman Schwarzkopf's extreme antipathy for irregular warfare and its special operations promulgators.[31] In particular the critical question to ask is why War College academic leadership selected this "traditional and winning" conventional case study and not a more likely but "messy and controversial" irregular and conventional warfare case like

Vietnam or Bosnia/Kosovo. Similar to Desert Storm, the latter examples provide insights into the strategic application of conventional landpower to include the "diplomatic, informational, and economic instruments of national power employed in conjunction with military activities together with a perspective on joint and multinational operations, including the importance of building coalitions."[32] Critically, Vietnam, Bosnia/Kosovo, Afghanistan, and Iraq offer irregular warfare components not inherent in the Persian Gulf War.

This syllabus assessment confirms a need to select a foundational campaign or series of cases for ISS study that adheres to Joint Chiefs of Staff directives on irregular warfare and requires senior students to assess and evaluate the intertwined nature of irregular warfare and conventional conflict. This assessment also aligns with that promulgated by irregular warfare scholar Heather Gregg in her article "Better Curricula, Better Strategic Outcomes: Irregular Warfare, Great Power Competition, and Professional Military Education," where she offers that studying a complex, amorphous case like the Great Game of the nineteenth and twentieth centuries provides senior officer students greater insights into irregular warfare and its symbiotic twin conventional war than the current World War II and Desert Storm investigations.[33] Fortunately, the Academic Year 2023 Foundations Course, which replaced ISS, now incorporates just such a case study focusing on the Democratic People's Republic of Korea (DPRK), which appears to better accommodate both irregular and conventional components of competition and warfare. This change is noteworthy and hopefully leads to further such adaptations in favor of better integrating irregular warfare into the War College curriculum.

Similarly, both War, Policy, & National Security I and II displayed a dearth of irregular warfare content. In War, Policy, & National Security I, a bedrock of the U.S. Army War College (USAWC) curriculum, a solitary, three-hour lesson, from a course of sixteen lessons, was dedicated to an overview of irregular warfare.[34] The subsequent War, Policy, & National Security II is essentially a graduate-level international relations curriculum on national security policy where warfighting in general is only obliquely addressed as an outcome of policy. A more specific discussion of irregular warfare at the strategic and policy level was not present in the syllabus.[35] This deficit was quite surprising on two counts. First, irregular warfare is not strategically different from conventional warfare.[36] Rhetorically, why would the subject be left out of professional military education on U.S. national security. Second, the United States, including its military, has performed poorly at developing strategy for irregular warfare engagements in Vietnam, Somalia, Bosnia, Kosovo, Iraq, and Afghanistan. Hence greater study of this challenge by the Army's future senior leaders in War, Policy, & National Security I and II would potentially contribute to major future improvements in crafting and integrating irregular warfare into successful national security policy. In fact, three potential case studies on conflict were subsequently added to the AY23 National Security Policy and Strategy course to start addressing this deficit: Bosnia, the DPRK nuclear program, and the Syrian intervention, all of which have significant irregular warfare components.

Correspondingly, the Military Strategy and Campaigning (MSC) course contained only limited irregular warfare content, even though the course was designed to "create strategically minded joint warfighters who exercise sound judgment in using military power to accomplish national security objectives."[37] Essentially two lessons out of thirty were dedicated to irregular warfare.[38] For some compensation, however, four lessons were devoted to an experiential learning event, Joint Overmatch, in which students develop and fight an operational approach which requires the integration of both conventional and irregular warfare techniques.[39]

The remaining two core courses assessed for irregular warfare content, Strategic Leadership and Defense Management, are generalist in nature, and therefore, while essential foundational courses for both forms of warfare, are neutral PME elements in evaluating the level of irregular warfare content in the War College curriculum. The Strategic Leadership course is intended to expose students to concepts that "enable effective leadership in the national security environment."[40] Its design is generic for leadership, critical thinking, and ethics, which are applicable topics for all warfare environments. Similarly, the Defense Management course affords an overview to Department of Defense and Army enterprise management, covering force structure, modernization, and readiness themes.[41] Like Strategic Leadership, its application is common for both forms of warfare.

Before suggesting a remedy to this apparent lack of depth on this subject, it is important to first understand why the curriculum is structured as it currently is. The Army War College curriculum is required to achieve joint learning outcomes as established by the Chairman of the Joint Chiefs of Staff, as well as other requirements of professional military education.[42] Based on this guidance, the Army War College develops program learning outcomes that seek to articulate what Army War College graduates are expected to be able to do upon graduation. Faculty members and subject matter experts view these outcomes from their own lens to determine the appropriate curriculum design, course content, and methods of evaluation. Therefore, there are limits to the amount of time that the curriculum can be devoted to any one topic, however important, and still meet the multiple requirements required for accreditation.

Of course, an analysis based on merely counting classes is not sufficient to make definitive judgements about the curriculum. If conflict is a unified whole, then you might expect that there would be limited coverage of distinct conventional and irregular warfare aspects. Therefore, a more thorough review of how irregular warfare is integrated with conventional warfare throughout the curriculum is in order. The main proposal to emerge out of this Army War College curriculum analysis is to properly institutionalize irregular warfare into professional military education by integrating it as a permeating subject like ethics or leadership into the entire curriculum. This step changes the "conventional warfare" discussion to a "warfare of all types" conversation. This technique would imply more irregular warfare readings, case studies, and examples combined into each lesson since irregular warfare dimensions characterize even the most conventional campaigns and subjects. Such an approach would mirror Mao Tse-Tung's thinking on irregular warfare, where guerrilla operations are not independent nor isolated from the conventional form of warfare.[43] Scholar-practitioner Mark Grdovic reinforces the pragmatic necessity of fully mixing irregular warfare into PME with the concrete example of the campaign planning for Operation Iraqi Freedom (OIF). According to Grdovic, U.S. Central Command (CENTCOM) was fully absorbed in conventional maneuver planning and did not show interest in the potential role of irregular warfare as part of the campaign. This cognitive deficit, likely derived from the collective military education of CENTCOM senior leadership, created opportunity costs for the overall coalition in both the southern and northern Iraq unconventional warfare efforts.[44]

So how do institutions like the USAWC go about reviewing and revising content to address the zero-sum nature of the curriculum? In his seminal book, *Leading Change*, John P. Kotter outlines an eight-step process to make important changes in an organization.[45] The first three steps can provide an initial focal point for improving professional military education. The poor outcomes in previous conflicts, most recently Afghanistan, should provide a sufficient sense of urgency in ensuring that irregular warfare is sufficiently examined and incorporated into the curriculum. His next step, creating a guiding coalition, requires identifying and empowering sufficient experts and advocates of irregular warfare to review the curriculum and advocate for change in the assessment and curriculum development process, not only identifying problems, but also proposing solutions based on the opportunity costs of removing other content. In addition to replacing existing content with irregular warfare topics, it is also important to examine ways to integrate irregular concepts into applicable lessons which also cover conventional warfare. Exercises like the Joint Overmatch lessons in MSC provide excellent opportunities to achieve the desired integration.

Naturally, such a process might raise both resistance and anxiety among War College faculty since curriculum planning is generally a zero-sum exchange where new content must replace existing material to remain within time and fiscal parameters. As Richard Hooker stated in "Taking the War Colleges from Good to Great," "proposals to modernize or transform the war colleges typically excite strong opposition from entrenched faculties."[46] Yet, consideration must be given as to why those faculty members have such views and what the costs might be of removing non-irregular warfare content. Although uncomfortable, such reform meets the Joint Chiefs of Staff vision that "to achieve deeper education on critical thinking, strategy, and warfighting, PME programs will have to ruthlessly reduce coverage of less important topics."[47] Of course, those less important topics are in the eyes of the beholder and require a deliberate process to identify. For example, in an era when the U.S. is grappling with the very definition of warfare, many other topics such as information, space, and cyber operations deserve thoughtful integration. That said, a PME change in favor of greater irregular warfare instruction and content at the War College level would be a substantial progression for inculcating the significance of irregular warfare into both the cognitive and cultural lobes of the U.S. Army's senior leaders.

ORGANIZATION

Finally, the U.S. Army requires irregular warfare–capable headquarters and units in order to achieve excellence in this type of warfighting, where the human domain as well as advising is so heavily weighted. This step goes well beyond special operations forces and requires a pragmatic degree of force structure specialization within the overall Army. The approach aligns closely with the spirit of the IW Annex, which advocates new ideas and innovative means to employ existing resources in creating better IW capabilities.[48] It also follows a key lesson of military history where superior organization and not technology has often been the path to military success.[49]

Pat Proctor in his book *Lessons Unlearned: The US Army's Role in Creating the Forever Wars in Afghanistan and Iraq* faulted the Army with failing to develop the capabilities to operate in irregular, or what he terms low-intensity conflict environments—skills like civil affairs, military police, psychological operations, and others.[50] This chapter would go a step further and offer that it is not the capabilities, which actually do exist, but their proper organization and employment. This perspective is supported by Charles Cleveland in *The American Way of Irregular Warfare*, where he elaborates on a U.S. Army not well organized for irregular warfare campaigns and with no designated operational level organization to properly orchestrate such activities.[51] This assertion is likely relevant for the other services as well. The Army's default and suboptimal solution to date has been to co-opt conventional, corps-level headquarters to conduct such operations with ad hoc augmentation from the reserve components and other active-duty units. The results of this methodology can be evaluated in the strategic outcomes found in the Iraq and Afghanistan experiences. Additionally, irregular warfare responsibility at the operational and campaign level is not just a SOF mission but includes the conventional Army. While the Security Force Assistance Brigades (SFABs) and the Asymmetric Warfare Group have been positive steps in the right direction for conventional organizational specificity for irregular warfare,[52] they remain insufficient in providing the U.S. Army a dedicated headquarters of action for irregular warfare matters, especially the all-important advisory component. These conventional elements also need to be integrated with relevant SOF formations for better irregular warfare cultural permeation between the two communities, for capability and capacity reasons, and to ensure their political survival in the regular army order of battle.

A case in point is the Asymmetric Warfare Group which the Army deactivated in March 2021. The Asymmetric Warfare Group's mandate was to seek material and nonmaterial solutions to operational challenges encountered during the irregular warfare operations in Afghanistan and Iraq and to "provide global operational advisory support to U.S. Army forces to rapidly transfer current threat based observations and solutions to tactical and operational commanders in order to defeat emerging asymmetric threats and enhance multi-domain effectiveness."[53] Yet the official Army statement justified this unit closure as a reprioritization in preparation for multidomain and near-peer threats like China and Russia.[54] Credible critics claim the decision was both shortsighted and bureaucratically driven, implying again a tendency for the U.S. Army to artificially bifurcate and group conventional warfare with state actors versus irregular warfare and nonstate entities, when reality shows the categories are strongly intermixed.[55] The deactivation of the Asymmetric Warfare Group was a premature abandonment of relevant expertise in irregular warfare amid ongoing conflicts with Russia, China, and violent extremist organizations.

For the remedy, and since the establishment of a new cabinet-level Office of Strategic Services–like organization, or a serious restructuring of SOCOM is bureaucratically, politically, and programmatically unlikely and unrealistic, the creation of a dedicated, operational-level irregular warfare–capable headquarters needs to originate from existing formations and capabilities and then be evaluated for its efficacy. Considering the proposed definition of irregular warfare as "armed conflict, even if it is below the traditional level of what is accepted as war, involving surrogate forces," then an irregular warfare capable headquarters must be able to conduct CT, COIN, UW, FID, and SO at a highly proficient level.

At the joint level, these capabilities can already be exercised through the existing, one- and two-star theater special operations commands (TSOCs). A TSOC is a subordinate unified command of U.S. Special Operations Command (SOCOM), with the respective geographic combatant commander (GCC) exercising operational control (OPCON) of the TSOC and its assigned or attached land, air, and maritime special operations tactical units. TSOCs perform campaign missions uniquely suited to SOF capabilities and integrate special operations into irregular warfare efforts that support GCC theater plans.[56]

Since the landpower mission capabilities for CT, COIN, UW, FID, and SO are currently found in Special Forces (SF), Psychological Operations (PSYOP), Civil Affairs (CA), and SFAB units, the pragmatic solution for solving this irregular warfare organizational deficit would be a careful study of the problem; followed by testing, experimentation, and evaluation of several organizational combinations of the aforementioned units of action; and then selection and implementation of one model to fill the operational IW headquarters void. Several broad variants could be considered. Two potential options for consideration include a modernized Special Action Force (SAF) or the establishment of a landpower component, Irregular Warfare Task Force (IWTF), one per GCC.[57]

The first option would be to replicate and update the 1969 SAF. This concept combined the unconventional warfare capabilities of a Special Forces Group with a CA group, PSYOP battalion, and engineer, medical, military police, and intelligence detachments.[58] Modernizing enhancements could include cyber teams and an extension of advisory capabilities with SFAB-like troops.

Similarly, a more radical option would be to create a one-star IWTF, incorporating a SF Group, fully integrated PSYOP and CA battalions (one each), and an allocated SFAB under the overall framework and leadership of the joint TSOC. Both models equal an irregular warfare land component command to be paired with its maritime and air equivalents. The billets for such organizations would come from reductions in the SOCOM Headquarters staff and the deactivation of the group and brigade level of command for the already regionally assigned PSYOP and CA battalions.

This restructuring would achieve several objectives. It would create a standing landpower operational headquarters with requisite capabilities to address the long missing and dedicated irregular warfare campaigning mechanism. The unit would also reduce SOCOM's proclivity to maintain a perpetual focus on direct action at the expense of indirect approaches.[59] The PSYOP and CA functions would be more tightly integrated in IW while eliminating unnecessary intermediate headquarters.

With the inclusion of an SFAB or SFAB-like conventional troops, the new formation would gain the extended and augmented advising expertise needed in the irregular warfare environment while incorporating the conventional Army system and culture into an overall irregular warfare construct. The advising function remains a significant component of irregular warfare success, and while military advising has historically been a core competence for U.S. special operations forces, the conventional military, with its greater resources, has continually been called upon to address this persistent mission mainly with ad hoc organizational and personnel solutions that often achieved suboptimal results.[60] In fact, as designed, SFABs can conduct the advise, support, liaise, and assess aspects of security force assistance for both regular and irregular partner forces, while alleviating conventional Army units from executing these operations in an ad hoc manner.[61] This capability provides a bridging function between conventional and irregular forces that could be useful in future and ambiguous irregular warfare campaigns. For example, in a hypothetical Ukraine scenario, the IWTF or SAF, with its constituent SFAB and SF assets, would be well positioned to advise the full spectrum of partner units—territorial defense forces, civilian irregulars, special forces, and even conventional formations. With six established SFABs, each aligned to a GCC, the Army could easily assign them into an Irregular Warfare Task Force or Special Action Force within the respective GCCs.

Such a move is also important for the long-term viability of the SFAB formation. This author, in his 2016 article "The Missing Lever: A Joint Military Advisory Command for Partner-Nation Engagement," reinforced the need for dedicated advisory units, especially given that the "mainstream military culture resists the strategic significance of military advisors and often relegates this mission to a second-tier status."[62] Similarly, already in 2008, John A. Nagl in his article "Institutionalizing Adaptation: It's Time for an Army Advisory Command," provided early advocacy for the establishment of a standing advisory organization for irregular warfare.[63] While the Army announced the creation of the aforementioned six SFABs in May 2018, their long-term survival in the force structure is subject to Army cultural, fiscal, and personality whims. Embedding them within a permanent irregular warfare construct, closely connected with SOF, could ensure the long-term sustainability of this important irregular warfare function.

While the premise of this essay is that an IWTF or SAF organization addresses the current gap of a dedicated IW campaigning headquarters, the concept needs to be tested for efficacy and to eliminate the guaranteed cultural and organizational resistance it will face from within the Department of the Army. Such organizational testing for new landpower formations has long historical precedent in the U.S. Army and offers a path for implementing needed organizational change and innovation in the face of Army traditionalism, while providing detailed refinement of future concepts. Two examples provide illustration. In World War II, General Leslie McNair and his Army Ground Forces headquarters tested all the divisional designs, resulting in the approval of some, like the airborne divisions, and the discarding of others, like the light divisions.[64] Similarly, the so-called Howze Board, named after General Hamilton Howze, resulted in the development of the airmobile concept and the subsequent creation of airmobile/air assault divisions and ultimately the U.S. Army Aviation branch.[65] The Secretary of Defense instructions for this Air Mobility Board provide useful guidance for the creation of an IWTF or SAF. "I shall be disappointed if the . . . reexamination merely produces more of the same, rather than a plan for implementing fresh and perhaps unorthodox concepts."[66] The IWTF or SAF, with one assigned to each existing joint TSOC, should undergo a similar testing process to create a refined theater-level formation for the orchestration and execution of irregular warfare land campaigns and operational level activities. This testing takes on greater importance for force validation since the SFABs were established without the experimentation and testing formalities normally associated with new unit designs.[67]

CONCLUSION

The twenty-first century U.S. landpower mindset is missing a key element. As Stephen Biddle lamented, "There is a widespread assumption that state and nonstate warfare are profoundly different phenomena."[68] In fact they are so closely intertwined that a land force must be prepared to shift fluidly between the two in all future conflicts. For the U.S. Army, simply increasing the number of troops committed and the money spent will not win an irregular conflict.[69] Rather, the landpower solution for irregular warfare requires something more nuanced, specifically in the cognitive and organizational spaces. For the cognitive dimension, a simplified definition of irregular warfare for the military element of national power coupled with Army professional military education reform could lead to better irregular warfare outcomes. The former change allows a better concentration on doctrine, organization, education, and training for those underlying military missions that deal with surrogates either as enemies (counterterrorism and counterinsurgency) or as allies (unconventional warfare, stabilization operations, and foreign internal defense). The broader population and legitimacy matters implied in the current definition, to include governance and economics, are best left to civilian agencies such as the Department of State and U.S. Agency for International Development to address. The latter professional military education reform needs to start at the War College level since "virtually every major military decision in time of war will be made by a war college graduate."[70] There, irregular warfare topics and spirit should be entwined throughout the entire curriculum. For as Heather Gregg wrote, "Foremost, PME needs to instill an irregular warfare mindset to prepare officers for strategic competition."[71] Additionally, future research should examine the role of irregular warfare in PME curriculum at other senior service colleges, and other levels, and in both resident and distance programs.

Finally, irregular warfare needs a dedicated organizational vessel for campaigning and operational activities. The proposed Irregular Warfare Task Force or Special Action Force, crafted from existing capabilities under the already existing joint theater special operations command, provides a pragmatic path to reform. For in the end, the desired outcome of this chapter is a more effective and ultimately victorious U.S. Army for the irregular operations of the future. For only by addressing current landpower deficits in the irregular warfare context can the United States maintain its strategic advantage in the twenty-first century and beyond.

NOTES

1. Department of Defense, *Joint Publication 1, Doctrine for the Armed Forces* (Washington, DC: Joint Chiefs of Staff, March 25, 2013, incorporating change 1, July 12, 2017.), I–5.

2. Department of Defense, *Summary of the Irregular Warfare Annex to the National Defense Strategy* (Washington, DC: Department of Defense, 2020), 1.

3. See for example Sandor Fabian, "Irregular Versus Conventional Warfare: A Dichotomous Misconception," *Modern Warfare Institute*, May 14, 2021, https://mwi.usma.edu/irregular-versus-conventional-warfare-a-dichotomous -misconception; James Derleth, "Failing to Train: Conventional Forces in Irregular Warfare," *Modern Warfare Institute*, May 5, 2021, https://mwi.usma.edu /failing-to-train-conventional-forces-in-irregular-warfare, accessed, January 17, 2022; David Ucko, "Nobody Puts IW in an Annex: It's Time to Embrace Irregular Warfare as a Strategic Priority," *Modern Warfare Institute*, October 14, 2021, https:// mwi.usma.edu/nobody-puts-iw-in-an-annex-its-time-to-embrace-irregular -warfare-as-a-strategic-priority; Kevin Bilms, "What's in a Name? Reimagining Irregular Warfare Activities for Competition," *War on the Rocks*, January 15, 2021, https://warontherocks.com/2021/01/whats-in-a-name-reimagining-irregular -warfare-activities-for-competition; John A. Pelleriti, Michael Maloney, David C. Cox, Heather J. Sullivan, J. Eric Piskura, and Montigo J. Hawkins, "The Insufficiency of US IW Doctrine," *Joint Force Quarterly* 93 (2nd Quarter, 2019), 104–110; Brian Petit, Steve Ferenzi, and Kevin Bilms, "An Irregular Upgrade to Operational Design," *War on the Rocks*, March 19, 2021; and Michael Noonan, "Not Just for SOF Anymore: Envisioning Irregular Warfare as a Joint Force Priority," *Modern War Institute* (April 21, 2021), https://mwi.usma.edu/not-just -for-sof-anymore-envisioning-irregular-warfare-as-a-joint-force-priority.

4. Sandor Fabian, "Irregular Versus Conventional Warfare: A Dichotomous Misconception," *Modern Warfare Institute*, May 14, 2021, https://mwi.usma.edu /irregular-versus-conventional-warfare-a-dichotomous-misconception.

5. Stephen Biddle, *Nonstate Warfare: The Military Methods of Guerrillas, Warlords, and Militias* (Princeton, NJ: Princeton University Press, 2021), 6–7.

6. Department of Defense, *Summary of the Irregular Warfare Annex to the National Defense Strategy* (Washington, DC: Department of Defense, 2020), 3.

7. Colin S. Gray, *Irregular Enemies and the Essence of Strategy: Can the American Way of War Adapt?* (Carlisle, PA: US Army War College Press, 2006), v, 8.

8. David Ucko, "Nobody Puts IW in an Annex: It's Time to Embrace Irregular Warfare as a Strategic Priority," *Modern Warfare Institute*, October 14, 2020, https://mwi.usma.edu/nobody-puts-iw-in-an-annex-its-time-to-embrace-irregular -warfare-as-a-strategic-priority.

9. Derek Jones, *Ending the Debate: Unconventional Warfare, Foreign Internal Defense, and Why Words Matter*, Thesis, U.S. Command and General Staff College, Ft. Leavenworth, KS (2006), 50.

10. Pat Proctor, *Lessons Unlearned: The US Army's Role in Creating the Forever Wars in Afghanistan and Iraq* (Columbia, MO: University of Missouri Press, 2020).

11. John Nagl, "In Era of Small Wars, U.S. Army Must Embrace Training Mission," *World Politics Review*, Tuesday, February 5, 2013, at https://www.worldpoliticsreview.com/articles/12693/in-era-of-small-wars-u-s-army-must-embrace-training-mission.

12. Charles Cleveland with Daniel Egel, *The American Way of Irregular War: An Analytical Memoir* (Santa Monica, CA: RAND Corporation, 2020), 12.

13. Charles Cleveland with Daniel Egel, *The American Way of Irregular War: An Analytical Memoir* (Santa Monica, CA: RAND Corporation, 2020), xii, xv, and xix.

14. Kevin Bilms, "What's in a Name? Reimagining Irregular Warfare Activities for Competition," *War on the Rocks*, January 15, 2021, https://warontherocks.com/2021/01/whats-in-a-name-reimagining-irregular-warfare-activities-for-competition.

15. Andrew Maher, "Theorising Rebellion: A Framework for Irregular Warfare," in *A Framework for Irregular Warfare: Irregular Warfare Essay Collection*, ed. Andrew Maher (Canberra: Australian Army Research Center, December 2021), 2–3.

16. After-action review (AAR), Special Operations Command Europe and Baltic Defence College Resistance Initiative Seminar, Baltic Defence College, Tartu, Estonia, 2014.

17. Department of Defense, *Summary of the Irregular Warfare Annex to the National Defense Strategy* (Washington, DC: Department of Defense, 2020), 2.

18. The definition derives from numerous conversations with CW2 (Retired) Tom Disburg, U.S Army Special Forces and Kelly H. Smith, "Surrogate Warfare for the 21st Century," in Richard Newton et al., eds., *Contemporary Security Challenges: Irregular Warfare and Indirect Approaches* (Tampa, FL: Joint Special Operations University, February 2009), 39–54.

19. *Field Manual 3-05.20 Special Forces Operations* (Washington, DC: Department of the Army, 2001), 2–5.

20. George C. Marshall European Center for Security Studies, Seminar in Irregular Warfare: Course Overview (Draft), January 2022.

21. Kelly H. Smith, "Surrogate Warfare for the 21st Century," in Richard Newton et al., eds., *Contemporary Security Challenges: Irregular Warfare and Indirect Approaches* (Tampa, FL: Joint Special Operations University, February 2009), 39–54.

22. Carl von Clausewitz, *On War*, ed. and trans. Michael Howard and Peter Paret, ind. ed. (Princeton: Princeton University Press, 1984), 89.

23. Charles Cleveland with Daniel Egel, *The American Way of Irregular War: An Analytical Memoir* (Santa Monica, CA: RAND Corporation, 2020), xix; and James Derleth, "Failing to Train: Conventional Forces in Irregular Warfare," *Modern Warfare Institute*, May 5, 2021, https://mwi.usma.edu/failing-to-train-conventional-forces-in-irregular-warfare, accessed January 17, 2022; and Heather S. Gregg, "Better Curricula, Better Strategic Outcomes: Irregular Warfare, Great Power Competition, and Professional Military Education," Modern Warfare Institute, March 15, 2022, at https://mwi.usma.edu/better

-curricula-better-strategic-outcomes-irregular-warfare-great-power-competition
-and-professional-military-education.

24. Sandor Fabian, "Irregular Versus Conventional Warfare: A Dichotomous Misconception," *Modern Warfare Institute*, May 14, 2021, https://mwi.usma.edu /irregular-versus-conventional-warfare-a-dichotomous-misconception.

25. James Derleth, "Failing to Train: Conventional Forces in Irregular Warfare," *Modern Warfare Institute*, May 5, 2021, https://mwi.usma.edu/failing-to -train-conventional-forces-in-irregular-warfare.

26. U.S. Joint Chiefs of Staff, Officer Professional Military Education Policy (OPMEP), Washington, DC, September 5, 2012.

27. Chairman of the Joint Chiefs of Staff Instruction 1800.01F, Officer Professional Military Education Policy, Washington, DC, May 15, 2020.

28. DoDD 3000.07: Irregular Warfare (IW), August 28, 2014 [Incorporating Change 1, May 12, 2017].

29. Jonathan Klug, Course Director, Academic Year 2022 Core Curriculum Introduction to Strategic Studies Directive, U.S. Army War College, 2021.

30. Jonathan Klug, Course Director, Academic Year 2022 Core Curriculum Introduction to Strategic Studies Directive, U.S. Army War College, 2021, 2.

31. Mark Moyer, *Oppose Any Foe: The Rise of America's Special Operations Forces* (New York: Basic Books, 2017), 186–191.

32. Jonathan Klug, Course Director, Academic Year 2022 Core Curriculum Introduction to Strategic Studies Directive, U.S. Army War College, 2021, 2.

33. Heather S. Gregg, "Better Curricula, Better Strategic Outcomes: Irregular Warfare, Great Power Competition, and Professional Military Education," *Modern Warfare Institute*, May 15, 2022, at https://mwi.usma.edu/better-curricula-better -strategic-outcomes-irregular-warfare-great-power-competition-and-professional -military-education, accessed March 25, 2022

34. See Paul Kan, Course Director, Department of National Security and Strategy, Academic Year 2022 War, Policy, & National Security I Course Directive, U.S. Army War College, 2021. Lesson 12 is entitled Irregular Warfare and conducted by Dr. Heather Gregg.

35. See Ronald J. Granieri, Course Director, Department of National Security and Strategy, Academic Year 2022 War, Policy, & National Security II Course Directive, U.S. Army War College, 2021.

36. Colin S. Gray, *Irregular Enemies and the Essence of Strategy: Can the American Way of War Adapt?* (Carlisle, PA: US Army War College Press, 2006), 4–5.

37. Michael Marra, Course Director, Department of Military Strategy, Planning, and Operations, Academic Year 2022 Military Strategy and Campaigning Course Directive, U.S. Army War College, 2021, 1.

38. See Michael Marra, Course Director, Department of Military Strategy, Planning, and Operations, Academic Year 2022 Military Strategy and Campaigning Course Directive, U.S. Army War College, 2021, LSN 13 and LSN 23. The required IW readings are Thomas A. Marks, "Counterinsurgency and Operational Art," *Low Intensity Conflict and Law Enforcement*; Kathleen H. Hicks and Alice Hunt Friend, *By Other Means, Part I: Campaign in the Gray Zone*, Center for Strategic and International Studies; and CJTF-OIR Historian, "History of CJTF-OIR," Operation INHERENT RESOLVE, September 9, 2020. The one suggested reading is Brian Petit, Steve

Ferenzi, and Kevin Bilms, "An Irregular Upgrade to Operational Design," *War on the Rocks*, March 19, 2021.

39. Michael Marra, Course Director, Department of Military Strategy, Planning, and Operations, Academic Year 2022 Military Strategy and Campaigning Course Directive, U.S. Army War College, 2021, 72–77.

40. Maurice L. Sipos, Course Director, Department of Command, Leadership, and Management, Academic Year 2022 Strategic Leadership Course Directive, U.S. Army War College, 2021

41. Douglas. E. Waters, Course Director, Department of Command, Leadership, and Management, Academic Year 2022 Defense Management Course Directive, U.S. Army War College, 2021.

42. U.S. Joint Chiefs of Staff, Officer Professional Military Education Policy (OPMEP), Washington, DC, May 15, 2020.

43. Mao Tse-Tung, *On Guerrilla Warfare*, trans. Samuel B. Griffith (New York, NY: Praeger, 1961), 55–56.

44. Mark Grdovic. Lessons from TF 103 in Northern Iraq 2002–2003, unpublished white paper, August 2015.

45. John P. Kotter, *Leading Change* (Boston, MA: Harvard Business Review Press, 1996), 21.

46. Richard D. Hooker Jr., "Taking the War Colleges from Good to Great," *Parameters* 49, 4 (2019), https://press.armywarcollege.edu/parameters/vol49/iss4/4.

47. Joint Chiefs of Staff, "Developing Today's Joint Officers for Tomorrow's Ways of War: The Joint Chiefs of Staff Vision and Guidance for Professional Military Education and Talent Management" (Washington, DC: Joint Chiefs of Staff, May 1, 2020), 6.

48. Department of Defense. *Summary of the Irregular Warfare Annex to the National Defense Strategy* (Washington, DC: Department of Defense, 2020), 1.

49. See Kevin D. Stringer, *Military Organizations for Homeland Defense and Smaller-Scale Contingencies* (Westport, CT: Praeger Security International, 2006), 6; and Martin van Creveld, *Command in War* (Cambridge, MA: Harvard University Press, 1985), 101.

50. Pat Proctor, *Lessons Unlearned: The US Army's Role in Creating the Forever Wars in Afghanistan and Iraq* (Columbia, MO: University of Missouri Press, 2020) 4.

51. Charles Cleveland with Daniel Egel, *The American Way of Irregular War: An Analytical Memoir* (Santa Monica, CA: RAND Corporation, 2020), xviii, 166.

52. Charles Cleveland with Daniel Egel, *The American Way of Irregular War: An Analytical Memoir* (Santa Monica, CA: RAND Corporation, 2020), 197.

53. Asymmetric Warfare Group website, at https://www.awg.army.mil/About-Us/Mission-Core-Functions-Priorities, accessed March 12, 2022, and U.S. Army, "Army to Discontinue Asymmetric Warfare Group and Rapid Equipping Force," accessed March 12, 2022, https://www.army.mil/article/239622/army_to_discontinue_asymmetric_warfare_group_and_rapid_equipping_force.

54. U.S. Army, "Army to Discontinue Asymmetric Warfare Group and Rapid Equipping Force," accessed March 12, 2022, https://www.army.mil/article/239622/army_to_discontinue_asymmetric_warfare_group_and_rapid_equipping_force.

55. Nolan Peterson, "Army's Vaunted Asymmetric Warfare Group is Closing Shop," *Coffee or Die*, October 5, 2020, https://coffeeordie.com/asymmetric-warfare-group-closing.

56. U.S. Joint Chiefs of Staff, *Joint Publication (JP) 3-05 Special Operations* (Washington, DC: Joint Chiefs of Staff, July 16, 2014), 1–3, 1–9.

57. Many thanks to LTG (R) Charles Cleveland for the seed of this idea. Charles Cleveland, Telephone discussion, March 10, 2022.

58. Thomas K. Adams, *US Special Operations Forces in Action: The Challenge of Unconventional Warfare* (London: Routledge, 1998), 100–101.

59. Charles Cleveland with Daniel Egel, *The American Way of Irregular War: An Analytical Memoir* (Santa Monica, CA: RAND Corporation, 2020), 203.

60. Kevin D. Stringer, "The Missing Lever: A Joint Military Advisory Command for Partner-Nation Engagement," *Joint Force Quarterly* 81 (4th Quarter, 2016).

61. Andrew Feickert, Army Security Force Assistance Brigades (SFABs), Report IF10675, Congressional Research Service, July 1, 2021.

62. Kevin D. Stringer, "The Missing Lever: A Joint Military Advisory Command for Partner-Nation Engagement," *Joint Force Quarterly* 81 (4th Quarter, 2016).

63. John A. Nagl, "Institutionalizing Adaptation: It's Time for an Army Advisory Command," *Military Review* (September–October 2008), 21–26.

64. Kent Roberts Greenfield, Robert R. Palmer, and Bell I. Wiley, *The Army Ground Forces: The Organization of Ground Combat Troops* (Washington, DC: Historical Division United States Army, 1947), 339–350.

65. See Shelby Stanton, *Anatomy of a Division: The 1st Cav in Vietnam* (New York, NY: Warner Books, 1987), 17; and Hamilton H. Howze, *A Cavalryman's Story* (Washington, DC: Smithsonian Institution, 1996), 233–257.

66. See "Memorandum for Mr. Stahr" quoted in Alain C. Enthoven and K. Wayne Smith, *How Much Is Enough? Shaping the Defense Program, 1961–1969* (New York, NY: Harper Colophon Books, Harper and Row, 1971), 103–104.

67. COL Scott Naumann, Security Force Assistance Proponent Office, Army War College Strategic Landpower Symposium, Carlisle, Pennsylvania, May 10, 2022.

68. Stephen Biddle, *Nonstate Warfare: The Military Methods of Guerrillas, Warlords, and Militias* (Princeton, NJ: Princeton University Press, 2021), xv.

69. Charles Cleveland with Daniel Egel, *The American Way of Irregular War: An Analytical Memoir* (Santa Monica, CA: RAND Corporation, 2020), 58.

70. Richard D. Hooker Jr., "Taking the War Colleges from Good to Great," *Parameters* 49, 4 (2019), https://press.armywarcollege.edu/parameters/vol49/iss4/4.

71. Heather S. Gregg, "Better Curricula, Better Strategic Outcomes: Irregular Warfare, Great Power Competition, and Professional Military Education," Modern Warfare Institute, May 15, 2021, at https://mwi.usma.edu/better-curricula-better-strategic-outcomes-irregular-warfare-great-power-competition-and-professional-military-education.

CHAPTER 14

Rethinking Security Cooperation in a Competitive World

Jerad I. Harper

Over the course of a year running from summer 2021 to summer 2022, the world stage showed two highly visible and dramatically different outcomes for U.S. security cooperation partners. First, over the summer of 2021, Afghanistan's military rapidly collapsed to the Taliban militia after the withdrawal of American military forces and supporting contract personnel. In stark contrast, however, was the much more highly proficient performance of Ukrainian armed forces during their 2022 resistance to invasion by the forces of the Russian Federation. Both of these were surprises for many analysts. While many had expected the Taliban to eventually win out against the Afghan government, most were surprised by the sheer speed of the Afghan military's collapse. Similarly, while many had expected the Ukrainian military to try hard, few expected that they would last long against what was seen as the competent military of a major world power. Clearly the opposing forces had a role in these circumstances, but in both of these cases, the *performance* of the defending government forces also appeared to be the significant factor.

The U.S. military is universally regarded as one of the most effective military forces in the world. Security cooperation is a major foreign policy tool for the United States—as it is for a number of other countries—and a major military contribution not only in competition but also in conflict. Despite this, however, the United States has experienced a wide variation in the outcome—the resulting quality of its security partners. To illustrate a small population of these varied outcomes: unsuccessful examples include the collapse of Iraqi Security Forces in 2014 and South

Vietnamese forces in 1975, while the Colombian military experienced significant improvement following Plan Colombia in the 1980s, and after more than a half-century partnership the Republic of Korea's military is recognized as a highly proficient force. So why does such variation occur?

With the United States entering a period of renewed great power competition, America should reconsider critical aspects of the way it approaches security cooperation. Rather than focusing primarily on *capabilities* and *capacity*, the U.S. military should instead seek to build *effectiveness* in its partners. Although there are some similarities to both approaches, a security cooperation model centered on building effectiveness requires a more detailed up-front and continuing analysis that is much more likely to produce more durable outcomes and proficient partners than our present approach. This is particularly true for the weak state partners with which the United States often finds itself deeply embroiled, but it also has important advantages for strong states as well. This chapter starts off with an exploration of security cooperation—what it is and why the United States pursues it—then examines the present model, introduces an *effectiveness model for security cooperation*, shows how this would achieve better outcomes for weak states, and finally, briefly shows its value for strong state partners in great power competition by applying the model to the Japanese Self Defense Forces.

With its large network of partners and allies, ensuring that the United States gets security cooperation right is a key concept for any discussion of sustaining America's competitive advantage. This is particularly important when considering the future of U.S. landpower. Many of the tasks that fall within security cooperation are common missions for ground forces and one for which they are uniquely suited. Their larger manpower and longer dwell time gives them the ability to execute a large number of the person-to-person exchanges within security cooperation necessary over a longer duration. And, of course, almost every country has a fairly sizable army, allowing extended contact between the two nations. With this being said, however, every service has a role in security cooperation.

UNDERSTANDING SECURITY COOPERATION
AND THE PRESENT U.S. APPROACH

What Is Security Cooperation and Why Does the United States Perform It?

Joint Publication 3-20, Security Cooperation, the U.S. military's doctrinal manual on this subject, defines security cooperation as "all [Department of Defense] interactions, programs, and activities with foreign security forces

(FSF) and their institutions to build relationships that help promote [U.S.] interests; enable [partner nations] to provide the [U.S.] access to territory, infrastructure, information and resources; and/or to build and apply their capacity and capabilities consistent with U.S. defense objectives."[1] Thus, security cooperation is a very broad set of tasks ranging from joint exercises, joint operations, and personnel exchanges all the way to the sale and transfer of weapons and training. This last set of tasks is actually administered by the U.S. military on behalf of the State Department, which oversees the foreign military sales program, foreign military financing, and the provision of training and education for foreign militaries under Title 22, U.S. Code. But while overseen by the State Department, this set of programs—known collectively as "security assistance"—is almost exclusively run by the military and thus falls under the broad umbrella of security cooperation.[2]

Many are often surprised by the fact that—despite the rhetoric on the "delivery" end of security cooperation—the top strategic consideration driving the U.S. provision of security cooperation is not always to make the partner nation better. As stated in the preceding definition, the United States pursues security cooperation for three purposes: (1) "to build security relationships," (2) "to gain or maintain operational access," and (3) to "build or apply their capability and capacity consistent with U.S. defense objectives."[3]

Thus, the overriding U.S. strategic goal driving military exercises, weapons sales, and even advisory efforts may primarily be to gain or deny a country's vote in the United Nations or otherwise support U.S. interests or to gain access to bases or the ability to deploy through a country. Making the partnered military better may be a peripheral goal but is not always the central priority. In other cases, however, security cooperation is very specifically aimed at making the partnered military more capable of accomplishing its required missions because that is important to advance U.S. interests. The United States may want that country to be able to stand on its own and stabilize the region—or it may seek to develop the country as a partner to integrate into future combined operations.

Regardless of whether it is a main effort or a peripheral goal, security cooperation does ultimately revolve around efforts to improve the partner nation. However, it's important for leaders and planners to note *where* the partner country falls in terms of these security cooperation purposes. This should drive the level of effort and the willingness of military and government officials to push certain objectives, as we shall see later. Even though the terms "capability" and "capacity" are used previously, there may very well be more to it than having a capability and the ability to use it. The United States may be interested in how *well* the partner nation uses its military forces—something which is more adequately approached

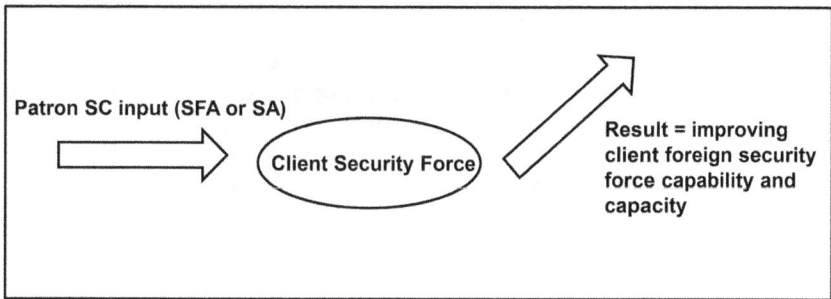

Figure 14.1 The "Input-Output" Model of Security Cooperation

with an effectiveness model for security cooperation as will be discussed subsequently.

The Present U.S. Approach to Security Cooperation

So, whether as the main effort goal or a peripheral outcome, how does the United States in general presently seek to make its partners better? Common U.S. doctrine and practice emphasizes finding the *overlap* between our own goals and those of our partner nation and meeting their needs.[4] Essentially this is a patron–client relationship, and ideally the relationship works like the model shown in Figure 14.1. Unfortunately, this is not always what happens.

The patron (the United States) provides a range of inputs, and together these result in improvements in the capability and capacity of the client's security forces. U.S. joint doctrine emphasizes that "multiple activities executed over time" are required to "make progress towards an objective," while admitting that "[o]ften, causal relationships are difficult to discern, making it difficult to determine the extent to which a [security cooperation] activity influenced a [partner nation]."[5]

And in some ideal cases this is actually how the process works. But why doesn't this work more often? Before we get there, we need to develop the concept of security force effectiveness and then examine the common explanations for variation in the field of security studies.

As currently structured, U.S. security cooperation is focusing on developing *capability* (can they do a given task?) and *capacity* (how much can they do, and can they sustain it?).[6] The problem with these terms for our purposes is that they don't lend themselves to explaining variations in performance. Instead, we'll use the term *effectiveness*. Since effectiveness gets to how well something is accomplished, this serves as better way of comparing the relative value or measurement of different security forces *regardless* of different characteristics or opponents.

AN EFFECTIVENESS MODEL FOR SECURITY COOPERATION

Although U.S. joint and military doctrine focus on capability and capacity rather than effectiveness, this term does find frequent use in doctrine. Although effectiveness, per se, is not defined, the Army *does* define a *measure of effectiveness*—the criteria "used to measure changes in system behavior, capability, or operational environment . . . tied to measuring the attainment of an endstate."[7] Measuring behaviors and capabilities is important, which is why this term is used more often in academic literature.

In their classic multivolume study, Millett and Murray define military effectiveness as "the process by which Armed Forces convert resources into fighting power."[8] They argue that military effectiveness exists at the political, strategic, operational, and tactical levels. For our purposes, however, it is easier to envision this as a political/strategic/organizational level dealing with resources and an operational/tactical level dealing with campaigns and battles. For Millett and Murray, a fully effective military is one deriving maximum combat power from the physical and political resources available. Combat power is "the ability to destroy the enemy while limiting the damage he can inflict in return."[9]

But conflict today is not simply the realm of the military; it involves close interaction between the military and intelligence and police forces—particularly if a nation is threatened on its own soil. Thus, we can expand the existing definitions of military effectiveness as similarly applying to "security force effectiveness," which also has a political/strategic component focused on acquiring and applying resources and an operational/tactical level focused on conducting campaigns.

Others have operationalized military effectiveness by examining the "outputs" of the security effort.[10] Building on this, the internal workings of the system at the political/strategic level can be seen as the "inputs" that emerge as the operational and tactical level "outputs" of the system. This is something that military analysts inherently recognize—that the structures and processes involved in managing organization, fielding, training, and educating (i.e., "the generating force")—produce trained and ready forces (i.e., the "operating force)." These outputs are the bottom-line requirements for any security force—they provide a country with the ability to apply a security effort at the operational/tactical level (i.e., versus a given opponent or to control or protect a given population).

We can further understand the concept of combat power (the output) by using Stephen Biddle's concept of "force employment" and modern warfare. In *Military Power*, Biddle argues that the possession of material capabilities alone is not an effective measure, but that instead the key factor is *how these capabilities are used*.[11] It is not simply developments in weapons technology but rather the integration of modern weapons with new tactics that have led to such tremendous increases in lethality over the course of the twentieth century. This increased lethality is available to those who can:

1. independently maneuver dispersed small units;
2. rapidly mass precision firepower; and
3. combine multispectrum capabilities (infantry, armor, airpower) through combined arms operations.[12]

The resulting "modern system" military can rapidly deliver tremendous casualties against exposed opponents while limiting their own casualties, although this can be mitigated by an opponent using defense in depth, defensive maneuver, and cover and concealment in the defense.[13]

Such operations are highly complex and require extensive training to accomplish. This system requires "extensive independent decision-making by junior officers and senior enlisted personnel," and Biddle acknowledges that "[a]mong the most serious drawbacks of the modern system is its tremendous complexity, and the high levels of skill it therefore demands in soldiers and officers. Not all armies can provide such skill."[14]

Talmadge expands further on Biddle's arguments and argues that we can evaluate "combat effectiveness" in militaries in two aspects: *basic tactics* (the ability to perform common tasks integral to an individual or small unit's job as infantry, armor, intelligence, etc.) and *complex operations* (the ability to integrate joint forces in operations, recognizing that such operations require "significant low level initiative and high-level coordination").[15] Rather than two separate variables (complex operations and basic tactics), these seem more like ends of a spectrum. We can thus envision security force effectiveness as a spectrum running from a high end of combined/joint operations to a mid-range of basic proficiency and a low end of lack of basic proficiency (Figure 14.2).

Combined arms/joint operations is the successful coordination and integration of multiple security elements (infantry, armor, airpower, intelligence, local and national police) within a given operation. Achieving combined/joint operations by a security force is optimal for meeting national political goals. The ability to conduct combined/joint operations inherently involves the ability to perform individual and small unit tasks for each of the security elements.

The middle range of the spectrum of variance in security force effectiveness is *basic proficiency*.

Basic proficiency is found where individual security force elements are successfully able to perform all or a majority of their basic individual and small unit skills (e.g., individual marksmanship, the ability to conduct local police operations, ability to maintain equipment and effectively gather and analyze intelligence). Though able to perform their particular function, security forces are unable to successfully coordinate and

Dependent Variable Conceptualization
Security Force Effectiveness

So if we envision "Combat Power" as the
"output" of security force organizational
structures ...

Security Force Effectiveness
(Operational and Tactical
 Performance – "Organizational
 Outputs")

Security Force
Effectiveness

Consists of both: Variance can be
1. Political/Strategic Level   measured
2. Operational/Tactical Level   by its outputs:

Spectrum of demonstrated performance

Combined/Joint
Operations *(optimal)*

Basic Proficiency
(suboptimal but may be
enough depending on the
threat environment)

Lack of Basic Proficiency
(ineffective)

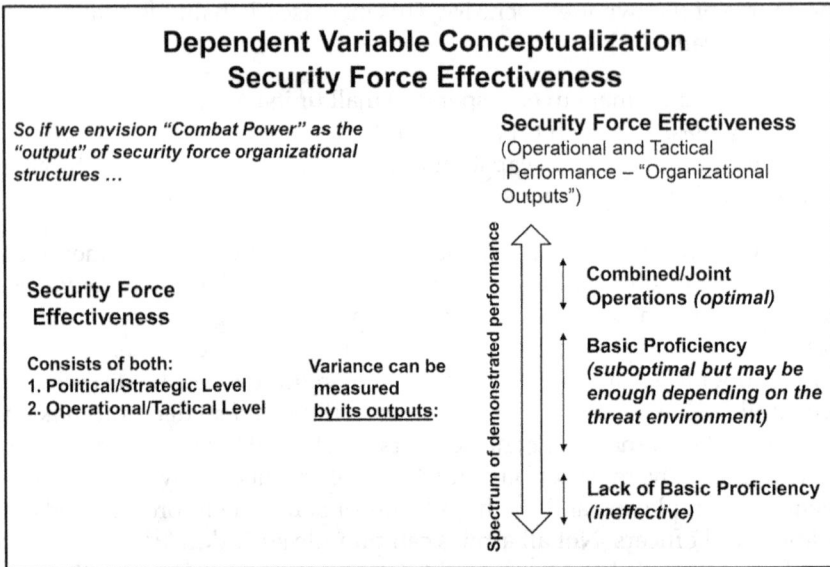

Figure 14.2   The Spectrum of Security Force Effectiveness

integrate with other elements during a given operation. Basic proficiency is a suboptimal outcome. However, depending on the nature of the threat environment, such suboptimal security forces may be equal to the requirements of the national political goals.

The bottom level in the spectrum of security effectiveness variation is *lack of basic proficiency,* which exists when security force elements are unable to achieve success in the majority of integral tasks during a given operation. At this level security forces are considered ineffective and incapable of meeting national political goals.

POSSIBLE EXPLANATIONS FOR VARIATIONS IN EFFECTIVENESS

So what are possible explanations for variations in security force effectiveness? A number of authors have provided explanations for such variation, although they overwhelmingly avoid discussing the impact of assistance from outside states.

Warrior versus Nonwarrior Cultures

One common explanation is that of warrior versus nonwarrior cultures. Cultural factors endemic within certain ethnic groups (e.g., Arabs) or dominant within a particular country (e.g., Israel) create dynamics that enhance or degrade tactical ability. No amount of training will overcome these inherent shortfalls.

Thus, in his comprehensive 2002 work, *Arabs at War: Military Effectiveness, 1948–1991*, Kenneth Pollack argues Arab armies from 1948 to 1991 suffered from inherent weaknesses in tactical leadership, information management, and technical incompetence not found in their Israeli and Western (1991 Gulf War) opponents. This set of "critical limitations on their combat effectiveness . . . proved to be insurmountable obstacles to their military operations" and was particularly "devastating in an age of warfare in which decentralized command, aggressive and innovative tactical leadership, accurate information flows, and advanced weaponry were the keys to victory."[16]

In his more recent (2019) *Armies of Sand: The Past, Present, and Future of Arab Military Effectiveness*, Pollack expands on how culture works to shape particular performances in Arab militaries. He argues that a reverence for tradition reinforced by educational practices emphasizing rote memorization rather than questioning produces a "dominant culture [that] consistently suppresses creativity, innovation, imagination and all similar divergences from established patterns," and favors the centralization of (and deference to) authority and information.

In both of his works, Pollack's description of the realities of modern combat where militaries combining small unit initiative with combined arms integration are dramatically more effective than those that do not runs parallel with Biddle's earlier argument. Additionally, his argument that certain security forces may exhibit a common set of flaws that inhibit their effectivess seems important. It is hard to discount the fact that education and environment do contribute to adult performance in at least some ways. But is culture the dominant reason for factor that shapes security force performance? Did these problems simply occur because they were Arabs—or is something else to blame?

If Arab militaries do indeed suffer from these particular shortfalls, then why have we seen variation in performance within this ethnic group— even within the military of a single country? For example, why did the Iraqi special forces that led operations to retake Mosul in 2016–2017 perform so much more effectively than their Shia Arab peers?[17] Additionally, why do we see *similar problems in performance across different cultures?* For example, why do we see similarities between the Iraqi military in 2014 and the army of the Republic of Vietnam during the Vietnam War?

Culture Lag

Another closely related explanation for security force ineffectiveness is culture lag or cultural resistance. The argument under this theory is that culture contributes to organizational shortfalls that impede military effectiveness. This is similar to Pollack's argument, but more generalized to third world countries rather than something specific to Arabs or other ethnic groups. Systemic impediments in many nations retard or frustrate attempts to translate military resources into improved military skill. These impediments are produced by governance shortfalls caused by cultural norms. Because they derive from norms, these impediments are extremely difficult to change by policy and thus tend to be stable over long periods of time. The observable indicators for this theory are problems occurring when it is difficult to train and utilize weapons systems designed for different (usually Western and non-Western) doctrines or cultures. Western doctrine and technology rely on initiative gained from Western educational systems and not found in most third world or non-Western (i.e., authoritarian) systems.[18]

This is taking Pollack's argument a step further. Rather than saying that some cultures fight better than others because of how they were raised, independent of organizational factors, this argument is saying that particular cultural factors contribute to organizational shortcomings that degrade military effectiveness. However, if this is the case, then why do we see similar problems occurring in different cultures? Unless there is such a thing as a "third world culture," we probably shouldn't see the same sorts of results happening in different cultures. Additionally, this approach is hard to quantify. However, we shouldn't discount this idea that organizational weaknesses within governments create impediments to security force effectiveness. This seems like a useful concept worth exploring. But as with Pollack's earlier argument—is this really because of culture or something else?

Democracy—or Not

A third commonly argued theory of effectiveness suggests that a multitude of inherent factors within democracies enhance military effectiveness: low power distances enhance communications, liberal culture encourages initiative, and democratic militaries are more likely to have merit-based promotions than authoritarian militaries.[19]

One key problem with this argument is in defining what is democracy or not. What about the large number of countries falling somewhere in the middle and defined as partial or hybrid democracies? Additionally, what mechanisms cause these various factors to impact security force effectiveness? Why do different democracies seem to have security forces with varying degrees of effectiveness? Or, instead, are some or all of these factors part of another, more comprehensive explanation?

Politicization

A final group of explanations is that variation in military effectiveness is caused by the degree of politicization of that military.

Politicization of security forces generally occurs when the central government seeks to increase its control over military forces to accomplish certain political goals and/or when the government seeks to protect itself from overt military influence or military takeover. This last dynamic has been labeled "coup-proofing" and is the most common form of politicization. To accomplish these goals, political leaders—particularly but not exclusively authoritarian ones—intervene in the organizational practices of security force organizations in ways that are clearly counterproductive, resulting in degraded performance but political loyalty.[20] This theoretical approach seems particularly useful, particularly with its close focus on the direct impact of political actions on particular organizational practices.

The commonality across politicization and many of the previous explanations is a focus on organizational practices as an intervening variable. Whatever the actual independent variable involved, variation in security force effectiveness occurs by first changing organizational practices, which then in turn contribute to poor battlefield performance. Arguments for politicization, however, generally provide a clear and well-defined explanation of the mechanisms leading to ineffective security forces. More importantly, it is easier to conceptualize such politicization occurring in many governments around the world, so this seems like a better answer for why we see similar shortfalls in security force effectiveness in different cultures.

CRITICAL SECURITY FORCE ORGANIZATIONAL PRACTICES AND SECURITY COOPERATION

We shall return to the concept of politicization further subsequently, but one clear commonality that stands out across all of the previous explanations is a link to battlefield performance stemming from three organizational practices that vary on a spectrum between professional and nonprofessional:

1. *Promotions and advancement*—Is the primary criteria to advance merit-based, or loyalty based?
2. *Command and control*—Is it a hierarchical structure with decentralized execution and lateral information sharing, or does the state rely on redundant and competing security services?
3. *Training and education*—Is it battle-focused and emphasizing initiative? Or is it constrained and scripted?[21]

If we envision combat power in the operating force as the output of security force organizational practices from the generating force, we can visualize this as in Figure 14.3.

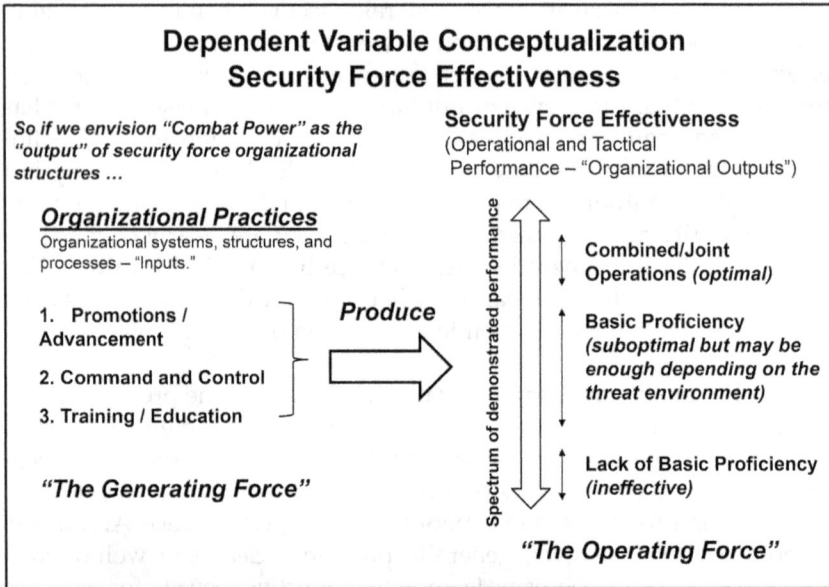

Figure 14.3   Security Force Organization Practices and Their Outputs

Security cooperation impacts organizational practices in the client state depending on the nature of the assistance rendered. There are two primary ways in which this works (as shown in Figure 14.4).

The first dynamic is the nature of patron control or influence. Is it oriented at the organizational practices of the client state? This is more than simply providing training, advising a partner on improvements to promotion or command and control systems, or providing hardware that allows these to function better. This is part of the dynamic, but this also speaks to the degree that this assistance is a professionalizing influence. The mechanism occurring here is *organizational intervention*. Despite the terminology, this doesn't mean that this is unwanted intervention—simply that it produces change. The impact of this intervention varies based on the number of practices targeted and the degree of effort expended to create change.

The second dynamic is the nature of the commitment of host nation partner combat forces. The mechanism for professionalizing influence here is that of *emulation*. Working with a more professional partner creates a degree of influence upon the client to adopt some of the best practices of their patron. The impact varies on the number, quality, and duration of patron state interactions with host nation tactical elements. These interactions produce changes—some small and some larger—which in turn filter up through the host nation system as the leaders in these tactical units advance or are seen to outperform their peers.

How does Security Cooperation impact client Organizational Practices?

1. Security Cooperation: Patron Control (Is it oriented at Organizational Practices or not?)

MECHANISM = *ORGANIZATIONAL INTERVENTION*
Impact varies based on the number of practices targeted and degree of effort expended to create change

Professionalizing Influence

2. Security Cooperation: Nature of Combat Force Commitment – What is the main effort for the partner? (see next page)

Professionalizing Influence

Organizational Practices
Organizational systems, structures, and processes – "Inputs."

1. **Promotions / Advancement**
2. **Command and Control**
3. **Training / Education**

*4. **Professional Ethos**
Norms reinforcing professional practices

MECHANISM = *EMULATION* – Impact varies based on the number, quality and duration of patron state interactions with host nation tactical elements, influencing emulation of patron state practices and host nation reforms to induce professional practices

3. *TIME (Duration applied) – AMPLIFIES THE IMPACT OF BOTH*

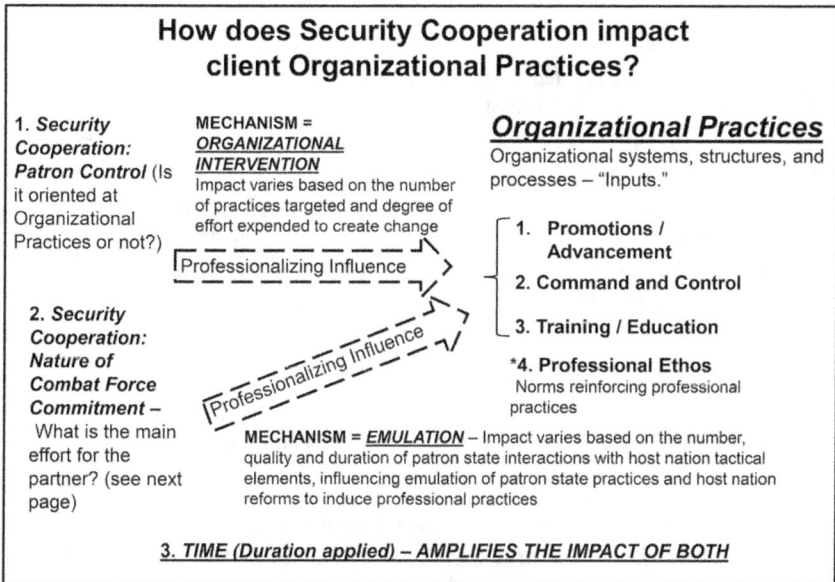

Figure 14.4 Security Cooperation Impacting Organizational Practices

The United States has an extensive history of security cooperation. Examination of this shows that the more extensive the interaction between the patron state and the client, the greater and faster the results. Thus, we can see a spectrum of impact based on the type of the commitment. An *advisory presence* produces gains based on the depth of this coverage. But advisory teams are small, and the number of individuals in direct contact with them is limited. A greater effect is produced by *partnered operations*. Working alongside military forces of the partner state produces a much greater number of opportunities for emulation. The greatest influence comes from *combined command*. When the two partners exist in the same headquarters, the opportunity for close interaction of current and future leaders is extensive and produces the greatest opportunity for rapid change. An example of this last dynamic would be Republic of Korea armed forces operating as part of the UN Command in the Korean War.

Alongside these options is a fourth type of patron state option for the commitment of combat forces, which is much less effective. When *independent combat operations* are the main effort for the patron state, this produces no gains in improvement within the client state and in fact often has a negative impact on effectiveness. Because the majority of patron state operations have no interaction with client forces, there is little opportunity for emulation. Additionally, this option often results in client forces sitting on the sidelines or being relegated to static and less intensive operations.

There is little potential for learning through doing, and this often leads to atrophy in the client forces, and the reliance on the patron state partner produces little incentive for improving organizational practices.

This is an effectiveness model for examining how security cooperation—in general—influences improvements in partnered security force effectiveness. The two means of impacting effectiveness do not have to both be pursued simultaneously. But these dynamics *are* mutually supporting. However, there is something missing. This model fails to account for a dynamic that has occurred in a number of the security cooperation relationships which the United States has entered since becoming a great power following World War II. This is the nature of security cooperation with weak states.

THE CHALLENGE: SECURITY COOPERATION IN WEAK STATES

The unstable nature of weak states provides a structural imperative that drives the relationship of weak state governments with their security forces. The foreign policy and development communities frequently use terms such as "weak" or "fragile" states to categorize states on the lower end of a spectrum assessing the administrative, security and economic capacity of a state. Weak states typically face a host of security challenges and are usually the focus of a significant amount of international foreign policy attention due to the inherent instability that they contribute to the international system.

All too often, weak state political leaders perceive that enabling their security forces with the ability to deal effectively with external or internal threats creates organizations that are equally capable of toppling the government. Given this danger, many weak state governments pursue what are labeled as "coup-proofing" strategies—generally understood by political scientists studying civil–military relations as a series of practices where the leadership of a particular country increases its direct and personal control over tactical and operational decision-making and decreases the coordination between individual units and services without central political oversight.

We can assume that most weak state governments will pursue policies that politicize their security force organizations because of the prevalence of these factors—government capacity shortfalls, potential for military coups, and turbulent societies where government supporters and political opponents are often at odds—present in weak state societies which are benefited in the short term by government intervention. Weak state governments will thus intervene directly upon the organizational practices of their security forces—generating a negative influence on security force effectiveness which outside patrons must work to counter if they seek to increase the capabilities of their partners.

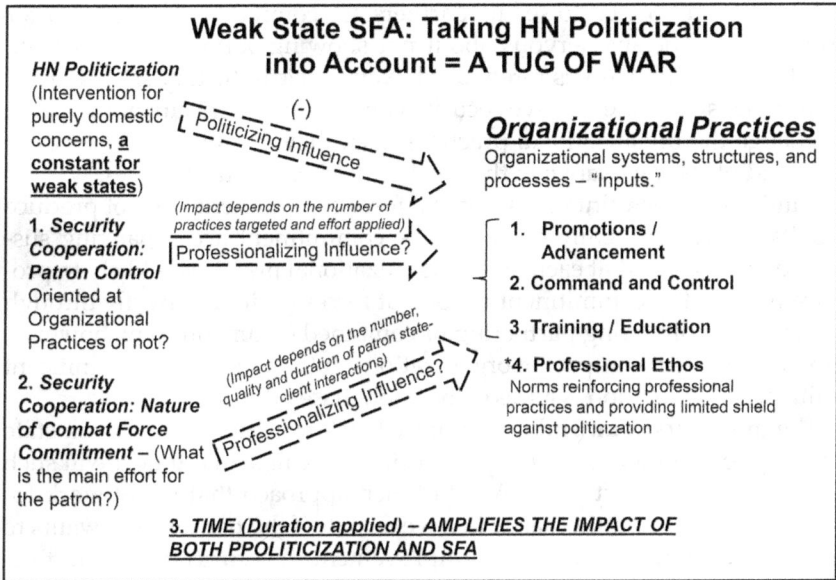

Figure 14.5   The "Tug of War" for Effectiveness Occurring in Security Cooperation to Weak States

The problem that the United States failed to account for in Iraq, Afghanistan, South Vietnam as well as numerous other cases is to recognize the unique nature of security cooperation in weak states. In such cases, the United States (or any other assisting country) is in a tug of war with a naturally occurring and corrosive drive to politicize the security forces of partnered weak state security forces. Positive change can only occur when appropriate and long-term strategies are implemented to counter this dynamic.

In such cases, it is critical that patron state organizational intervention targets multiple practices with a commitment of extensive resources and time. Whereas in cases of *non*-weak state security cooperation any interventions targeting organizational practices have a good chance of sticking, this is different in weak states where security force organizational practices are often under sustained pressure from the forces of politicization. This leads to a discussion of *how* the patron state intervention occurs.

Patron states can seek to influence their clients through either *inducement* (giving clients assistance and gifts and hoping that this will influence them to change) or through use of quid pro quo *conditionality* (assistance will only come *after* the client has made changes or at least begun movement along such a path). An in-depth 2017 analysis of weak state security cooperation by Walter Ladwig argues inducement will rarely produce results and that conditionality is almost always necessary to produce lasting change.[22] This is very challenging for the U.S. military, which is culturally conditioned to see its partners as "comrades in arms" and for American politicians nervous about not showing support for a partner, and thus conditionality is rarely used. Nevertheless, history shows that if comprehensive and extensive security cooperation efforts are not targeted to change the organizational practices, then they are unlikely to produce any kind of lasting change in the weak state client security forces.

Similarly, in such situations, combat force commitment will not produce results on its own. Only combat force commitment accompanying sustained efforts aimed at each of the organizational practices is likely to produce results. The commitment of combat forces (if done correctly through strategies emphasizing partnering or combined command) will, however, greatly amplify the impact of organizational intervention as the significant influence of emulation is added alongside these efforts.

The most important and critical initial task for situations of weak state security cooperation is for the patron state to realize that they are in such a situation in the first place. A light touch approach that might work for stronger states has little chance in a weak state if the patron truly wants to achieve meaningful and lasting improvement in security force effectiveness. However, a more comprehensive, targeted (and almost certainly long duration) approach has the chance of producing the kind of changes that will result not only in improvements to the three primary organizational practices. When applied alongside more comprehensive capacity building efforts that target other sectors of government capacity, the chance for moving the client from a weak state to a stronger state without the same drive to politicize its security forces also increases.

KEY TAKEAWAYS FROM APPLYING THE EFFECTIVENESS MODEL TO SECURITY COOPERATION IN WEAK STATES

Capability Substitution

When U.S. forces (or other patrons) insert their own capabilities, systems, or processes, it doesn't necessarily mean client forces have gotten better, particularly for the long term. In many cases, the use of advisors with enhanced communications and intelligence support or partnered patron state forces alongside client forces will produce more optimal battlefield outcomes. This is the result of *capability substitution*—the short-term substitution of patron state capabilities in order to achieve immediate impact upon the battlefield rather than developing these capabilities over the long term. What is occurring on the battlefield is a *combination* of patron and client effectiveness, *not* a true picture of the effectiveness of host nation forces.

This is a way for the United States to produce rapid and temporary increases in the security force effectiveness of the client state forces with which we work. However, overreliance on patron systems such as command and control, intelligence, transportation, and precision firepower can create a crutch for the client state, preventing their own development of systems and processes to coordinate and deliver their own capabilities or even to coordinate those of their partners.

This takes us back to the security cooperation purposes introduced at the beginning of this chapter. If the strategic goals of security cooperation with a client are short term, then capability substitution can be an acceptable course of action. But policymakers and senior leaders need to understand that such results will likely be fleeting.

If, however, the long-term effectiveness of the client is the goal, then the use of capability substitution needs to be carefully considered and modulated. When client state security forces are actively involved in planning and coordination, a degree of improvement rubs off on them through emulation. However, when these functions are exclusively performed by the patron, there is little chance of improvement. In some cases, when advisors substitute themselves and exclusively perform roles that should normally be accomplished by client state personnel—such as planning or coordination between adjacent or higher units—they may actually impede the normal development that would occur through repeated performance of such tasks. Even better for long-term effectiveness improvement is when client forces are forced to use their own capabilities to accomplish required missions and forced to develop their own systems and processes to command and control them.

Achieving Long-Term Effectiveness in a Weak State Requires a Major and Likely Long-Term Commitment

The tug of war with a weak state client requires significant intervention by the patron in order to achieve sustainable results in security force effectiveness. This intervention must be targeted and leaders must be willing to commit to intervene with our clients' organizational practices in order to achieve results. Simply providing training and education—standard tools in security cooperation—is not enough in these cases—promotions and advancement as well as command and control must be addressed. Combat force commitment requires a close partnership. Although extended duration may not be absolutely necessary, it is likely. It can help counteract previous shortfalls in assistance and ensure that necessary pressure for change is exerted over time. The reverse requires the major commitment of resources applied across the spectrum of intervention (the U.S.–ROK partnership in the Korean War from 1950 to 1953 produced major strides in South Korean effectiveness, although these gains were *still* cemented further by a long-running U.S. presence).

If this is not politically feasible or if leaders do not judge this worth the commitment of resources, then serious thought must be given to avoiding such an endeavor altogether—unless commitment to a status quo, unending, and unimproving quagmire is an acceptable foreign policy option.

Dedicated Roles and Missions for Security Cooperation

The effectiveness model for security cooperation shows the importance of targeting partnered organizations—not just their tactical assets. The U.S. military currently pursues extensive training and education efforts with a number of partners around the world. Such engagements remain highly useful tools from the standpoint of relationship building or maintenance. *However*, this is unlikely to result in substantive improvements in the effectiveness of weak state partners. To achieve this requires additional investment targeting their command and control and personnel practices.

The U.S. Army's Security Force Assistance Brigades (SFABs) are a critical capability for future security cooperation efforts. Having forces trained and focused on security cooperation will be vital to sustaining and improving partnerships in an increasingly competitive world. While other U.S. military forces are capable of performing combined exercises and operations, the ability to perform advisory and partnering missions needs to be resourced. These are also the forces to lead assistance on the ground to those more challenging weak state relationships, whether to achieve short-term goals or long-term stability. However, where is the matching organizational structure to impact the client's generating force and executive functions? Absent this capability, our organizational structure aimed at providing lasting change in weak states seems limited. Additionally, SFABs seem much more suited for partnering at the division level and below, and the division level is a stretch. The Army should consider establishing a Security Forces Assistance Command (SFAC) with the capability to fill two critically needed gaps as well as serving as a proponent for the existing SFABs. First, such a command needs to be able to field the assistance efforts for partnered division and corps level headquarters. Second, it needs to be able to field or support the training of qualified personnel to advise the generating forces of partnered nations.

The U.S. military's long-running State Partnership Program pairs state-level National Guard organizations with other nations. These partnerships benefit from extended relationships with recurring engagements over a long duration, although the number of contacts varies by the particular relationship. These are important tools, but relationships generally are focused at the partnership level and lack the ability to incentivize organizational change. On its own this program may produce resulting improvements in effectiveness for moderate or strong state security forces, but it appears unlikely to be a tool to improve weak state security force effectiveness. State partnerships could, however, be tools in a more comprehensive security cooperation approach to these more challenging missions as one of multiple assets used in combination with each other.

Combat force commitment in weak states requires close partnership. As the U.S. military looks to embed lessons learned from Iraq and Afghanistan, it is critical to integrate these into its doctrine and institutional memory rather than forgetting them as it did with the lessons of South Vietnam. Partnered operations are critical, and combined rather than separate command relationships should be sought wherever possible. This, however, requires significant political commitment on the part of the patron, possibly including a great deal of arm twisting of the client to achieve a combined command arrangement. The U.S. military's "By, With, and Through" policy appears to be a positive outcome resulting from lessons learned from U.S. involvement in Iraq and Afghanistan, where coalition forces spent extensive time pursuing independent operations during much of these campaigns that did not aid in building partner capacity. To date, such efforts have been focused on more short-term capability substitution efforts such as retaking Mosul and other occupied areas of Iraq from ISIS. However, this sort of approach is also important to the sort of long-term partnering necessary to achieve success in future long-term commitments, when—not if—the United States again finds itself developing the security forces of a weak state in the future.

Up to this point, this chapter has shown how the effectiveness model is a useful way of examining security cooperation in general as well as very specifically in understanding the unique challenges of assistance to weak states. But it is also very useful in cases where the United States seeks to develop the capabilities of a strong state, as we can see from applying the model to examining the Japanese Self Defense Forces, a critical partner to deterring—and, if necessary, in fighting—any future conflict with the People's Republic of China (PRC).

APPLYING THE MODEL TO A STRONG STATE—JAPANESE SELF DEFENSE FORCES

Japan is a critical American ally in the Indo-Pacific region, a member of the Quadrilateral Security Dialogue, and increasingly part of combined planning efforts with the United States for potential operations to defend against a PRC invasion of Taiwan.[23] Despite this, however, numerous concerns have been raised about the ability of Japanese Self-Defense Forces (JSDF) to conduct joint operations. The JSDF not only faces constitutional constraints against offensive operations, it is part of a defense system not designed for jointness, with the heritage of a military culture suffering from extensive interservice infighting. The JSDF continues to suffer from institutional stovepipes, interservice stovepipes, and the lack of joint integration.[24]

Applying the effectiveness model to look at the potential for targeted security cooperation to improve the JSDF provides some key insights. The JSDF already has merit-based promotions and advancement and extensive battle-focused training. Their problems lie in the area of command and control, where they have significant shortfalls in the joint command and control (C2) necessary to fight and win a major high-end conflict as part of combined operations with the United States.

Using the model to develop a prescription for targeted security cooperation would suggest increased combined–joint operations with the United States and Australia to allow opportunities for emulation of more advanced joint integration. Additionally, the United States should focus its partnering efforts on helping the Japanese improve their organizational practices in joint C2.

Outlook for a Competitive World

As it contemplated an increasingly competitive world, the Biden administration's initial national security documents prioritized efforts to "reinvigorate and modernize [U.S.] alliances and partnerships" with an awareness that such alliances and partnerships are "our greatest asymmetric strength" in its Interim National Security Strategic Guidance of 2021 and Indo-Pacific Strategy of 2022.[25] Alliances and partnerships have long been a U.S. strength, and these goals will be important for future administrations as well. Security cooperation will be a critical tool to accomplish this.

Despite being such a long-running and important tool for U.S. foreign policy, security cooperation has not always achieved the results that it could have. Reorientating toward great power competition deserves rethinking the model used to conceptualize security cooperation. Rather than the present input–output model of capability and capacity development, the U.S. military needs to shift to a broader and more holistic model focusing on effectiveness.

An effectiveness model would allow the United States to better analyze the necessary requirements for performing security cooperation in weak states—a challenge that has stymied U.S. foreign policy efforts from South Vietnam to Iraq and Afghanistan. Such failures could be preventable—or at least avoided. But an effectiveness model *equally* allows improved analysis of the particular challenges of strong state security cooperation and to more efficiently deploy U.S. resources and achieve more positive results. As the United States advances into a world where it faces adaptive and determined adversaries, the efficient and effective deployment of American resources will be critically important.

Focusing on effectiveness should be grounded in the understanding that the United States pursues security cooperation for several reasons. Ultimately, however, the United States is usually trying to at least secondarily produce improvements in effectiveness, even though this may not be the primary reason for the relationship. A more holistic way of examining particular challenges allows U.S. policymakers and strategists a better means to judge how and when to pursue particular strategies. This will be critical to accomplishing U.S. strategic goals in the twenty-first century.

NOTES

1. *Joint Publication 3-20: Security Cooperation*, May 23, 2017, I-1.
2. *Joint Publication 3-20*, vii.
3. *Joint Publication 3-20*, v.; Joint Center for International Security Force Assistance, *Security Force Assistance Planner's Guide*, January 1, 2016, 1-1.
4. *Joint Publication 3-20*, III-4.
5. *Joint Publication 3-20*, III-5.
6. *Joint Publication 3-20*, I-2.
7. U.S. Army, *Army Doctrinal Reference Publication 3-07: Stability* (Washington, DC: Headquarters, Department of the Army, August 31, 2012), 4–12.
8. Allan R. Millett, Williamson Murray, and Kenneth Watman, "The Effectiveness of Military Organizations," in *Military Effectiveness: Volume 1: The First World War* (Cambridge, UK: Cambridge University Press, 1988), 2.
9. *Ibid.*
10. Caitlin Talmadge, *The Dictator's Army: Battlefield Effectiveness in Authoritarian Regimes* (Ithaca, NY: Cornell University Press, 2015), 13.
11. Stephen Biddle, *Military Power: Explaining Victory and Defeat in Modern Warfare* (Princeton University Press, Princeton, New Jersey, 2004), 191–192.
12. Biddle, 35–39.
13. Biddle, 44–51, 51.
14. Biddle, 49–50.
15. Talmadge, 6.
16. Kenneth Pollack, *Arabs at War: Military Effectiveness, 1948–1991* (Lincoln: University of Nebraska Press, 2004), 582.
17. Jahara Matisek and William Reno, "Getting American Security Force Assistance Right: Political Context Matters," *Joint Forces Quarterly* 92 (1st Quarter, 2019), 70.
18. Wade Hinkle, Michael Fischerkeller, Matthew Diascro, and Rafael Bonoan, "Why Nations Differ in Military Skill—and How That Should Affect U.S. Defense Planning," Institute for Defense Analysis, October 1999, es-2–3.
19. Dan Reiter and Allan Stamm, "Democracy and Battlefield Military Effectiveness," *Journal of Conflict Resolution* 2, 3 (June 1998), 265; Stephen Biddle and Stephen Long, "Democracy and Military Effectiveness: A Deeper Look" *Journal of Conflict Resolution* 48, 4 (2004), 541.
20. Joel Migdal, *State in Society: Studying How States and Societies Transform and Constitute One Another* (New York: Cambridge University Press, 2001), 80; Talmadge, 15.

21. Talmadge, 13–17; Kenneth Pollack, *Armies of Sand: The Past, Present, and Future of Arab Military Effectiveness* (New York: Oxford University Press, 2019), 112–128.

22. Walter Ladwig, *The Forgotten Front: Patron–Client Relationships in Counterinsurgency*, (Cambridge, UK: Cambridge University Press, 2017), 292.

23. "Japan, U.S. Draw Up Plan for Any Taiwan Emergency, Kyodo," *Reuters*, December 23, 2021.

24. John Wright, "Solving Japan's Joint Operations Problem," *The Diplomat*, January 31, 2018.

25. "Interim National Security Strategic Guidance" (Washington, DC: The White House), March 2021, 10; "Indo-Pacific Strategy (Washington, DC: The White House), February 2022, 12.

About the Editors
and Contributors

Editors

Dr. Joel R. Hillison, the lead editor, is an associate professor and the General Colin Powell Chair of Military and Strategic Studies at the U.S. Army War College. Dr. Hillison has published many articles and podcasts on national security issues. In 2019, he contributed a chapter to *Landpower in the Long War* (The University of Kentucky Press) and in 2014 authored the book *Stepping Up: Burden Sharing by NATO's Newest Members* (U.S. Army War College Press). He retired after thirty years of service in the U.S. Army, serving in Operations Desert Storm and Iraqi Freedom.

Dr. Jerad I. Harper, co-editor, is an assistant professor at the U.S. Army War College. A former U.S. Army strategic intelligence officer, he teaches courses in national security, the Indo-Pacific, the Middle East, and building partner capacity. He has been published in *Foreign Policy, Parameters,* FPRI *e-notes*, and the *Journal of Military Learning*. He retired after thirty years of service in the U.S. Army, serving in Operations Enduring Freedom and Iraqi Freedom.

Dr. Christopher J. Bolan, co-editor, is an associate professor at the U.S. Army War College and a senior fellow in the Middle East Program at the Foreign Policy Research Institute. He has served as a foreign policy advisor on Middle East and South Asia affairs for Vice Presidents Gore and Cheney. Dr. Bolan has published over fifty articles and book chapters on national security issues. He retired after thirty years of service in the U.S. Army and twelve years of civil service.

Contributors

Colonel Thomas J. Bouchillon is a faculty member at the U.S. Army War College. A specialist in Southeast Asia security affairs, he has served in multiple U.S. Embassy posts in the region, including Indonesia, Laos, and the U.S. Mission to ASEAN. His operational work and research focuses on the Mekong subregion, the South China Sea, and transnational and environmental security issues.

Professor Maryann F. Foster is a faculty instructor at the U.S. Army War College. She has also served as a U.S. Army Reserve military intelligence officer, a U.S. government civilian overseas, a U.S. federal contractor, a U.S. Department of Defense contractor, and a U.S. Department of State contractor, most recently at the U.S. Mission to the North Atlantic Treaty Organization in Brussels, where she served as a speechwriter to the U.S. ambassador. She is the founding senior editor of *War Room*, the online journal of the U.S. Army War College.

Dr. James A. Frick is an assistant professor and director of the Second Year Studies Distant Education Program at the U.S. Army War College. As a strategist within Indo-PACOM he served as a planner for major contingency operations while also collaborating on land power posture and strategies for the U.S. Army in the Asia Pacific region. He currently teaches courses on contemporary security issues, defense management, Asia Pacific studies, and theater strategy.

Colonel Heather Levy is an active-duty Army officer currently serving as the commander of the Far East District of the U.S. Army Corps of Engineers. She is a former Carlisle Scholar at the Army War College. Her published articles cover a range of topics from military structure to senior military education.

Dr. Craig Morrow retired from the faculty of the U.S. Army War College in 2021 as an associate professor of strategic leadership. Dr. Morrow previously served as director of the General Psychology for Leaders program at the U.S. Military Academy and as a research fellow at the NATO Defense College in Rome. His published articles and book chapters span a variety of topics including critical thinking development, strategic intelligence, and pedagogy.

Dr. John A. Mowchan is an assistant professor at the U.S. Army War College. As a strategic intelligence officer at the Army War College, he has taught courses in national security, strategic leadership, cyber warfare, and Russia–Eurasia. While on faculty at the National Intelligence University, he taught classes on joint intelligence doctrine, the organization and management of the intelligence community, intelligence analysis, and asymmetric warfare. His previous publications include "Don't Draw the (Red) Line," *Proceedings*, October 2011; "The Black Sea Threat," *Proceedings*, February 2011, and the "The Militarization of the Collective Security Treaty Organization," *Collins Hall Update*, June 2009.

Dr. John A. Nagl is associate professor of warfighting studies at the U.S. Army War College. A retired Army officer, he served in both Iraq wars. Nagl is the author of *Learning to Eat Soup with a Knife: Counterinsurgency Lessons from Malaya and Vietnam* (University of Chicago Press, 2005) and of *Knife Fights: A Memoir of Modern War* (Penguin Press, 2013).

Commander Thomas P. Newman is an active-duty U.S. Navy information warfare officer currently assigned as the force oceanographer and deputy director of the Maritime Intelligence Operations Center on the staff of Commander, Submarine Forces, U.S. Pacific Fleet. A distinguished graduate of the U.S. Army War College, his contribution is adapted from a 2022 original research requirement.

Dr. Kevin D. Stringer is an associate professor at the General Jonas Žemaitis Military Academy of Lithuania and lecturer at the University of Northwestern Switzerland. As a retired foreign area officer, his research areas are irregular warfare, special operations, and Russian indirect action. Dr. Stringer has published in *Naval War College Review, Joint Force Quarterly, Military Review, Special Operations Journal*, and the *Journal on Baltic Security*. He is the author of the book *Military Organizations for Homeland Defense and Smaller-Scale Contingencies* (Praeger, 2006).

Dr. Brett D. Weigle is an associate professor and the General John J. Pershing Chair of Military Planning and Operations at the U.S. Army War College. He teaches courses on operational design and energy's role in national security. Weigle is the coauthor with Charles D. Allen of "Keeping David from Bathsheba: The Four-Star General's Staff as Nathan" in the *Journal of Military Ethics*.

Dr. Kevin J. Weddle is professor of military theory and strategy and the Elihu Root Chair of Military Studies at the U.S. Army War College. He served over twenty-eight years as a combat engineer officer, including command of a combat engineer battalion. He also served in Operations Desert Storm and Enduring Freedom. He has written numerous articles for scholarly and popular publications and has authored two award-winning books, *Lincoln's Tragic Admiral: The Life of Samuel Francis Du Pont* (University of Virginia Press, 2005) and *The Compleat Victory: Saratoga and the American Revolution* (Oxford University Press, 2021).

Index

Note: Page numbers followed by *t* indicate tables and *f* indicate figures.

www.ingramcontent.com/pod-product-compliance
Lightning Source LLC
Chambersburg PA
CBHW060139280326
41932CB00012B/1564